Communications and Control Engineering

Series editors

Alberto Isidori, Roma, Italy
Jan H. van Schuppen, Centre for Mathematics and Computer Science, Amsterdam,
The Netherlands
Eduardo D. Sontag, Piscataway, NY, USA
Miroslav Krstic, La Jolla, CA, USA

Communications and Control Engineering is a high-level academic monograph series publishing research in control and systems theory, control engineering and communications. It has worldwide distribution to engineers, researchers, educators (several of the titles in this series find use as advanced textbooks although that is not their primary purpose), and libraries.

The series reflects the major technological and mathematical advances that have a great impact in the fields of communication and control. The range of areas to which control and systems theory is applied is broadening rapidly with particular growth being noticeable in the fields of finance and biologically-inspired control. Books in this series generally pull together many related research threads in more mature areas of the subject than the highly-specialised volumes of *Lecture Notes in Control and Information Sciences*. This series' mathematical and control-theoretic emphasis is complemented by *Advances in Industrial Control* which provides a much more applied, engineering-oriented outlook.

More information about this series at http://www.springer.com/series/61

Arjan van der Schaft

L_2-Gain and Passivity Techniques in Nonlinear Control

Third Edition

 Springer

Arjan van der Schaft
Johann Bernoulli Institute for Mathematics
  and Computer Science
University of Groningen
Groningen
The Netherlands

ISSN 0178-5354   ISSN 2197-7119   (electronic)
Communications and Control Engineering
ISBN 978-3-319-84294-3   ISBN 978-3-319-49992-5   (eBook)
DOI 10.1007/978-3-319-49992-5

Printed on acid-free paper

This Springer imprint is published by Springer Nature
The registered company is Springer International Publishing AG
The registered company address is: Gewerbestrasse 11, 6330 Cham, Switzerland

Preface

The third edition of this book differs substantially from the second edition that came more than fifteen years ago. Approximately one-third of the book is new material, while existing parts have undergone major rewritings and extensions.

On the other hand, the spirit of the third edition as compared with the second and first edition has remained the same: to provide a compact presentation of the basic ideas in the theory of L_2-gain and passivity of nonlinear systems, starting from a brief summary of classical results on input–output maps, to a broad range of analysis and control theories for nonlinear state space systems, regarded from the unifying perspective of dissipative systems theory.

A major change with respect to the second edition is the splitting, as well as substantial extension, of the old Chap. 3 on dissipative systems, formerly also including passivity and L_2-gain theory, into three separate chapters on dissipative systems theory (Chap. 3), passive systems (Chap. 4), and L_2-gain theory (Chap. 8). Furthermore, the old Chapter 4 on port-Hamiltonian systems has been reworked and extended into two chapters on port-Hamiltonian systems theory (Chap. 6) and on control theory of port-Hamiltonian systems (Chap. 7). Also, the theory of all-pass factorizations (new Chap. 9) has been augmented with a treatment of nonlinear state space "spectral factorization" theory.

Apart from all the rewritings and extensions, a relative novelty from a conceptual point of view is the increased attention towards network dynamics and large-scale systems. The general concept of an interconnected system already was at the core of the passivity and small-gain theories, and even more so of the overarching theory of dissipative systems, having their origin in network theory and closed-loop stability, and emphasizing the "systems point of view." But the recent developments in dynamics on networks, including the use of algebraic graph theory, have certainly been influential in extending classical results, which is reflected in this new edition of the book.

Acknowledgements In addition to the acknowledgements of the previous editions (see the Preface to the 2nd edition), I would like to thank everybody who has contributed to the development of the new material in the third edition. Apart from the continuing collaborations with Bernhard Maschke, Romeo Ortega, Jacquelien

Scherpen, Alessandro Macchelli, Peter Breedveld, Rodolphe Sepulchre, and Stefano Stramigioli, since 2000 the collaborations with in particular Noboru Sakamoto, Dimitri Jeltsema, Jorge Cortes, Hans Zwart, Gjerrit Meinsma, Claudio De Persis, Bayu Jayawardhana, Kanat Camlibel, and Shodhan Rao have been very stimulating. Also I thank the Ph.D. students and postdocs since 2000 working on these and related topics: Guido Blankenstein, Goran Golo, Viswanath Talasila, Agung Julius, Ramkrishna Pasumarthy, Javier Villegas, Norbert Ligterink, Damien Eberard, Rostyslav Polyuga, Aneesh Venkatraman, Florian Kerber, Harsh Vinjamoor, Marko Seslija, Shaik Fiaz, Ewoud Vos, Jieqiang Wei, Geert Folkertsma, Nima Monshizadeh, Tjerk Stegink, Filip Koerts, Pooya Monshizadeh, and Rodolfo Reyes Baez, as well as the short-term visiting Ph.D. students and postdocs Hugo Rodrigues, Joaquin Cervera, Luca Gentili, Fabio Gomez-Estern, Joaquin Carrasco, Marius Zainea, Ioannis Sarras, Daniele Zonetti, and Luis Pablo Borja for numerous discussions and feedback. Furthermore, I thank the Johann Bernoulli Institute for Mathematics and Computer Science of the University of Groningen and my colleagues of the Jan C. Willems Center for Systems and Control for providing a stimulating working environment. Finally, once more I would like to acknowledge the influence of my former thesis advisor and inspirator Jan Willems. Especially this book would not have been possible without his fundamental contributions to systems and control, and thus is rightfully dedicated to him.

Groningen, The Netherlands   Arjan van der Schaft
September 2016

Preface to the First Edition

The first version of these lecture notes were prepared for part of a graduate course taught for the Dutch Graduate School of Systems and Control (DISC) in the spring trimester of 1994.

My main goal was to provide a synthesis between the classical theory of input-output and closed-loop stability on the one hand, and recent work on nonlinear $\mathcal{H}_\infty$ control and passivity-based control on the other hand. Apart from my own research interests in nonlinear $\mathcal{H}_\infty$ control and in passive and Hamiltonian systems, this motivation was further triggered by some discussions with David Hill (Sydney, Australia), Romeo Ortega, Rogelio Lozano (both Compiègne, France) and Olav Egeland (Trondheim, Norway), at a meeting of the GR Automatique du CNRS in Compiègne, November 1993, devoted to passivity-based and H_∞ control. During these discussions also the idea came up to organize a pre-CDC tutorial workshop on passivity-based and nonlinear H_∞ control, which took place at the 1994 CDC under the title "Nonlinear Controller Design using Passivity and Small-Gain techniques". Some improvements of the contents and presentation of Chapter 2 of the final version of these lecture notes are directly due to the lecture [122] presented at this workshop, and the set of handwritten lecture notes [183].

As said before, the main aim of the lecture notes is to provide a synthesis between classical input-output and closed-loop stability theory, in particular the small-gain and passivity theorems, and the recent developments in passivity-based and nonlinear H_∞ control. From my point of view the trait d'union between these two areas is the theory of dissipative systems, as laid down by Willems in the fundamental paper [350], and further developed by Hill and Moylan in a series of papers [123, 124, 125, 126]. Strangely enough, this theory has never found its place in any textbook or research monograph; in fact I have the impression that the paper [350] is still relatively unknown. Therefore I have devoted Chapter 3 to a detailed treatment of the theory of dissipative systems, although primarily geared towards L_2-gain and passivity supply rates. One of the nice aspects of classical input-output and closed-loop stability theory, as well as of dissipative systems theory, is their firm rooting in electrical network analysis, with the physical notions of passivity, internal energy and supplied power. Furthermore, using the scattering transformation a direct link is established with the finite gain property. Passivity-based control, on the other hand, used these same physical notions but draws its motivation primarily from the control of mechanical systems, especially robotics. Indeed, a usual approach is via the Euler–Lagrange equations of mechanical systems. In Chapter 4 of the lecture notes my aim is to show that the passivity properties of electrical networks, of mechanical systems described by Euler–Lagrange equations, and of constrained mechanical systems, all can be unified within a (generalized) Hamiltonian framework. This leaves open, and provokes, the question how other properties inherent in the generalized Hamiltonian structure, may be exploited in stability analysis and design. The rest of the lecture notes is mainly devoted to the use of L_2-gain techniques in nonlinear control, with an emphasis on nonlinear H_∞ control. The approach mimics to a large extent similar developments in robust linear control theory, while the specific choice of topics is biased by my own recent research interests and recent collaborations, in particular with Joe Ball and Andrew Paice. The application of these L_2-gain techniques relies on solving (stationary) Hamilton–Jacobi inequalities, and sometimes on their nonlinear factorization. This constitutes a main bottleneck in the application of the theory, which is similar to the problems classically encountered in nonlinear optimal control theory (solving Hamilton–Jacobi–Bellman equtions), and, more generally, in nonlinear state space stability analysis (the construction of Lyapunov functions). On the other hand, a first-order approach (linearization) may already yield useful information about the local solvability of Hamilton–Jacobi inequalities.

Enschede   Arjan van der Schaft
January 1996

Preface to the Second Edition

With respect to the first edition as Volume 218 in the Lecture Notes in Control and Information Sciences series the basic idea of the second edition has remained the same: to provide a compact presentation of some basic ideas in the classical theory of input-output and closed-loop stability, together with a choice of contributions to the recent theory of nonlinear robust and $\mathcal{H}_\infty$ control and passivity-based control. Nevertheless, some parts of the book have been thoroughly revised and/or expanded, in order to have a more balanced presentation of the theory and to include some of the new developments which have been taken place since the appearance of the first edition. I soon realized, however, that it is not possible to give a broad exposition of the existing literature in this area without affecting the spirit of the book, which is precisely aimed at a compact presentation. So as a result the second edition still reflects very much my personal taste and research interests. I trust that others will write books emphasizing different aspects. Major changes with respect to the first edition are the following: A new section has been added in Chapter 2 relating L_2-gain and passivity via scattering, emphasizing a coordinate-free, geometric, treatment. The section on stability in Chapter 3 has been thoroughly expanded, also incorporating some recent results presented in [312]. Chapter 4 has been largely rewritten and expanded, incorporating new developments. A new Chapter 5 has been added on the topic of feedback equivalence to a passive system, based on the paper [56].

Acknowledgements Many people have contributed to the genesis of this book. Chapter 3 is based on the work of my former thesis advisor Jan C. Willems, who also otherwise has shaped my scientific attitude and taste in a deep way. Chapter 4 owes a lot to an inspiring and fruitful cooperation with Bernhard Maschke, as well as with Romeo Ortega and Morten Dalsmo, while Chapter 6 is based on joint research with Andrew Paice, Joe Ball and Jacquelien Scherpen. Also I acknowledge useful and stimulating discussions with many other people, including Peter Crouch, Bill Helton, David Hill, Alberto Isidori, Gjerrit Meinsma, Carsten Scherer, Hans Schumacher, Rodolphe Sepulchre, Stefano Stramigioli, and my colleague Henk Nijmeijer. I thank the graduate students of the spring trimester of 1994 for being an attentive audience. Gjerrit Meinsma is furthermore gratefully acknowledged for his patient way of handling all sorts of LaTeX problems. Finally, I thank the Faculty of Mathematical Sciences, University of Twente, for providing me with secretarial support for the preparation of this book; in particular I sincerely thank Marja Langkamp, and for the first edition Marjo Mulder, for their great efforts in bringing the manuscript to its present form.

Enschede   Arjan van der Schaft
August 1999

Contents

Introduction

L_2-gain and passivity are fundamental concepts in the stability analysis and control of dynamical systems. Their traditional scenario of application concerns the feedback interconnection of two systems. By computing the L_2-gain of one of the two systems closed-loop stability is guaranteed once the second system has an L_2-gain less than the reciprocal of the first one. Similarly, by verifying passivity of one system, closed-loop stability results for any second system that is also passive. Thus passivity or L_2-gain provides a stability "certificate" for interconnection with an unknown system. At the same time it can be regarded as a robustness guarantee. As a result, many control problems can be cast into the L_2-gain or passivity framework; from stabilization, adaptive control, to robustness optimization and disturbance attenuation.

Furthermore, the L_2-gain and passivity theories are firmly rooted in the mathematical modeling of physical systems. Passivity can be considered as an abstraction of the common property that the time-derivative of the energy stored in a physical system is less than or equal than the externally supplied power. Port-based network modeling of physical systems takes the same point of view, leading to the further structured class of port-Hamiltonian systems. At the same time, (inverse) scattering of a system with L_2-gain ≤ 1 defines a passive system. Moreover passivity naturally extends to a cyber-physical context by emphasizing Lyapunov stability theory and physical analogies. On the other hand, the small-gain theory connects to contraction theorems.

Not only the basic reasoning in L_2-gain and passivity theory is similar, but both theories can be regarded as parts of the overarching theory of dissipative systems, as was formulated by Willems in the early 1970s. This theory provides a unified framework for a compositional theory of dynamical systems, directly applicable to large-scale network systems. Furthermore, it reveals, through the dissipation inequality, the close connection with optimal control.

The focus of this book is on the use of passivity and L_2-gain techniques for *nonlinear* systems and control. In particular, no frequency domain conditions are treated. Solutions to the dynamical analysis and control problems are expressed in terms of solvability of nonlinear (partial differential) equations and inequalities, mostly in Hamilton–Jacobi form.

The contents of the third edition of the book are organized as follows:

Chapter 1 summarizes the classical notions of input–output and closed-loop stability. The presentation is based on (and is sometimes almost literally taken from) Vidyasagar's excellent "Nonlinear Systems Analysis" [343].

Chapter 2 also largely follows the treatment of nonlinear small-gain and passivity theorems from Vidyasagar [343], with additions from e.g. Desoer and Vidyasagar [83], as well as further extensions. Section 2.4 provides a geometric treatment of scattering in this context.

Chapter 3 gives a detailed treatment of the theory of dissipative systems starting from the fundamental paper by Willems [350], thereby laying the conceptual foundation for much of the rest of the book. Main parts include the description of the solution set of the dissipation inequality, and the close link with stability and stabilization theory using Lyapunov functions, as well as with optimal control. Furthermore, dissipative systems theory as a tool for the analysis of large-scale interconnected systems is introduced.

Chapter 4 focusses on dissipative systems with respect to the passivity supply rate, and the resulting analysis and control theory of nonlinear passive state space systems. The passivity theorems of Chap. 2 are revisited from the state space point of view, including converse passivity theorems and extensions to networks of passive systems. Euler–Lagrange equations and second-order systems are dealt with as a special class of passive systems, emphasizing a geometric point of view. Also the notions of incremental and shifted passivity are introduced.

Chapter 5 deals with the problem of rendering a nonlinear system passive by the use of state feedback. This is employed for the stabilization of cascaded systems, and in particular the "backstepping" approach.

Chapter 6 is devoted to the theory of port-Hamiltonian systems. While in the previous chapters passivity is employed as a tool for analysis and control, in port-Hamiltonian systems theory (cyclo-)passivity follows from the mathematical modeling. A broad range of examples from mechanical systems and electrome-chanical systems is provided. Key properties of port-Hamiltonian systems are discussed, including the availability of Casimirs and shifted passivity. Furthermore, from a network modeling perspective physical systems are naturally described as differential-algebraic equation (DAE) systems, leading to the introduction of the geometric notion of Dirac structures and port-Hamiltonian DAE systems. This Dirac structure is partly determined by the incidence structure of the underlying graph. Finally, the notion of scattering is revisited within the port-Hamiltonian context.

Chapter 7 considers control strategies exploiting the port-Hamiltonian structure. The notion of control by interconnection (with a controller port-Hamiltonian system) for stabilization is emphasized. This is extended to passivity-based control including energy and interconnection structure shaping.

Chapter 8 treats the notion of L_2-gain within a state space context, revisiting the small-gain theorems and extending this to a network setting. It also briefly discusses connections with input-to-state stability theory.

Chapter 9 deals with the nonlinear analogs of the notions of left- and right factorization of transfer matrices for linear systems, and with nonlinear all-pass (inner–outer) factorization. These are used for constructing nonlinear uncertainty models, for obtaining a nonlinear Youla–Kucera parametrization of stabilizing controllers, and for deriving the minimum-phase factor of nonlinear systems.

Chapter 10 treats the theory of nonlinear state feedback H_∞ control, and derives necessary conditions for the output feedback H_∞ problem.

Finally, Chapter 11 is devoted to checking (local) solvability of Hamilton–Jacobi inequalities and to the structure of their solution set. Emphasis is on the relations between nonlinear dissipation and Hamilton–Jacobi inequalities on the one hand and linearized dissipation and Riccati inequalities on the other hand, with applications towards nonlinear optimal and H_∞ control.

Each chapter is ended by a section containing notes on related developments not discussed in the book, and a few pointers to the literature. It should be emphasized that these references are primarily based on personal, biased, interests and insights, and are certainly not meant to be an accurate reflection of the existing literature.

The relation between the chapters can be explained by the following diagram:

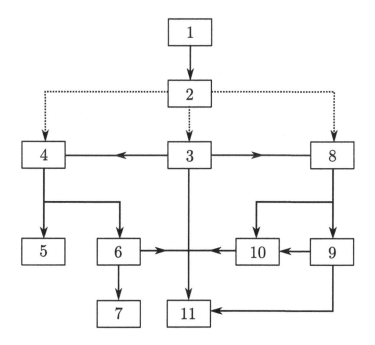

Here dashed lines indicate a motivational background from nonlinear input–output map theory. In particular, a self-contained path through the book is to start in Chap. 3, and then to continue either to Chap. 4 and further (the passivity route), or to continue to Chap. 8 and further (the L_2-gain route). Both routes finally meet in Chap. 11.

Chapter 1
Nonlinear Input–Output Stability

In this chapter, we briefly discuss the basic notions of input–output stability for nonlinear systems described by input–output maps. Also the stability of input–output systems in standard feedback closed-loop configuration is treated.

1.1 Input–Output Maps on Extended L_q-Spaces

The signal spaces under consideration are L_q, $q = 1, 2, \ldots, \infty$, and their extensions:

Definition 1.1.1 (L_q-spaces) For each positive integer $q \in \{1, 2, \ldots\}$, the set $L_q[0, \infty) = L_q$ consists of all functions[1] $f : \mathbb{R}^+ \to \mathbb{R}$ ($\mathbb{R}^+ = [0, \infty)$), which are measurable[2] and satisfy

$$\int_0^\infty |f(t)|^q dt < \infty \qquad (1.1)$$

The set $L_\infty[0, \infty) = L_\infty$ consists of all measurable functions $f : \mathbb{R}^+ \to \mathbb{R}$ which are bounded, i.e.,

$$\sup_{t \in \mathbb{R}^+} |f(t)| < \infty \qquad (1.2)$$

It is well known that L_q are *Banach spaces* (i.e., complete normed linear spaces) with respect to the norms

---

[1] Throughout we will *identify* functions which are equal except for a set of Lebesgue measure zero. Thus conditions imposed on functions are always to be understood in the sense of being valid for all $t \in \mathbb{R}^+$ except for a set of measure zero.

[2] A function $f : \mathbb{R}^+ \to \mathbb{R}$ is measurable if it is the pointwise limit (except for a set of measure zero) of a sequence of piecewise constant functions on $\mathbb{R}^+$.

© Springer International Publishing AG 2017

A. van der Schaft, *L2-Gain and Passivity Techniques in Nonlinear Control*,
Communications and Control Engineering, DOI 10.1007/978-3-319-49992-5_1

$$\|f\|_q = \left(\int_0^\infty |f(t)|^q dt\right)^{\frac{1}{q}} \qquad q = 1, 2, \ldots \qquad (1.3)$$

$$\|f\|_\infty = \sup_{t \in \mathbb{R}^+} |f(t)|$$

Definition 1.1.2 (*Extended L_q-spaces*) Let $f : \mathbb{R}^+ \to \mathbb{R}$. Then for each $T \in \mathbb{R}^+$, the function $f_T : \mathbb{R}^+ \to \mathbb{R}$ is defined by

$$f_T(t) = \begin{cases} f(t), & 0 \le t < T \\ 0, & t \ge T \end{cases} \qquad (1.4)$$

and is called the *truncation* of f to the interval $[0, T]$. For each $q = 1, 2, \ldots, \infty$, the set L_{qe} consists of all measurable functions $f : \mathbb{R}^+ \to \mathbb{R}$ such that $f_T \in L_q$ for all T with $0 \le T < \infty$. L_{qe} is called the *extension* of L_q or the *extended L_q-space*.

Trivially $L_q \subset L_{qe}$. Note that L_{qe} is a linear space but *not* a normed space like L_q. Note also that $\|f_T\|_q$ is an increasing function of T, and that

$$\|f\|_q = \lim_{T \to \infty} \|f_T\|_q \qquad (1.5)$$

whenever $f \in L_q$.

In order to deal with *multi-input multi-output* systems, we consider instead of the one-dimensional space $\mathbb{R}$ any finite-dimensional *linear space* $\mathcal{V}$ endowed with a *norm* $\|\ \|_\mathcal{V}$. Then $L_q(\mathcal{V})$ consists of all measurable functions $f : \mathbb{R}^+ \to \mathcal{V}$ such that

$$\int_0^\infty \|f(t)\|_\mathcal{V}^q dt < \infty, \quad q = 1, 2, \ldots, \infty \qquad (1.6)$$

By defining the norm

$$\|f\|_q := \left(\int_0^\infty \|f(t)\|_\mathcal{V}^q dt\right)^{\frac{1}{q}}, \|f\|_\infty = \sup_{t \in \mathbb{R}^+} \|f\|_\mathcal{V} \qquad (1.7)$$

$L_q(\mathcal{V})$ becomes a Banach space, for any $q = 1, 2, \ldots, \infty$.

The extended space $L_{qe}(\mathcal{V})$ is defined similar to Definition 1.1.2, that is, for $f : \mathbb{R}^+ \to \mathcal{V}$ we define the truncation $f_T : \mathbb{R}^+ \to \mathcal{V}$, and $f \in L_{qe}(\mathcal{V})$ if $f_T \in L_q(\mathcal{V})$ for all $0 \le T \le \infty$.

The cases L_2 and $L_2(\mathcal{V})$ are special. Indeed, in the first case the norm $\|f\|_2$ given in (1.3) is associated with the *inner product*

$$< f, g > = \int_0^\infty f(t)g(t)dt, \ f, g \in L_2$$
$$\|f\|_2 = < f, f >^{\frac{1}{2}}, \qquad f \in L_2 \qquad (1.8)$$

Note that by the Cauchy–Schwartz inequality $< f, g > \leq ||f||_2 ||g||_2$ and thus $< f, g >$ is well-defined. Hence L_2 is a *Hilbert space* (complete linear space with inner product).

Similarly, let $\mathcal{V}$ be a finite-dimensional linear space with an inner product $<, >_\mathcal{V}$. Then $L_2(\mathcal{V})$ becomes a Hilbert space with respect to the inner product

$$< f, g > = \int_0^\infty < f(t), g(t) >_\mathcal{V} dt, \quad f, g \in L_2(\mathcal{V}) \tag{1.9}$$

$$||f||_2 = < f, f >^{\frac{1}{2}}, \quad f \in L_2(\mathcal{V})$$

Let now U be an m-dimensional linear space with norm $||\ ||_U$, and Y be a p-dimensional linear space with norm $||\ ||_Y$. Consider the *input signal space* $L_{qe}(U)$ and the *output signal space* $L_{qe}(Y)$, together with an *input–output mapping*

$$\begin{aligned} G : L_{qe}(U) &\to L_{qe}(Y) \\ u &\mapsto y = G(u) \end{aligned} \tag{1.10}$$

Definition 1.1.3 (*Causal input–output maps*) A mapping $G : L_{qe}(U) \to L_{qe}(Y)$ is said to be *causal* (or *nonanticipating*) if

$$(G(u))_T = (G(u_T))_T, \quad \forall\, T \geq 0,\ u \in L_{qe}(U) \tag{1.11}$$

Lemma 1.1.4 $G : L_{qe}(U) \to L_{qe}(Y)$ *is causal if and only if*

$$u, v \in L_{qe}(U),\ u_T = v_T \Rightarrow (G(u))_T = (G(v))_T,\ \forall\, T \geq 0 \tag{1.12}$$

Lemma 1.1.4 states that G is causal if, whenever two inputs u and v are equal over any interval $[0, T]$, the corresponding outputs are also equal over the same interval.

Definition 1.1.5 (*Time-invariant input–output maps*) A mapping $G : L_{qe}(U) \to L_{qe}(Y)$ is said to be *time invariant* if

$$S_\tau G = G S_\tau, \quad \forall\, \tau \geq 0, \tag{1.13}$$

where $S_\tau : L_{qe}(U) \to L_{qe}(U)$ and $S_\tau : L_{qe}(Y) \to L_{qe}(Y)$ is the shift operator defined as $(S_\tau u)(t) = u(t + \tau)$, $(S_\tau y)(t) = y(t + \tau)$.

Example 1.1.6 Consider the *linear* operator $G : L_{qe} \to L_{qe}$ of convolution type

$$(G(u))(t) = \int_0^\infty h(t, \tau) u(\tau) d\tau \tag{1.14}$$

for some kernel $h(\cdot, \cdot)$. Then G is causal if and only if

$$h(t, \tau) = 0, \qquad t < \tau \tag{1.15}$$

Furthermore, G is time invariant if and only if $h(t, \tau)$ only depends on the difference $t - \tau$.

1.2 L_q-Stability and L_q-Gain; Closed-Loop Stability

The basic definitions of input–output stability are as follows.

Definition 1.2.1 (L_q-stability and finite L_q-gain) Consider $G : L_{qe}(U) \to L_{qe}(Y)$. Then G is said to be L_q-stable if

$$u \in L_q(U) \;\Rightarrow\; G(u) \in L_q(Y), \tag{1.16}$$

i.e., G maps the subset $L_q(U) \subset L_{qe}(U)$ into the subset $L_q(Y) \subset L_{qe}(Y)$.

The map G is said to have *finite L_q-gain* if there exist nonnegative constants γ_q and b_q such that

$$\|(G(u))_T\|_q \;\leq\; \gamma_q \|u_T\|_q + b_q, \qquad \forall\, T \geq 0, \; u \in L_{qe}(U) \tag{1.17}$$

G is said to have *finite L_q-gain with zero bias* if b_q in (1.17) can be taken equal to zero.

Remark 1.2.2 *Linear* input–output maps G with finite L_q-gain have zero bias. Indeed, given (1.17), then for any $\lambda > 0$

$$\lambda\|(G(u))_T\|_q = \|(G(\lambda u))_T\|_q \;\leq\; \gamma_q\|\lambda u_T\|_q + b_q = \lambda\gamma_q\|u_T\|_q + b_q$$

implying $b_q = 0$.

Note that if G has finite L_q-gain then it is automatically L_q-stable. Indeed, taking $u \in L_q(U)$ and letting $T \to \infty$ in (1.17) we obtain

$$\|G(u)\|_q \leq \gamma_q\|u\|_q + b_q, \qquad \forall\, u \in L_q(U) \tag{1.18}$$

implying that $G(u) \in L_q(Y)$ for all $u \in L_q(U)$.

Conversely, for *causal* maps (1.18) implies (1.17):

Proposition 1.2.3 *Let $G : L_{qe}(U) \to L_{qe}(Y)$ be causal and satisfy (1.18). Then G satisfies (1.17) for the same b_q, and thus has finite L_q-gain.*

Proof Let $u \in L_{qe}(U)$, then $u_T \in L_q(U)$ and by (1.18)

$$\|G(u_T)\|_q \leq \gamma_q \|u_T\|_q + b_q$$

Since G is causal, $(G(u))_T = (G(u_T))_T$, and thus

$$\|(G(u))_T\|_q = \|(G(u_T))_T\|_q \leq \|G(u_T)\|_q \leq \gamma_q \|u_T\|_q + b_q$$

$\square$

Definition 1.2.4 (L_q-*gain*) Let $G : L_{qe}(U) \to L_{qe}(Y)$ have finite L_q-gain. Then the L_q-gain of G is defined as

$$\gamma_q(G) := \inf\{\gamma_q \mid \exists\, b_q \text{ such that (1.17) holds}\} \tag{1.19}$$

Remark 1.2.5 Since all norms on a finite-dimensional linear space are equivalent the property of finite L_q-gain is independent of the choice of the norms on U and Y. Of course, the value of the L_q-gain *does* depend on these norms.

Remark 1.2.6 In the linear case the L_2-gain is equal to the induced norm of the input–output map.

A slightly different formulation of finite L_2-gain, to be used later on in Chaps. 2, 3, and 9, can be given as follows. By Definition 1.2.1, an input–output map $G : L_{2e}(U) \to L_{2e}(Y)$ has finite L_2-gain if there exist constants γ and b such that

$$\|(G(u))_T\|_2 \leq \gamma \|u_T\|_2 + b, \quad \forall u \in L_{2e}(U), \ \forall T \geq 0 \tag{1.20}$$

Alternatively, we may consider the inequality

$$\|(G(u))_T\|_2^2 \leq \tilde{\gamma}^2 \|u_T\|_2^2 + c, \quad \forall u \in L_{2e}(U), \ \forall T \geq 0 \tag{1.21}$$

for some nonnegative constants $\tilde{\gamma}$ and c.

Proposition 1.2.7 *If (1.20) holds, then for every $\tilde{\gamma} > \gamma$ there exists c such that (1.21) holds, and conversely, if (1.21) holds, then for every $\gamma > \tilde{\gamma}$ there exists b such that (1.20) holds. In particular, the L_2-gain $\gamma(G)$ defined in Definition 1.2.4, cf. (1.19), is alternatively given as*

$$\gamma(G) = \inf\{\tilde{\gamma} \mid \exists c \text{ such that (1.21) holds }\} \tag{1.22}$$

Proof Recall that (1.20) amounts to (with $y = G(u)$)

$$\left(\int_0^T ||y(t)||^2 dt \right)^{\frac{1}{2}} \leq \gamma \left(\int_0^T ||u(t)||^2 dt \right)^{\frac{1}{2}} + b,$$

while (1.21) amounts to

$$\int_0^T ||y(t)||^2 dt \leq \tilde{\gamma}^2 \int_0^T ||u(t)||^2 dt + c$$

Denote $Y = \int_0^T ||y(t)||^2 dt$, $U = \int_0^T ||u(t)||^2 dt$, and start from the inequality

$$Y^{\frac{1}{2}} \leq \gamma U^{\frac{1}{2}} + b$$

Then by quadrature $Y \leq \gamma^2 U + 2\gamma b U^{\frac{1}{2}} + b^2$. Let now $\tilde{\gamma} > \gamma$, then $(\gamma^2 - \tilde{\gamma}^2)U + 2\gamma b U^{\frac{1}{2}} + b^2$ as a function of U is bounded from above by some constant c, and thus

$$Y \leq \tilde{\gamma}^2 U + c$$

Conversely, if $Y \leq \tilde{\gamma}^2 U + c$, then $Y^{\frac{1}{2}} \leq (\tilde{\gamma}^2 U + c)^{\frac{1}{2}}$, and for any $\gamma > \tilde{\gamma}$ there exists b such that $(\tilde{\gamma}^2 U + c)^{\frac{1}{2}} \leq \gamma U^{\frac{1}{2}} + b$, whence $Y^{\frac{1}{2}} \leq \gamma U^{\frac{1}{2}} + b$.   □

A special case of finite L_2-gain is defined as follows.

Definition 1.2.8 (*Inner map*) A map $G : L_{2e}(U) \rightarrow L_{2e}(Y)$ is called *inner* if it has L_2-gain ≤ 1, and furthermore

$$||G(u)||_2^2 = ||u||_2^2 + c, \quad \text{for all } u \in L_2(U) \tag{1.23}$$

for some constant c.

Note that by Proposition 1.2.3 a *causal* map $G : L_{2e}(U) \rightarrow L_{2e}(Y)$ is inner if and only if it satisfies (1.23). Also note that (1.23) implies that $c = ||G(0)||_2^2$ with 0 the zero input function. We obtain the following characterization of inner causal maps, which will be used in Chaps. 8 and 9.

Proposition 1.2.9 *A causal L_2-stable map $G : L_{2e}(U) \rightarrow L_{2e}(Y)$ is inner if and only if*

$$(DG(u))^* \circ G(u) = u, \text{ for all } u \in L_2(U), \tag{1.24}$$

where $DG(u)$ denotes the Fréchet derivative of G at $u \in L_2(U)$, and $(DG(u))^*$ its adjoint map with respect to the L_2 inner product.

Proof (Only if) Differentiation of (1.23) with respect to u in the direction $h \in L_2(U)$ yields

$$< DG(u)^* \circ G(u), h > = < u, h >, \tag{1.25}$$

which by arbitrariness of h implies (1.24). (If) Note that

$$
\begin{aligned}
&< u, u > - < G(u), G(u) > + < G(0), G(0) > \\
&= \int_0^1 \tfrac{d}{ds}[< su, su > - < G(u), G(u) >]\,ds \\
&= 2 \int_0^1 [< su, u > - < DG(su)^* \circ G(su), u >]\,ds
\end{aligned}
\tag{1.26}
$$

Hence (1.23) follows from (1.24).   □

It is sometimes useful to generalize input–output *maps* to *relations*. Let R be a subset of $L_{qe}(U) \times L_{qe}(Y)$. Then we say that $u \in L_{qe}(U)$ is *related* to $y \in L_{qe}(Y)$ via the *relation* R if $(u, y) \in R \subset L_{qe}(U) \times L_{qe}(Y)$.

Definition 1.2.10 (*L_q-stable relations*) The relation $R \subset L_{qe}(U) \times L_{qe}(Y)$ is said to be *L_q-stable* if

$$
(u, y) \in R, \quad u \in L_q(U) \quad \Rightarrow \quad y \in L_q(Y),
\tag{1.27}
$$

while R is said to have *finite L_q-gain* if $\exists\, \gamma_q, b_q$ such that for all $T \geq 0$

$$
(u, y) \in R, \quad u \in L_{qe}(U) \quad \Rightarrow \quad \|y_T\|_q \leq \gamma_q \|u_T\|_q + b_q,
\tag{1.28}
$$

and its L_q-gain is denoted as $\gamma_q(R)$.

Any map $G : L_{qe}(U) \to L_{qe}(Y)$ defines a relation R_G, namely

$$
R_G = \{(u, G(u)) \mid u \in L_{qe}(U)\}
\tag{1.29}
$$

The converse, however, need not to be true; for a particular $u \in L_{qe}(U)$ there may not exist an $y \in L_{qe}(Y)$ such that $(u, y) \in R$, or, alternatively, there may exist *many* such y.

So far we have discussed *open-loop* stability. For *closed-loop* stability we look at the standard negative feedback configuration of Fig. 1.1, denoted by $G_1 \|_f G_2$, where $G_1 : L_{qe}(U_1) \to L_{qe}(Y_1)$, $G_2 : L_{qe}(U_2) \to L_{qe}(Y_2)$ are input–output maps,

Fig. 1.1 Standard feedback configuration $G_1 \|_f G_2$

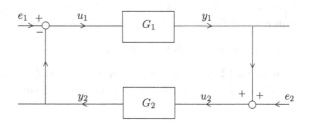

and $U_1 = Y_2 =: E_1$, $U_2 = Y_1 =: E_2$. Furthermore, $e_1 \in L_{qe}(E_1)$, $e_2 \in L_{qe}(E_2)$ represent *external* input signals injected in the closed-loop configuration. The closed-loop system $G_1 \|_f G_2$ is thus described by the equations

$$u_1 = e_1 - y_2 , \quad u_2 = e_2 + y_1$$
$$y_1 = G_1(u_1) , \quad y_2 = G_2(u_2) \tag{1.30}$$

or, more compactly,

$$u = e - Fy, \quad y = G(u), \tag{1.31}$$

with

$$u = \begin{bmatrix} u_1 \\ u_2 \end{bmatrix}, y = \begin{bmatrix} y_1 \\ y_2 \end{bmatrix}, e = \begin{bmatrix} e_1 \\ e_2 \end{bmatrix},$$
$$F = \begin{bmatrix} 0 & I_{m_1} \\ -I_{m_2} & 0 \end{bmatrix}, G = \begin{bmatrix} G_1 & 0 \\ 0 & G_2 \end{bmatrix} \tag{1.32}$$

($\dim U_i = m_i$, $\dim Y_i = p_i$, $i = 1, 2$, where $m_2 = p_1, m_1 = p_2$).

The closed-loop system $G_1 \|_f G_2$ defines two relations. Indeed, eliminate y from (1.31) to obtain

$$u = e - FG(u) \tag{1.33}$$

leading to the relation

$$R_{eu} = \{(e, u) \in L_{qe}(E_1 \times E_2) \times L_{qe}(U_1 \times U_2) \mid u + FG(u) = e\} \tag{1.34}$$

Alternatively, eliminate u from (1.31) to obtain

$$y = G(e - Fy) \tag{1.35}$$

and the relation

$$R_{ey} = \{(e, y) \in L_{qe}(E_1 \times E_2) \times L_{qe}(Y_1 \times Y_2) \mid y = G(e - Fy)\} \tag{1.36}$$

Definition 1.2.11 (*Closed-loop L_q-stability*) The closed-loop system $G_1 \|_f G_2$ is L_q-*stable* if both R_{eu} and R_{ey} are L_q-stable relations. $G_1 \|_f G_2$ has finite L_q-gain if both R_{eu} and R_{ey} have finite L_q-gain.

Actually, the situation is more simple:

Lemma 1.2.12 (a) R_{eu} is L_q-stable $\iff$ R_{ey} is L_q-stable.
(b) R_{eu} has finite L_q-gain $\iff$ R_{ey} has finite L_q-gain.

Proof (a) Suppose R_{eu} is L_q-stable. Let $(e, y) \in R_{ey}$. By (1.31)

$$(e, e - Fy) \in R_{eu} \tag{1.37}$$

Let $e \in L_q(E_1 \times E_2)$. Then, since R_{eu} is L_q-stable, $e - Fy \in L_q(E_1 \times E_2)$. Since F is a nonsingular matrix this implies that $y \in L_q(Y_1 \times Y_2)$.

Conversely, suppose R_{ey} is L_q-stable. Let $(e, u) \in R_{eu}$, and take $e \in L_q(E_1 \times E_2)$. Then $(e, y = G(u)) \in R_{ey}$, and by L_q-stability of R_{ey}, $y = G(u) \in L_q(Y_1 \times Y_2)$. Since F is a constant matrix this implies that $u = e - Fy \in L_q(U_1 \times U_2)$. Part (b) follows similarly using the triangle inequality $||a + b||_q \leq ||a||_q + ||b||_q$ for any norm $|| \ ||_q$. □

Remark 1.2.13 Note that the implication ($\Rightarrow$) hinges upon the nonsingularity of the interconnection matrix F. This may not be valid anymore for more general feedback configurations. In particular, for $e_2 = 0$ the L_q-stability or finite L_q-gain of $R_{e_1 y_1}$ and of $R_{e_1 u_1}$ are *not* equivalent; see also Remark 2.2.17.

With regard to causality we have the following simple observation.

Proposition 1.2.14 *Let G_1 and G_2 be causal input–output mappings. Then also $G_1 \|_f G_2$ is causal in the sense that for $(e, u) \in R_{eu}$, u_T only depends on e_T, and for $(e, y) \in R_{ey}$, y_T only depends on e_T, $\forall T \geq 0$.*

Proof By causality of G_1 and G_2, y_{1T} only depends on e_{1T} and y_{2T}, while y_{2T} only depends on e_{2T} and y_{1T}. Thus y_{1T}, y_{2T} only depend on e_{1T}, e_{2T}. □

The relations R_{eu} and R_{ey} as defined above do not necessarily correspond to mappings from e to u, respectively, from e to y. Indeed, solving (1.33) for u would correspond to

$$u = (I + FG)^{-1}e, \tag{1.38}$$

but the inverse of $I + FG$ need not exist. Thus for arbitrary external signals $e_1 \in L_{qe}(E_1)$, $e_2 \in L_{qe}(E_2)$ there do not necessarily exist internal signals $u_1 \in L_{qe}(U_1)$, $u_2 \in L_{qe}(U_2)$, as well as $y_1 \in L_{qe}(Y_1)$, $y_2 \in L_{qe}(Y_2)$. This constitutes a main bottleneck of the theory; see also the developments in the next Chap. 2 around an incremental form of the small-gain theorem.

Remark 1.2.15 If R_{ey} defines an input–output mapping G from $e = (e_1, e_2)$ to (y_1, y_2), and G_1 and G_2 are causal, then by Proposition 1.2.14 G is a causal map. The same holds for R_{eu}.

1.3 Input–Output Maps from State Space Models

Let us now take a different point of view (to be continued in Chap. 3 and all subsequent chapters), by starting from *state space systems*

$$\Sigma \ : \ \begin{matrix} \dot{x} = f(x, u) \ , \ u \in U \\[4pt] y = h(x, u) \ , \ y \in Y \end{matrix} \qquad (1.39)$$

with U, Y normed finite-dimensional linear spaces (of dimension m and p, respectively), and with $x = (x_1, \ldots, x_n)$ local coordinates for some n-dimensional state space manifold $\mathcal{X}$. Furthermore, f, h are sufficiently smooth mappings. For every initial condition $x_0 \in \mathcal{X}$ this defines, *in principle*, an input–output map $G_{x_0} : L_{qe}(U) \rightarrow L_{qe}(Y)$, by substituting any input function $u(\cdot) \in L_{qe}(U)$ in $\dot{x} = f(x, u)$, solving these differential equations for the initial condition $x(0) = x_0$, and substituting $u(\cdot)$ and the resulting state space trajectory $x(\cdot)$ in $y = h(x, u)$ in order to obtain the output function $y(\cdot)$. However, the differential equations may have finite escape time, and in general additional conditions are necessary to ensure that[3] $y \in L_{qe}(Y)$ for every $u \in L_{qe}(U)$.

In the following it will be assumed that for every x_0 the input–output map $G_{x_0} : L_{qe}(U) \rightarrow L_{qe}(Y)$ is indeed well-defined. Definition 1.2.1 then extends as follows.

Definition 1.3.1 (L_q-*stability of state space system*) The state space system Σ is L_q-*stable* if for all initial conditions $x_0 \in \mathcal{X}$ the input–output map G_{x_0} maps $L_q(U)$ into $L_q(Y)$. The system is Σ is said to have finite L_q-*gain* if there exists a finite constant γ_q such that for every initial condition x_0 there exists a finite constant $b_q(x_0)$ with the property that

$$||(G_{x_0}(u))_T||_q \leq \gamma_q ||u_T||_q + b_q(x_0) \,, \quad \forall T \geq 0, \ u \in L_{qe}(U) \qquad (1.40)$$

Furthermore, let us replace in the feedback configuration $G_1 \|_f G_2$ of Fig. 1.1 the input–output maps $G_i, i = 1, 2$, by state space systems

$$\Sigma_i \ : \ \begin{matrix} \dot{x}_i = f_i(x_i, u_i) \ , \ u_i \in U_i \\[4pt] y_i = h_i(x_i, u_i) \ , \ y_i \in Y_i \end{matrix} \ , \quad x_i \in \mathcal{X}_i, \ i = 1, 2 \qquad (1.41)$$

with $U_1 = Y_2, U_2 = Y_1$, and consider again the feedback interconnection

$$u_1 = -y_2 + e_1, \ u_2 = y_1 + e_2 \qquad (1.42)$$

The closed-loop system with inputs e_1, e_2 and outputs y_1, y_2 will be denoted by $\Sigma_1 \|_f \Sigma_2$. For every pair of initial conditions $x_{0i} \in \mathcal{X}_i, i = 1, 2$, we can define the relations $R_{eu}^{x_{01}, x_{02}}$ and $R_{ey}^{x_{01}, x_{02}}$ as in (1.34), respectively (1.36).

Definition 1.3.2 The closed-loop system $\Sigma_1 \|_f \Sigma_2$ is L_q-*stable* if for every pair $(x_{01}, x_{02}) \in \mathcal{X}_1 \times \mathcal{X}_2$ the relations $R_{eu}^{x_{01}, x_{02}}$ and $R_{ey}^{x_{01}, x_{02}}$ (or equivalently, see Lemma 1.2.12, one of them) are L_q-stable.

---

[3]Note the abuse of notation, with, e.g., $u \in U$ denoting the value of the input and on the other hand $u \in L_{qe}(U)$ denoting a time function $u : \mathbb{R}^+ \rightarrow U$.

In the state space context it is relatively easy to ensure that for any pair of initial conditions $x_{0i} \in \mathcal{X}_i$, $i = 1, 2$, the relations $R_{eu}^{x_{01}, x_{02}}$ and $R_{ey}^{x_{01}, x_{02}}$ correspond to *mappings* from e to u, respectively, from e to y. Indeed, the closed-loop system $\Sigma_1 \|_f \Sigma_2$ is described by the equations

$$
\begin{cases}
\dot{x}_1 = f_1(x_1, u_1) \\
\dot{x}_2 = f_2(x_2, u_2)
\end{cases}
$$
$$
\begin{cases}
y_1 = h_1(x_1, u_1) \\
y_2 = h_2(x_2, u_2)
\end{cases}
\tag{1.43}
$$
$$
\begin{cases}
e_1 = h_2(x_2, u_2) + u_1 \\
e_2 = -h_1(x_1, u_1) + u_2
\end{cases}
$$

If at least one of the two mappings $h_i(x_i, u_i)$, $i = 1, 2$, does not depend on u_i, then we may immediately eliminate u_1, u_2. For instance, if h_1 does not depend on u_1, then (1.43) can be rewritten as

$$
\begin{cases}
\dot{x}_1 = f_1(x_1, e_1 - h_2(x_2, e_2 + h_1(x_1))) \\
\dot{x}_2 = f_2(x_2, e_2 + h_1(x_1))
\end{cases}
$$
$$
\begin{cases}
y_1 = h_1(x_1) \\
y_2 = h_2(x_2, e_2 + h_1(x_1))
\end{cases}
\tag{1.44}
$$
$$
\begin{cases}
u_1 = e_1 - h_2(x_2, e_2 + h_1(x_1)) \\
u_2 = e_2 + h_1(x_1)
\end{cases}
$$

and, under suitable technical conditions as alluded to above, this will define input–output mappings from e to u, and from e to y. If both h_1 and h_2 depend on u_1, respectively u_2, then conditions have to be imposed in order that the static map from u to e given by the last two equations of (1.43) has (at least locally) an inverse. This can be done by considering the Jacobian matrix of the map from u to e, given by

$$
\begin{bmatrix}
I_{m_1} & \frac{\partial h_2}{\partial u_2}(x_2, u_2) \\
-\frac{\partial h_1}{\partial u_1}(x_1, u_1) & I_{m_2}
\end{bmatrix}
\tag{1.45}
$$

Invertibility of this Jacobian for all x_1, x_2, u_1, u_2 ensures by the Inverse Function theorem that locally u can be expressed as a function of e. Finally, it is easy to see that invertibility of the Jacobian in (1.45) is equivalent to invertibility of the matrices $I_{m_2} + \frac{\partial h_1}{\partial u_1}(x_1, u_1)\frac{\partial h_2}{\partial u_2}(x_2, u_2)$ or $I_{m_1} + \frac{\partial h_2}{\partial u_2}(x_2, u_2)\frac{\partial h_1}{\partial u_1}(x_1, u_1)$, for all x_1, x_2, u_1, u_2.

1.4 Notes for Chapter 1

1. The exposition of the classical material in this chapter is largely based on Vidyasagar [343].

2. Proposition 1.2.9 is stated in Ball & van der Schaft [24]. See also Scherpen & Gray [309] and Fujimoto, Scherpen & Gray [108] for a study of the closely related notion of the nonlinear Hilbert adjoint of an input–output map, and the application toward the analysis of the nonlinear Hankel singular values. In Chap. 9 we will approach the adjoint of the Fréchet derivative from the state space point of view, based on the Hamiltonian extension introduced in Crouch & van der Schaft [73].

3. Various theory has been developed for shedding light on the invertibility of the map $I + FG$ in (1.38); see, e.g., Willems [348].

4. In Chaps. 3, 4 and 8 a synthesis will be provided between the stability of the input–output maps G_{x_0} of a state space system Σ, and Lyapunov stability of the state space system $\dot{x} = f(x, u)$ from the vantage point of dissipative systems theory.

5. The relation to (integral) Input-to-State Stability theory developed by Sontag and coworkers will be addressed in Chap. 8, Sect. 8.5.

Chapter 2
Small-Gain and Passivity for Input–Output Maps

In this chapter we give the basic versions of the classical *small-gain* (Sect. 2.1) and *passivity* theorems (Sect. 2.2) in the study of closed-loop stability. Section 2.3 briefly touches upon the "loop transformations" which can be used to expand the domain of applicability of the small-gain and passivity theorems. Finally, Sect. 2.4 deals with the close relation between passivity and L_2-gain via the scattering representation.

2.1 The Small-Gain Theorem

A straightforward, but very important, theorem is as follows.

Theorem 2.1.1 (Small-gain theorem) *Consider the closed-loop system $G_1 \|_f G_2$ given in Fig. 1.1, and let $q \in \{1, 2, \ldots, \infty\}$. Suppose that G_1 and G_2 have L_q-gains $\gamma_q(G_1)$, respectively $\gamma_q(G_2)$. Then the closed-loop system $G_1 \|_f G_2$ has finite L_q-gain (see Definition 1.2.11) if*

$$\gamma_q(G_1) \cdot \gamma_q(G_2) \; < \; 1 \tag{2.1}$$

Remark 2.1.2 Inequality (2.1) is known as the *small-gain condition*. Two stable systems G_1 and G_2 which are interconnected as in Fig. 1.1 result in a stable closed-loop system provided the *"loop gain"* is "small" (i.e., less than 1). Note that the small-gain theorem implies an inherent *robustness* property: the closed-loop system remains stable for all *perturbed* input–output maps, as long as the small-gain condition remains satisfied.

Proof By the definition of $\gamma_q(G_1)$, $\gamma_q(G_2)$ and (2.1) there exist constants $\gamma_{1q}, \gamma_{2q}, b_{1q}, b_{2q}$ with $\gamma_{1q} \cdot \gamma_{2q} < 1$, such that for all $T \geq 0$

© Springer International Publishing AG 2017   13
A. van der Schaft, *L₂-Gain and Passivity Techniques in Nonlinear Control*,
Communications and Control Engineering, DOI 10.1007/978-3-319-49992-5_2

$$\|(G_1(u_1))_T\|_q \leq \gamma_{1q}\|u_{1T}\|_q + b_{1q} , \qquad \forall u_1 \in L_{qe}(U_1)$$
$$\|(G_2(u_2))_T\|_q \leq \gamma_{2q}\|u_{2T}\|_q + b_{2q} , \qquad \forall u_2 \in L_{qe}(U_2) \qquad (2.2)$$

For simplicity of notation we will drop the subscripts "q." Since $u_{1T} = e_{1T} - (G_2(u_2))_T$

$$\|u_{1T}\| \leq \|e_{1T}\| + \|(G_2(u_2))_T\| \leq \|e_{1T}\| + \gamma_2\|u_{2T}\| + b_2$$
$$\|u_{2T}\| \leq \|e_{2T}\| + \|(G_1(u_1))_T\| \leq \|e_{2T}\| + \gamma_1\|u_{1T}\| + b_1.$$

Combining these two inequalities, using the fact that $\gamma_2 \geq 0$, yields

$$\|u_{1T}\| \leq \gamma_1\gamma_2\|u_{1T}\| + (\|e_{1T}\| + \gamma_2\|e_{2T}\| + b_2 + \gamma_2 b_1).$$

Since $\gamma_1\gamma_2 < 1$ this implies

$$\|u_{1T}\| \leq (1 - \gamma_1\gamma_2)^{-1}(\|e_{1T}\| + \gamma_2\|e_{2T}\| + b_2 + \gamma_2 b_1). \qquad (2.3)$$

Similarly we derive

$$\|u_{2T}\| \leq (1 - \gamma_1\gamma_2)^{-1}(\|e_{2T}\| + \gamma_1\|e_{1T}\| + b_1 + \gamma_1 b_2). \qquad (2.4)$$

This proves finite L_q-gain of the relation R_{eu}, and thus by Lemma 1.2.12 finite L_q-gain of $G_1\|_f G_2$. □

Remark 2.1.3 Note that in (2.3) and (2.4) we have actually derived a bound on the L_q-gain of the relation R_{eu}. Substituting $y_1 = G_1(u_1)$, $y_2 = G_2(u_2)$, and combining (2.2) with (2.3) and (2.4), we also obtain the following bound on the L_q-gain of the relation R_{ey}:

$$\|y_{1T}\| \leq (1 - \gamma_1\gamma_2)^{-1}\gamma_1(\|e_{1T}\| + \gamma_2\|e_{2T}\| + b_2 + \gamma_2 b_1) + b_1$$
$$\|y_{2T}\| \leq (1 - \gamma_1\gamma_2)^{-1}\gamma_2(\|e_{2T}\| + \gamma_1\|e_{1T}\| + b_1 + \gamma_1 b_2) + b_2. \qquad (2.5)$$

Remark 2.1.4 Theorem 2.1.1 remains valid for *relations* $R_{u_1 y_1}$ and $R_{u_2 y_2}$, instead of maps G_1 and G_2.

Note that in many situations, e_1 and e_2 are *given* and u_1, u_2 (as well as y_1, y_2) are *derived*. The above formulation of the small-gain theorem (as well as the definition of L_q-stability of the closed-loop system $G_1\|_f G_2$, cf. Definition 1.2.11) *avoids* the question of *existence* of solutions $u_1 \in L_{qe}(U_1)$, $u_2 \in L_{qe}(U_2)$ to $e_1 = u_1 + G_2(u_2)$, $e_2 = u_2 - G_1(u_1)$ for given $e_1 \in L_{qe}(E_1)$, $e_2 \in L_{qe}(E_2)$. As we will see, a *stronger* version of the small-gain theorem does also answer this question, as well as some other issues. First, we extend the definition of L_q-gain to its *incremental* version.

Definition 2.1.5 *(Incremental L_q-gain)* The input–output map $G : L_{qe}(U) \to L_{qe}(Y)$ is said to have *finite incremental L_q-gain* if there exists a constant $\Gamma_q \geq 0$ such that

$$\|(G(u))_T - (G(v))_T\|_q \leq \Gamma_q \|u_T - v_T\|_q , \quad \forall T \geq 0, \ u, v \in L_{qe}(U) \qquad (2.6)$$

Furthermore, its *incremental L_q-gain* $\Gamma_q(G)$ is defined as the infimum over all such Γ_q.

The property of finite incremental L_q-gain is seen to imply *causality*.

Proposition 2.1.6 *Let $G : L_{qe}(U) \to L_{qe}(Y)$ have finite incremental L_q-gain. Then it is causal.*

Proof Let $u, v \in L_{qe}(U)$ be such that $u_T = v_T$. Then by (2.6)

$$\|(G(u))_T - (G(v))_T\|_q \ \leq \ \Gamma_q \|u_T - v_T\|_q = 0,$$

and thus $(G(u))_T = (G(v))_T$, implying by Lemma 1.1.4 causality of G. $\square$

Remark 2.1.7 Hence, finite incremental L_q-gain for causal maps is the same as requiring that for all $T \geq 0$

$$\|(G(u_T))_T - (G(v_T))_T\|_q \leq \Gamma_q \|u_T - v_T\|_q , \quad \forall u, v \in L_{qe}(U) \qquad (2.7)$$

Theorem 2.1.8 (Incremental form of small-gain theorem) *Let $G_1 : L_{qe}(U_1) \to L_{qe}(Y_1)$, $G_2 : L_{qe}(U_2) \to L_{qe}(Y_2)$ be input–output maps with incremental L_q-gains $\Gamma_q(G_1)$, respectively $\Gamma_q(G_2)$. Consider the closed-loop system $G_1 \|_f G_2$. Then, if $\Gamma_q(G_1) \cdot \Gamma_q(G_2) < 1$,*

(i) *For all $(e_1, e_2) \in L_{qe}(E_1 \times E_2)$ there exists a unique solution $(u_1, u_2, y_1, y_2) \in L_{qe}(U_1 \times U_2 \times Y_1 \times Y_2)$.*

(ii) *The map $(e_1, e_2) \mapsto (u_1, u_2)$ is uniformly continuous on the space $L_{qe}(E_1 \times E_2)$.*

(iii) *If the solution (u_1, u_2) to $e_1 = e_2 = 0$ is in $L_q(U_1 \times U_2)$, then $(e_1, e_2) \in L_q(E_1 \times E_2)$ implies that $(u_1, u_2) \in L_q(U_1 \times U_2)$.*

Proof First we note that since $\Gamma_q(G_1) \cdot \Gamma_q(G_2) < 1$, there exist constants Γ_{1q}, Γ_{2q} with $\Gamma_{1q} \cdot \Gamma_{2q} < 1$ such that for all $T \geq 0$ and for all $u_1, v_1 \in L_{qe}(U_1), u_2, v_2 \in L_{qe}(U_2)$

$$\begin{aligned} \|(G_1(u_1))_T - (G_1(v_1))_T\|_q &\leq \Gamma_{1q} \|u_{1T} - v_{1T}\|_q \\ \|(G_2(u_2))_T - (G_2(v_2))_T\|_q &\leq \Gamma_{2q} \|u_{2T} - v_{2T}\|_q \end{aligned} \qquad (2.8)$$

Furthermore, by Proposition 2.1.6 G_1, G_2 are causal. The statements (i), (ii) and (iii) are now proved as follows.

(i) Since $u_2 = e_2 + G_1(e_1 - G_2(u_2))$ it follows that

$$u_{2T} = e_{2T} + [G_1(e_1 - G_2(u_2))]_T$$

Using causality of G_1 and G_2 this yields

$$u_{2T} = e_{2T} + \{G_1[e_{1T} - (G_2(u_{2T}))_T]\}_T \tag{2.9}$$

For every e_1, e_2 this is an equation of the form $u_{2T} = C(u_{2T})$. We claim that C is a *contraction* on $L_{q,[0,T]}(U_2)$ (the space of L_q-functions on $[0, T]$). Indeed for all $u_{2T}, v_{2T} \in L_{q,[0,T]}(U_2)$

$$\begin{aligned}
&\|G_1[e_{1T} - (G_2(u_{2T}))_T] - G_1[e_{1T} - (G_2(v_{2T}))_T]\|_{q,[0,T]} \\
&\leq \Gamma_{1q}\|(G_2(v_{2T}))_T - (G_2(u_{2T}))_T\|_q \leq \Gamma_{1q} \cdot \Gamma_{2q}\|u_{2T} - v_{2T}\|_q
\end{aligned}$$

by (2.8). By assumption $\Gamma_{1q} \cdot \Gamma_{2q} < 1$, and thus C is a contraction. Therefore, for all $T \geq 0$, and all $(e_1, e_2) \in L_{qe}(E_1 \times E_2)$, there is a uniquely defined element of $u_{2T} \in L_{q,[0,T]}(U_2)$ solving $u_{2T} = C(u_{2T})$. The same holds trivially for u_{1T} since

$$u_{1T} = e_{1T} - (G_2(u_{2T}))_T$$

Thus for all $(e_1, e_2) \in L_{qe}(E_1 \times E_2)$ there exists a unique solution $(u_1, u_2) \in L_{qe}(U_1 \times U_2)$ to (1.30).

(ii) Since $u_{1T} = e_{1T} - (G_2(u_{2T}))_T$, $u'_{1T} = e'_{1T} - (G_2(u'_{2T}))_T$ we obtain by subtraction and the triangle inequality

$$\|u_{1T} - u'_{1T}\| \leq \|e_{1T} - e'_{1T}\| + \Gamma_{2q}\|u_{2T} - u'_{2T}\|$$

for all (e_1, e_2), (e'_1, e'_2) and corresponding solutions (u_1, u_2), (u'_1, u'_2). Similarly

$$\|u_{2T} - u'_{2T}\| \leq \|e_{2T} - e'_{2T}\| + \Gamma_{1q}\|u_{1T} - u'_{1T}\|$$

and thus

$$\|u_{1T} - u'_{1T}\| \leq (1 - \Gamma_{1q}\Gamma_{2q})^{-1}(\|e_{1T} - e'_{1T}\| + \Gamma_{2q}\|e_{2T} - e'_{2T}\|), \tag{2.10}$$

and similarly for $\|u_{2T} - u'_{2T}\|$. This yields (ii).

(iii) Insert $e'_1 = e'_2 = 0$ in (2.10) and in the same inequality for the expression $\|u_{2T} - u'_{2T}\|$. $\qquad\square$

Remark 2.1.9 For a *linear* map G, property (2.6) is equivalent to

$$\|(G(u))_T\|_q \leq \Gamma_q\|u_T\|_q$$

and thus to the property that G has L_q-gain $\leq \Gamma_q$ (with zero bias). Note also that in this case the solution to $e_1 = e_2 = 0$ is $u_1 = u_2 = 0$, and thus (iii) is always satisfied.

2.2 Passivity and the Passivity Theorems

While the small-gain theorem is naturally concerned with *normed* (finite-dimensional) linear spaces $\mathcal{V}$ and the corresponding Banach spaces $L_q(\mathcal{V})$ for every $q = 1, 2, \ldots, \infty$, passivity is, *at least in first instance*, independent of any norm, but, at the same time, requires a duality between the input and output space.

Indeed, let us consider any finite-dimensional linear input space U (of dimension m), and let the output space Y be the *dual space* U^* (the set of linear functions on U). Denote the duality product between U and $U^* = Y$ by $< y \mid u >$ for $y \in U^*, u \in U$. (That is, $< y \mid u >$ is the linear function $y : U \to \mathbb{R}$ evaluated at $u \in U$.) Furthermore, take any linear space of functions $u : \mathbb{R}^+ \to U$, denoted by $L(U)$, and any linear space of functions $y : \mathbb{R}^+ \to Y = U^*$, denoted by $L(U^*)$. Define the extended spaces $L_e(U)$, respectively $L_e(U^*)$, similar to Definition 1.1.2, that is, $u \in L_e(U)$ if $u_T \in L(U)$ for all $T \geq 0$ and $y \in L_e(U^*)$ if $y_T \in L(U^*)$ for all $T \geq 0$. Define a duality pairing between $L_e(U)$ and $L_e(U^*)$ by defining for $u \in L_e(U), y \in L_e(U^*)$

$$< y \mid u >_T := \int_0^T < y(t) \mid u(t) > dt, \tag{2.11}$$

assuming that integral on the right-hand side exists. In examples, the duality product $< y(t) \mid u(t) >$ usually is the (instantaneous) *power* (electrical power if the components of u, y are voltages and currents, or mechanical power if the components of u, y are forces and velocities). In these cases, $< y \mid u >_T$ will denote the externally *supplied energy* during the time interval $[0, T]$.

Definition 2.2.1 *(Passive input–output maps)* Let $G : L_e(U) \to L_e(U^*)$. Then G is *passive* if there exists some constant β such that

$$< G(u) \mid u >_T \geq -\beta, \quad \forall u \in L_e(U), \quad \forall T \geq 0, \tag{2.12}$$

where additionally it is assumed that the left-hand side of (2.12) is well defined.

Note that (2.12) can be rewritten as

$$- < G(u) \mid u >_T \leq \beta, \quad \forall u \in L_e(U), \quad \forall T \geq 0, \tag{2.13}$$

with the interpretation that the *maximally extractable energy* is bounded by a finite constant β. Hence, G is passive iff only a *finite amount* of energy can be extracted from the system defined by G. This interpretation, together with its ramifications, will become more clear in Chaps. 3 and 4.

Definition 2.2.1 directly extends to relations.

Definition 2.2.2 *(Passive relation)* A relation $R \subset L_e(U) \times L_e(U^*)$ is said to be passive if $< y \mid u >_T \geq -\beta$, for all $(u, y) \in R$ and $T \geq 0$, assuming that $< y \mid u >_T$ is well defined for all $(u, y) \in R$ and all $T \geq 0$.

Remark 2.2.3 In many applications $L_e(U)$ will be defined as $L_{2e}(U)$ for some norm $\|\ \|_U$ on U. Then $L_e(U^*)$ can be taken to be $L_{2e}(U^*)$, with $\|\ \|_{U^*}$ the norm on U^* canonically induced by $\|\ \|_U$, that is,

$$\|y\|_{U^*} := \max_{u \neq 0} \frac{< y \mid u >}{\|u\|_U}.$$

This implies $| < y \mid u > | \leq \|y\|_{U^*} \cdot \|u\|_U$, yielding

$$| < G(u) \mid u >_T | = | \int_0^T < G(u)(t) \mid u(t) > dt | \leq \\ \left(\int_0^T \|G(u)(t)\|_{U^*}^2 dt \right)^{\frac{1}{2}} \cdot \left(\int_0^T \|u(t)\|_U^2 dt \right)^{\frac{1}{2}}. \tag{2.14}$$

Hence, in this case the left-hand side of (2.12) is automatically well defined. The same holds for a passive relation $R \subset L_{2e}(U) \times L_{2e}(U^*)$

Remark 2.2.4 For a *linear* single-input single-output map the property of passivity is equivalent to the *phase shift* of an input sinusoid being always less than or equal to 90° (see e.g., [343]). This should be contrasted with the L_q-gain of a linear input–output map, which deals with the amplification of the input signal.

Similarly to Proposition 1.2.3 we have the following alternative formulation of passivity for *causal* maps G.

Proposition 2.2.5 *Let $G : L_e(U) \to L_e(U^*)$ satisfy (2.12). Then also*

$$< G(u) \mid u > \ \geq \ -\beta \ , \quad \forall u \in L(U), \tag{2.15}$$

if the left-hand side of (2.15) is well defined. Conversely, if G is causal, then (2.15) implies (2.12).

Proof Suppose (2.12) holds. By letting $T \to \infty$ we obtain (2.15) for $u \in L(U)$. Conversely, suppose (2.15) holds and G is causal. Then for $u \in L_e(U)$

$$< G(u) \mid u >_T = < (G(u))_T \mid u_T > = < (G(u_T))_T \mid u_T >$$

$$= < G(u_T) \mid u_T > \ \geq -\beta.$$

$\square$

We are ready to state the first version of the Passivity theorem.

Theorem 2.2.6 (Passivity theorem; first version) *Consider the closed-loop system $G_1 \|_f G_2$ in Fig. 1.1, with $G_1 : L_e(U_1) \to L_e(U_1^*)$ and $G_2 : L_e(U_2) \to L_e(U_2^*)$ passive, and $E_1 = U_2^* = U_1$, $E_2 = U_1^* = U_2$.*

(a) *Assume that for any $e_1 \in L_e(U_1)$, $e_2 \in L_e(U_2)$ there are solutions $u_1 \in L_e(U_1)$ and $u_2 \in L_e(U_2)$. Then $G_1 \|_f G_2$ with inputs (e_1, e_2) and outputs (y_1, y_2) is passive.*

(b) *Assume that for any* $e_1 \in L_e(U_1)$ *and* $e_2 = 0$ *there are solutions* $u_1 \in L_e(U_1)$, $u_2 \in L_e(U_2)$. *Then* $G_1 \|_f G_2$ *with* $e_2 = 0$ *and input* e_1 *and output* y_1 *is passive.*

Proof The definition of standard negative feedback, cf. (1.30), implies the key property

$$
\begin{aligned}
&< y_1 \mid u_1 >_T + < y_2 \mid u_2 >_T \\
&= < y_1 \mid e_1 - y_2 >_T + < y_2 \mid e_2 + y_1 >_T \\
&= < y_1 \mid e_1 >_T + < y_2 \mid e_2 >_T,
\end{aligned}
\tag{2.16}
$$

and thus for any $e_1 \in L_e(U_1)$, $e_2 \in L_e(U_2)$ and any $T \geq 0$

$$
\begin{aligned}
&< y_1 \mid u_1 >_T + < y_2 \mid u_2 >_T \\
&= < y_1 \mid e_1 >_T + < y_2 \mid e_2 >_T
\end{aligned}
\tag{2.17}
$$

with $y_1 = G_1(u_1)$, $y_2 = G_2(u_2)$. By passivity of G_1 and G_2, $< y_1 \mid u_1 >_T \geq -\beta_1$, $< y_2 \mid u_2 >_T \geq -\beta_2$, and thus by (2.17)

$$
< y_1 \mid e_1 >_T + < y_2 \mid e_2 >_T \geq -\beta_1 - \beta_2
\tag{2.18}
$$

implying part (a). For part (b) take $e_2 = 0$ in (2.17). □

Remark 2.2.7 Theorem 2.2.6 expresses an inherent *robustness* property of passive systems: the closed-loop system $G_1 \|_f G_2$ remains passive for all perturbations of the input–output maps G_1, G_2, as long as they *remain passive* (compare with Remark 2.1.2).

In order to state a stronger version of the Passivity theorem we need stronger notions of passivity. First of all, we will assume that the input space U is equipped with an *inner product* $<, >$. Using the linear bijection

$$
u \in U \longmapsto < u, \cdot > \in U^*,
\tag{2.19}
$$

we may then identify $Y = U^*$ with U. That is, $Y = U^* = U$, and $< y \mid u >=< y, u >$. Furthermore, for any input function $u \in L_{2e}(U)$ and corresponding output function $y = G(u) \in L_{2e}(U)$ we will have $< y \mid u >_T = \int_0^T < y(t), u(t) > dt$, which will be throughout denoted by $< y, u >_T$.

Definition 2.2.8 *(Output and input strict passivity)* Let $U = Y$ be a linear space with inner product $<, >$ and corresponding norm $\| \cdot \|$. Let $G : L_{2e}(U) \to L_{2e}(Y)$ be an input–output map. Then G is *input strictly passive* if there exists β and $\delta > 0$ such that

$$
< G(u), u >_T \geq \delta \|u_T\|_2^2 - \beta, \quad \forall u \in L_{2e}(U), \forall T \geq 0,
\tag{2.20}
$$

and *output strictly passive* if there exists β and $\varepsilon > 0$ such that

$$
< G(u), u >_T \geq \varepsilon \|(G(u))_T\|_2^2 - \beta, \quad \forall u \in L_{2e}(U), \forall T \geq 0.
\tag{2.21}
$$

Furthermore, $G : L_{2e}(U) \to L_{2e}(Y)$ is merely *passive* if there exists β such that (2.21) holds for $\varepsilon = 0$ (or equivalent (2.20) for $\delta = 0$). Whenever we want to emphasize the role of the constants δ, ε we will say that G is δ-*input strictly passive* or ε-*output strictly passive*. In the same way we define (δ-)input and (ε-)output strict passivity for *relations* $R \subset L_{2e}(U) \times L_{2e}(Y)$.

Remark 2.2.9 Note that by Remark 2.2.3 the left-hand sides of (2.20) and (2.21) are well defined.

Remark 2.2.10 Proposition 2.2.5 immediately generalizes to input, respectively, output strict passivity.

We obtain the following extension of Theorem 2.2.6.

Theorem 2.2.11 (Passivity theorem; second version) *Consider the closed-loop system* $G_1\|_f G_2$ *in Fig. 1.1, with* $G_1 : L_{2e}(U_1) \to L_{2e}(U_1)$, $G_2 : L_{2e}(U_2) \to L_{2e}(U_2)$, *and* $E_1 = U_1 = U_2 = E_2 =: U$ *an inner product space.*

(a) *Assume that for any* $e_1, e_2 \in L_{2e}(U)$ *there are solutions* $u_1, u_2 \in L_{2e}(U)$. *If* G_1 *and* G_2 *are respectively* ε_1- *and* ε_2-*output strictly passive, then* $G_1\|_f G_2$ *with inputs* (e_1, e_2) *and outputs* (y_1, y_2) *is* ε-*output strictly passive, with* $\varepsilon = \min(\varepsilon_1, \varepsilon_2)$.

(b) *Assume that for any* $e_1 \in L_{2e}(U)$ *and* $e_2 = 0$ *there are solutions* $u_1, u_2 \in L_{2e}(U)$. *If* G_1 *is passive and* G_2 *is* δ_2-*input strictly passive, or if* G_1 *is* ε_1-*output strictly passive and* G_2 *is passive, then* $G_1\|_f G_2$ *for* $e_2 = 0$, *with input* e_1 *and output* y_1, *is* δ_2-*input, respectively* ε_1-*output strictly passive.*

Proof Equation (2.17) becomes

$$< y_1, u_1 >_T \, + \, < y_2, u_2 >_T \, = \, < y_1, e_1 >_T \, + \, < y_2, e_2 >_T \qquad (2.22)$$

(a) Since G_1 and G_2 are output strictly passive (2.22) implies

$$< y_1, e_1 >_T \, + \, < y_2, e_2 >_T \, = \, < y_1, u_1 >_T \, + \, < y_2, u_2 >_T$$

$$\geq \varepsilon_1 \|y_{1T}\|_2^2 + \varepsilon_2 \|y_{2T}\|_2^2 - \beta_1 - \beta_2$$

$$\geq \varepsilon(\|y_{1T}\|_2^2 + \|y_{2T}\|_2^2) - \beta_1 - \beta_2$$

for $\varepsilon = \min(\varepsilon_1, \varepsilon_2) > 0$.

(b) Let G_1 be passive and G_2 be δ_2-input strictly passive. By (2.22) with $e_2 = 0$

$$< y_1, e_1 >_T \, = \, < y_1, u_1 >_T \, + \, < y_2, u_2 >_T$$

$$\geq -\beta_1 + \delta_2 \|u_{2T}\|_2^2 - \beta_2 = \delta_2 \|y_{1T}\|_2^2 - \beta_1 - \beta_2$$

If G_1 is ε_1-output strictly passive and G_2 is passive, then the same inequality holds with δ_2 replaced by ε_1. $\square$

Remark 2.2.12 A similar theorem can be stated for *relations* R_1 and R_2.

For statements regarding the L_2-stability of the feedback interconnection of passive systems a key observation will be the fact that *output strict passivity* implies *finite L_2-gain.*

Theorem 2.2.13 *Let* $G : L_{2e}(U) \to L_{2e}(U)$ *be* ε-*output strictly passive. Then* G *has* L_2-*gain* $\leq \frac{1}{\varepsilon}$.

Proof Since G is ε-output strictly passive there exists β such that $y = G(u)$ satisfies

$$\varepsilon \|y_T\|_2^2 \leq \; <y, u>_T + \beta$$

$$\leq \; <y, u>_T + \beta + \tfrac{1}{2}\|\tfrac{1}{\sqrt{\varepsilon}} u_T - \sqrt{\varepsilon} y_T\|_2^2 \qquad (2.23)$$

$$= \beta + \tfrac{1}{2\varepsilon}\|u_T\|_2^2 + \tfrac{\varepsilon}{2}\|y_T\|_2^2,$$

whence $\frac{\varepsilon}{2}\|y_T\|_2^2 \leq \frac{1}{2\varepsilon}\|u_T\|_2^2 + \beta$, proving that $\gamma_2(G) \leq \frac{1}{\varepsilon}$. $\qquad \square$

Remark 2.2.14 As a partial converse statement, note that if G is δ-*input* strictly passive *and* has L_2-gain $\leq \gamma$, then

$$<G(u), u> \; \geq \delta\|u\|_2^2 - \beta \geq \frac{\delta}{\gamma}\|G(u)\|_2^2 - \beta,$$

implying that G is $\frac{\delta}{\gamma}$-output strictly passive.

Combining Theorems 2.2.11 and 2.2.13 one directly obtains the following.

Theorem 2.2.15 (Passivity theorem; third version) *Consider the closed-loop system* $G_1\|_f G_2$ *in Fig. 1.1, with* $G_1 : L_{2e}(U_1) \to L_{2e}(U_1)$, $G_2 : L_{2e}(U_2) \to L_{2e}(U_2)$, *and* $E_1 = E_2 = U_1 = U_2 =: U$ *an inner product space.*

(a) *Assume that for any* $e_1, e_2 \in L_{2e}(U)$ *there exist solutions* $u_1, u_2 \in L_{2e}(U)$. *If* G_i *is* ε_i-*output strictly passive,* $i = 1, 2$, *then* $G_1\|_f G_2$ *with inputs* (e_1, e_2) *and outputs* (y_1, y_2) *has* L_2-*gain* $\leq \frac{1}{\varepsilon}$ *with* $\varepsilon = \min(\varepsilon_1, \varepsilon_2)$. *For* $e_1, e_2 \in L_2(U)$ *it follows that* $u_1, u_2, y_1, y_2 \in L_2(U)$.
(b) *Assume that for any* $e_1 \in L_{2e}(U)$ *and* $e_2 = 0$ *there are solutions* $u_1, u_2 \in L_{2e}(U)$. *If* G_1 *is passive and* G_2 *is* δ_2-*input strictly passive, or if* G_1 *is* ε_1-*output strictly passive and* G_2 *is passive, then* $G_1\|_f G_2$ *for* $e_2 = 0$ *with input* e_1 *and output* y_1 *has* L_2-*gain* $\leq \frac{1}{\delta_2}$, *respectively* $\leq \frac{1}{\varepsilon_1}$. *Furthermore, if* $e_1 \in L_2(U)$ *then also* $y_1 = u_2 \in L_2(U)$.

Remark 2.2.16 Suppose G_1 and G_2 are causal. Then by Propositions 2.2.5 and 1.2.14 we can relax the assumption in (a) to assuming that for any $e_1, e_2 \in L_2(U)$ there exist solutions $u_1, u_2 \in L_{2e}(U)$. Similarly, we can relax the assumption in (b) to assuming that for any $e_1 \in L_2(U)$ and $e_2 = 0$ there exist solutions $u_1, u_2 \in L_{2e}(U)$. If G_1 and/or G_2 are not causal, then this relaxation of assumptions will guarantee at least L_2-stability.

Example 2.2.17 Note that in Theorem 2.2.15 (b) it is *not* claimed that u_1 and $y_2 = G_2(u_2)$ are in $L_2(U)$. In fact, a *physical counterexample* to such a claim can be given as follows. Consider a mass moving in one-dimensional space. Let the mass be subject to a friction force which is the sum of an ideal Coulomb friction and a linear damping. Furthermore, let the mass be actuated by a force $u_1 = e_1 - y_2$, where e_1 is an external force and y_2 is the force delivered by a linear spring. Defining y_1 as the velocity of the mass, the input–output map G_1 from u_1 to y_1 for zero initial condition (velocity zero) is output strictly passive, as follows from the definition of the friction force. Furthermore, let G_2 be the passive input–output map defined by the linear spring for zero initial extension, with the spring attached at one end to a wall and with the velocity of the other end being its input u_2 and with output y_2 being the spring force (acting on the mass). Now let $e_1(\cdot)$ be an external force time function with the shape of a pulse, of magnitude h and width w. Then by taking h large enough the force e_1 will overcome the total friction force (in particular the Coulomb friction force), resulting in a motion of the mass and thus of the free end of the spring. On the other hand by taking the width w of the pulse small enough the extension of the spring will be such that the spring force does not overcome the Coulomb friction force. As a result, the velocity of the mass y_1 will converge to zero, while the spring force y_2 will converge to a nonzero constant value (smaller than the Coulomb friction constant). Hence, y_2 and u_1 will *not* be in $L_2(\mathbb{R})$.

A useful generalization of the Passivity Theorems 2.2.11 (a) and 2.2.15 (a), where we do *not* necessarily require passivity of G_1 and G_2 separately, can be stated as follows.

Theorem 2.2.18 *Suppose there exist constants $\varepsilon_i, \delta_i, \beta_i, i = 1, 2$, satisfying*

$$\varepsilon_1 + \delta_2 > 0, \ \varepsilon_2 + \delta_1 > 0 \tag{2.24}$$

such that
$$< G_i(u_i), u_i >_T \ \geq \ \varepsilon_i ||(G_i(u_i))_T||_2^2 + \delta_i ||u_{iT}||_2^2 - \beta_i , \tag{2.25}$$

for all $u_i \in L_{2e}(U_i)$ and all $T \geq 0, i = 1, 2$. Then $G_1 \|_f G_2$ has finite L_2-gain from (e_1, e_2) to (y_1, y_2).

Proof Addition of (2.25) with $y_i = G_i(u_i)$ for $i = 1, 2$ yields

$$< y_1, u_1 >_T + < y_2, u_2 >_T$$
$$\geq \varepsilon_1 ||y_{1T}||_2^2 + \delta_1 ||u_{1T}||_2^2 + \varepsilon_2 ||y_{2T}||^2 + \delta_2 ||u_{2T}||^2 - \beta_1 - \beta_2. \tag{2.26}$$

Substitution of the negative feedback $u_1 = e_1 - y_2$, $u_2 = e_2 + y_1$ results in

$$
\begin{aligned}
& < y_1, e_1 >_T + < y_2, e_2 >_T + \beta_1 + \beta_2 \\
& \geq \varepsilon_1 \|y_{1T}\|_2^2 + \delta_1 \|e_1 - y_2\|_2^2 + \varepsilon_2 \|y_2\|_2^2 + \delta_2 \|e_2 + y_1\|_2^2.
\end{aligned}
\tag{2.27}
$$

Writing out and rearranging terms leads to

$$
\begin{aligned}
& -\delta_1 \|e_{1T}\|_2^2 - \delta_2 \|e_{2T}\|_2^2 + \beta_1 + \beta_2 \\
& \geq (\varepsilon_1 + \delta_2)\|y_{1T}\|_2^2 + (\varepsilon_2 + \delta_1)\|y_{2T}\|_2^2 \\
& \quad -2\delta_1 < y_2, e_1 >_T -2\delta_2 < y_1, e_2 >_T - < y_1, e_1 >_T - < y_2, e_2 >_T .
\end{aligned}
$$

By the positivity assumption on $\alpha_1^2 := \varepsilon_1 + \delta_2$, $\alpha_2^2 := \varepsilon_2 + \delta_1$ we can perform "completion of the squares" on the right-hand side of this inequality, to obtain an expression of the form

$$
\left\| \begin{bmatrix} \alpha_1 y_{1T} \\ \alpha_2 y_{2T} \end{bmatrix} - A \begin{bmatrix} e_{1T} \\ e_{2T} \end{bmatrix} \right\|_2^2 \leq c^2 \left\| \begin{bmatrix} e_{1T} \\ e_{2T} \end{bmatrix} \right\|_2^2 + \beta_1 + \beta_2,
\tag{2.28}
$$

for a certain 2×2 matrix A and constant c. In combination with the triangle inequality

$$
\left\| \begin{bmatrix} \alpha_1 y_{1T} \\ \alpha_2 y_{2T} \end{bmatrix} \right\|_2 \leq \left\| \begin{bmatrix} \alpha_1 y_{1T} \\ \alpha_2 y_{2T} \end{bmatrix} - A \begin{bmatrix} e_{1T} \\ e_{2T} \end{bmatrix} \right\|_2 + \left\| A \begin{bmatrix} e_{1T} \\ e_{2T} \end{bmatrix} \right\|_2,
\tag{2.29}
$$

this yields finite L_2-gain from (e_1, e_2) to (y_1, y_2). $\qquad\square$

Remark 2.2.19 Clearly, Theorem 2.2.18 includes Part (a) of Theorems 2.2.11 and 2.2.15 by taking $\delta_1 = \delta_2 = 0$. Importantly, it shows that $\varepsilon_1, \varepsilon_2, \delta_1, \delta_2$ need not all be nonnegative. Negativity of ε_1 ("lack of passivity" of G_1) can be "compensated" by a sufficiently large positive δ_2 ("surplus of passivity" of G_2).

Notice that the last version of the Passivity Theorem 2.2.15 still assumes the *existence of solutions* $u_1, u_2 \in L_{2e}(U)$. In the small-gain case this was remedied, cf. Theorem 2.1.8, by replacing finite L_q-gain and the small-gain condition by their *incremental versions*. Similarly this can be done by invoking a notion of *incremental passivity* defined as follows.

Definition 2.2.20 (*Incremental passivity*) An input–output map $G : L_{2e}(U) \to L_{2e}(Y)$ is $\mathfrak{E}$-*output strictly incrementally passive* for some $\mathfrak{E} > 0$ if there exists β such that

$$
\mathfrak{E}\|y_T - z_T\|_2^2 \leq \; < y - z, u - v >_T \; + \beta
\tag{2.30}
$$

for all $u, v \in L_{2e}(U)$ and corresponding outputs $y = G(u), z = G(v)$. If $\mathfrak{E} = 0$ then G is *incrementally passive*.

Furthermore, G is called Δ-*input strictly incrementally passive* for some $\Delta > 0$ if there exists β such that

$$
\Delta\|u_T - v_T\|_2^2 \leq \; < y - z, u - v >_T \; + \beta
\tag{2.31}
$$

for all $u, v \in L_{2e}(U)$ and corresponding outputs $y = G(u), z = G(v)$.

We immediately obtain the following incremental version of Theorem 2.2.15.

Proposition 2.2.21 *Consider the closed-loop system $G_1 \|_f G_2$ in Fig. 1.1, with G_1 : $L_{2e}(U_1) \to L_{2e}(U_1)$, $G_2 : L_{2e}(U_2) \to L_{2e}(U_2)$, and $E_1 = U_1 = U_2 = E_2 =: U$ an inner product space.*

(a) *Assume that for any $e_1, e_2 \in L_{2e}(U)$ there are solutions $u_1, u_2 \in L_{2e}(U)$. If G_1 and G_2 are respectively $\mathfrak{E}_1$- and $\mathfrak{E}_2$-output strictly incrementally passive, then $G_1 \|_f G_2$ with inputs (e_1, e_2) and outputs (y_1, y_2) is $\mathfrak{E}$-output strictly incrementally passive, with $\mathfrak{E} = \min(\mathfrak{E}_1, \mathfrak{E}_2)$.*

(b) *Assume that for any $e_1 \in L_{2e}(U)$ and $e_2 = 0$ there are solutions $u_1, u_2 \in L_{2e}(U)$. If G_1 is incrementally passive and G_2 is Δ_2-input strictly incrementally passive, or if G_1 is $\mathfrak{E}_1$-output strictly incrementally passive and G_2 is incrementally passive, then $G_1 \|_f G_2$ with $e_2 = 0$ and input e_1 and output y_1 is $\mathfrak{E}$-output strictly incrementally passive, with $\mathfrak{E}$ equal to Δ_2 respectively $\mathfrak{E}_1$.*

The following crucial step is the observation that output strict incremental passivity implies finite incremental L_2-gain in the same way as output strict passivity implies finite L_2-gain, cf. Theorem 2.2.13.

Proposition 2.2.22 *Let $G : L_{2e}(U) \to L_{2e}(U)$ be $\mathfrak{E}$-output strictly incrementally passive. Then G has incremental L_2-gain $\leq \frac{1}{\mathfrak{E}}$.*

Proof Repeat the same argument as in the proof of Theorem 2.2.13, but now in the incremental setting, to conclude that

$$\mathfrak{E}\|y_T - z_T\|_2^2 \leq \beta + \frac{1}{2\mathfrak{E}}\|u_T - v_T\|_2^2 + \frac{\mathfrak{E}}{2}\|y_T - z_T\|_2^2,$$

where $y = G(u), z = G(v)$. This proves that the incremental L_2-gain of G is $\leq \frac{1}{\mathfrak{E}}$. $\qquad\square$

By combining Propositions 2.2.21 and 2.2.22 with Theorem 2.1.8 we immediately obtain the following corollary.

Corollary 2.2.23 *Consider the closed-loop system $G_1 \|_f G_2$ in Fig. 1.1, with G_1 : $L_{2e}(U_1) \to L_{2e}(U_1)$, $G_2 : L_{2e}(U_2) \to L_{2e}(U_2)$, and $E_1 = E_2 = U_1 = U_2 =: U$ an inner product space.*

Assume that G_1 and G_2 are $\mathfrak{E}_1$-, respectively $\mathfrak{E}_2$-, output strictly incrementally passive, and that

$$\mathfrak{E}_1 \cdot \mathfrak{E}_2 > 1. \tag{2.32}$$

Then

(i) *For all $(e_1, e_2) \in L_{2e}(E_1 \times E_2)$ there exists a unique solution $(u_1, u_2, y_1, y_2) \in L_{2e}(U_1 \times U_2 \times Y_1 \times Y_2)$.*

(ii) The map $(e_1, e_2) \mapsto (u_1, u_2)$ is uniformly continuous on the domain $L_{2e}(E_1 \times E_2)$.

(iii) If the solution (u_1, u_2) to $e_1 = e_2 = 0$ is in $L_2(U_1 \times U_2)$, then $(e_1, e_2) \in L_2(E_1 \times E_2)$ implies that $(u_1, u_2) \in L_2(U_1 \times U_2)$.

Remark 2.2.24 (General power-conserving interconnections) All the derived passivity theorems can be generalized to interconnections which are more general than the standard feedback interconnection of Fig. 1.1. This relies on the observation that the essential requirement in the proof of Theorem 2.2.6 is the identity (2.16), expressing the fact that the feedback interconnection $u_1 = -y_2 + e_1$, $u_2 = y_1 + e_2$ is *power-conserving*. Many other interconnections share this property, and as a result the interconnected systems share the same passivity properties as the closed-loop systems arising from standard feedback interconnection. As an example, consider the following system (taken from [355]) given in Fig. 2.1. Here R represents a robotic system and C is a controller, while E represents the environment interacting with the controlled robotic mechanism. The external signal e denotes a velocity command. We assume R and E to be passive, and C to be a output strictly passive controller. By the interconnection constraints $u_C = y_E + e$, $u_R = y_E$ and $u_E = -y_R - y_C$ we obtain

$$< y_C \mid u_C > + < y_R \mid u_R > + < y_E \mid u_E > \;=\; < y_C \mid e >$$

and hence, as in Theorem 2.2.15 part (b), the interconnected system with input e and output y_C is output strictly passive, and therefore has finite L_2-gain.

This idea will be further developed in the subsequent chapters, especially in Chaps. 4, 6 and 7 in the passive and port-Hamiltonian systems context.

Fig. 2.1 An alternative power-conserving interconnection

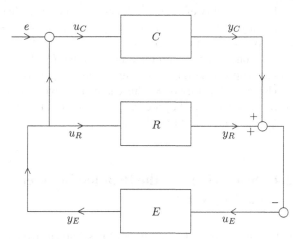

Fig. 2.2 Feedback system
with multipliers

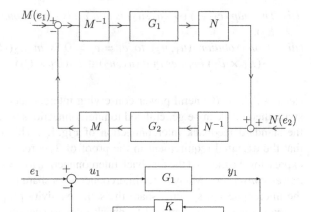

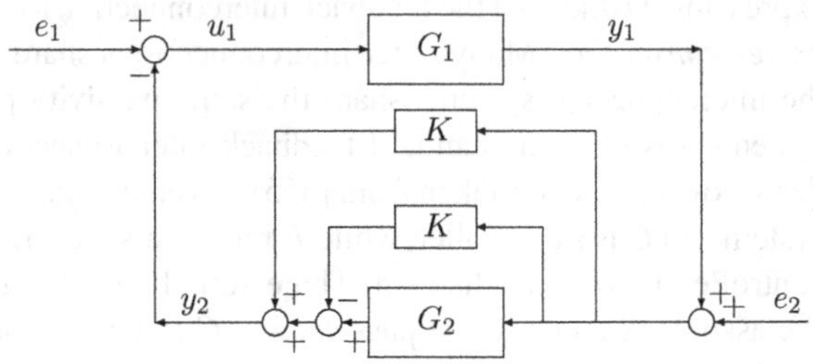

Fig. 2.3 Transformed
closed-loop configuration

2.3   Loop Transformations

The range of applicability of the small-gain and passivity theorems can be considerably enlarged using *loop transformations*. We will only indicate two basic ideas.

The first possibility is to insert *multipliers* in Fig. 1.1 by pre- and post-multiplying G_1 and G_2 by L_q-stable input–output mappings M and N and their inverses M^{-1} and N^{-1}, which are also assumed to be L_q-stable input–output mappings, see Fig. 2.2.

By L_q-stability of M, M^{-1}, N and N^{-1} it follows that $e_1 \in L_q(E_1)$, $e_2 \in L_q(E_2)$ if and only if $M(e_1) \in L_q(E_1)$, $M(e_2) \in L_q(E_2)$. Thus stability of $G_1 \|_f G_2$ is equivalent to stability of $G_1' \|_f G_2'$, with $G_1' = NG_1M^{-1}$, $G_2' = MG_2N^{-1}$.

A second idea is to introduce an additional L_q-stable and *linear* operator K in the closed-loop system $G_1 \|_f G_2$ by first subtracting and then adding to G_2 (see Fig. 2.3).

Using the linearity of K, this can be redrawn as in Fig. 2.4. Clearly, by stability of K, $e_1 - K(e_2)$ and e_2 are in L_q if and only if e_1, e_2 are in L_q. Thus stability of $G_1 \|_f G_2$ is equivalent to stability of $G_1' \|_f G_2'$.

2.4   Scattering and the Relation Between Passivity
and L_2-Gain

Let us return to the basic setting of passivity, as exposed in Sect. 2.2, starting with a finite-dimensional linear input space U (without any additional structure such as inner product or norm) and its dual space $Y := U^*$ defining the space of outputs.

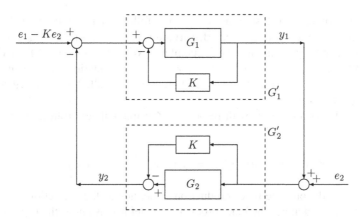

Fig. 2.4 Redrawn transformed closed-loop configuration

On the product space $U \times Y$ of inputs and outputs there exists a canonically defined symmetric bilinear form $\ll, \gg$, given as

$$\ll (u_1, y_1), (u_2, y_2) \gg := < y_1 \mid u_2 > + < y_2 \mid u_1 > \qquad (2.33)$$

with $u_i \in U$, $y_i \in Y$, $i = 1, 2$, and $< \mid >$ denoting the duality pairing between $Y = U^*$ and U. With respect to a basis $e_1, \ldots, e_m$ of U (where $m = \dim U$), and the corresponding dual basis $e_1^*, \ldots, e_m^*$ of $Y = U^*$, the bilinear form $\ll, \gg$ has the matrix representation

$$\begin{bmatrix} 0 & I_m \\ I_m & 0 \end{bmatrix} \qquad (2.34)$$

It immediately follows that $\ll, \gg$ has singular values $+1$ (with multiplicity m) and -1 (also with multiplicity m), and thus defines an *indefinite* inner product on the space $U \times Y$ of inputs and outputs. *Scattering* is based on decomposing the combined vector $(u, y) \in U \times Y$ with respect to the positive and negative singular values of this indefinite inner product. More precisely, we obtain the following definition.

Definition 2.4.1 Any pair (V, Z) of subspaces $V, Z \subset U \times Y$ is called a pair of *scattering subspaces* if

(i) $V \oplus Z = U \times Y$
(ii) $\ll v_1, v_2 \gg > 0$, for all $v_1, v_2 \in V$ unequal to 0,
     $\ll z_1, z_2 \gg < 0$, for all $z_1, z_2 \in Z$ unequal to 0
(iii) $\ll v, z \gg = 0$, for all $v \in V, z \in Z$.

It follows from (2.34) that any pair of scattering subspaces (V, Z) satisfies

$$\dim V = \dim Z = m$$

Given a pair of scattering subspaces (V, Z) it follows that any combined vector $(u, y) \in U \times Y$ also can be represented, in a unique manner, as a pair $v \oplus z \in V \oplus Z$, where v is the projection along Z of the combined vector $(u, y) \in U \times Y$ on V, and z is the projection of (u, y) along V on Z. The representation $(u, y) = v \oplus z$ is called a *scattering* representation of (u, y), and v, z are called the *wave vectors* of the combined vector (u, y).

Using orthogonality of V with respect to Z it immediately follows that for all $(u_i, y_i) = v_i \oplus z_i, i = 1, 2$,

$$\ll (u_1, y_1), (u_2, y_2) \gg = < v_1, v_2 >_V - < z_1, z_2 >_Z \qquad (2.35)$$

where $< , >_V$ denotes the inner product on V defined as the restriction of $\ll , \gg$ to V, and $< , >_Z$ denotes the inner product on Z defined as *minus* the restriction of $\ll , \gg$ to Z.

In particular, taking $(u_1, y_1) = (u_2, y_2) = (u, y)$, we obtain for any $(u, y) = v \oplus z$ the following fundamental relation between (u, y) and its wave vectors v, z

$$< y \mid u > = \frac{1}{2} \ll (u, y), (u, y) \gg = \frac{1}{2}\|v\|_V^2 - \frac{1}{2}\|z\|_Z^2, \qquad (2.36)$$

where $\| \ \|_V$, $\| \ \|_Z$ are the norms on V, Z, defined by $< , >_V$, respectively $< , >_Z$.

Identifying as before $< y \mid u >$ with *power*, the vector v thus can be regarded as the *incoming* wave vector, with half times its norm being the *incoming power*, and the vector z is the *outgoing* wave vector, with half times its norm being the *outgoing power*.

Now let $G : L_e(U) \to L_e(Y)$, with $Y = U^*$, be an input–output map as before. Expressing $(u, y) \in U \times Y$ in a scattering representation as $v \oplus z \in V \oplus Z$, it follows that G transforms into the *relation*

$$\begin{aligned} R_{vz} = \{v \oplus z \in L_e(V) \oplus L_e(Z) \mid \\ v(t) \oplus z(t) = (u(t), y(t)), \ t \in \mathbb{R}^+, y = G(u)\}, \end{aligned} \qquad (2.37)$$

with the function spaces $L_e(V)$ and $L_e(Z)$ yet to be defined. As a direct consequence of (2.36) we obtain the following relation between G and R_{vz}:

$$< G(u) \mid u >_T = \frac{1}{2}\|v_T\|_V^2 - \frac{1}{2}\|z_T\|_Z^2, \quad T \geq 0. \qquad (2.38)$$

In particular, if u and $y = G(u)$ are such that $v \in L_{2e}(V)$ and $z \in L_{2e}(Z)$ then, since the right-hand side of (2.38) is well defined, also the expression $< G(u) \mid u >_T$ is well defined for all $T \geq 0$.

We obtain from (2.38) the following key relation between passivity of G and the L_2-gain of R_{vz}.

Proposition 2.4.2 *Consider the relation* $R_{vz} \subset L_{2e}(V) \oplus L_{2e}(Z)$ *as defined in (2.37), with L_e replaced by L_{2e}. Then G is passive if and only if R_{vz} has L_2-gain ≤ 1.*

Proof By (2.38), $||z_T||_Z^2 \leq ||v_T||_V^2 + c$ if and only if $< G(u) \mid u >_T \geq -\frac{c}{2}$. $\square$

If the relation R_{vz} can be written as the graph of an input–output map

$$S : L_{2e}(V) \rightarrow L_{2e}(Z), \tag{2.39}$$

(with respect to the intrinsically defined norms $|| \ ||_V$ and $|| \ ||_Z$) then we call S the *scattering operator* of the input–output map G. We obtain the following fundamental relation between passivity and L_2-gain.

Corollary 2.4.3 *The scattering operator S has L_2-gain ≤ 1 if and only if G is passive.*

As noted before, the choice of scattering subspaces V, Z, and therefore of the scattering representation, is not unique. Particular choices of scattering subspaces are given as follows. Take any basis $e_1, \ldots, e_m$ for U, with dual basis $e_1^*, \ldots, e_m^*$ for $U^* = Y$. Then it can be directly checked that the pair (V, Z) given as

$$V = \text{span}\left\{ \left(\frac{e_i}{\sqrt{2}}, \frac{e_i^*}{\sqrt{2}} \right), i = 1, \ldots, m \right\}$$
$$Z = \text{span}\left\{ \left(\frac{-e_i}{\sqrt{2}}, \frac{e_i^*}{\sqrt{2}} \right), i = 1, \ldots, m \right\} \tag{2.40}$$

defines a pair of scattering subspaces. (In the above the factors $\frac{1}{\sqrt{2}}$ were inserted in order that the vectors spanning V, respectively Z, are *orthonormal* with respect to the intrinsically defined inner products $<, >_V$ and $<, >_Z$.) In these bases for U, Y and V, Z the relation between (u, y) and its scattering representation (v, z) is given as

$$v = \frac{1}{\sqrt{2}}(u + u^*)$$
$$z = \frac{1}{\sqrt{2}}(-u + u^*). \tag{2.41}$$

Hence, with $y = G(u)$, the relation R_{vz} has the coordinate expression

$$R_{vz} = \{(v, z) : \mathbb{R}^+ \rightarrow V \times Z \mid$$
$$v(t) = \frac{1}{\sqrt{2}}(G + I)(u)(t), \ z(t) = \frac{1}{\sqrt{2}}(G - I)(u)(t)\}, \tag{2.42}$$

where I denotes the identity operator. In particular, R_{vz} can be expressed as the graph of a scattering operator S if and only if the operator $G + I : L(U) \rightarrow L(V)$ is *invertible*, in which case S takes the standard form

$$S = (G - I)(G + I)^{-1}. \tag{2.43}$$

In case U is equipped with an inner product $<, >_U$, and U^* can be identified with U (see Sect. 2.2), we obtain the following relation between passivity of G and L_2-gain of R_{vz}.

Proposition 2.4.4 *Let U be endowed with an inner product $<, >_U$. Consider an input–output mapping $G : L_{2e}(U) \to L_{2e}(U)$ and the corresponding relation $R_{vz} \subset L_{2e}(V) \times L_{2e}(Z)$. Then G is input and output strictly passive if and only if the L_2-gain of R_{vz} (or, if $G + I$ is invertible, the L_2-gain of the scattering operator S) is strictly less than 1.*

Proof Let the L_2-gain of R_{vz} be $\leq 1 - \delta$, with $1 \geq \delta > 0$. Then $||z_T||_2^2 \leq (1 - \delta)||v_T||_2^2 + c$, and thus by (2.38)

$$2 < G(u) \mid u > \geq \delta||v_T||_2^2 - c$$

Since $||v_T||_2^2 = ||u_T + (G(u))_T||_2^2 = ||u_T||_2^2 + ||G(u)_T||_2^2 + 2 < G(u) \mid u >$, this implies for some $\epsilon > 0$ and β

$$< G(u) \mid u > \geq \varepsilon||G(u)||_2^2 + \varepsilon||u||_2^2 - \beta$$

The converse statement follows similarly.   □

Remark 2.4.5 Since "input strict passivity" plus "finite L_2-gain" implies output strict passivity, cf. Remark 2.2.14, and conversely output strict passivity implies finite L_2-gain, the condition of input and output strict passivity in the above proposition can be replaced by input strict passivity and finite L_2-gain.

2.5 Notes for Chapter 2

1. The treatment of Sects. 2.1 and 2.2 is largely based on Vidyasagar [343], with extensions from Desoer & Vidyasagar [83]. We have emphasized a "coordinate-free" treatment of the theory, which in particular has some impact on the formulation of passivity. See also Sastry [267] and Khalil [160] for expositions. The developments regarding incremental passivity, in particular Corollary 2.2.23, seem to be relatively new.

2. The small-gain theorem is usually attributed to Zames [362, 363], and in its turn is closely related to the Nyquist stability criterion. See also Willems [348]. A classical treatise on passivity and its implications for stability is Popov [255].

3. Theorem 2.2.18 is treated in Sastry [267], Vidyasagar [343].

4. An interesting generalization of the small-gain theorem (Theorem 2.1.1) is obtained by considering input–output maps G_1 and G_2 that have a finite "*nonlinear gain*" in the following sense. Suppose there exist functions $\gamma_i : \mathbb{R}^+ \to \mathbb{R}^+$

of class[1] $\mathcal{K}$ and constants b_i, $i = 1, 2$, such that

$$\|G_i(u_T)\| \leq \gamma_i(\|u_T\|) + b_i, \quad T \geq 0, \tag{2.44}$$

for $i = 1, 2$, where $\|\ \|$ denotes some L_q-norm. Note that by taking *linear* functions $\gamma_i(z) = \gamma_i z$, with constant $\gamma_i > 0$, we recover the usual definition of finite gain. Then, similar to the proof of Theorem 2.1.1, we derive the following inequalities for the closed-loop system $G_1 \|_f G_2$:

$$\begin{aligned}
\|u_{1T}\| &\leq \|y_{2T}\| + \|e_{1T}\| \\
\|u_{2T}\| &\leq \|y_{1T}\| + \|e_{2T}\|
\end{aligned} \tag{2.45}$$

and thus by (2.44)

$$\begin{aligned}
\|y_{1T}\| &\leq \gamma_1(\|y_{2T}\| + \|e_{1T}\|) + b_1 \\
\|y_{2T}\| &\leq \gamma_2(\|y_{1T}\| + \|e_{2T}\|) + b_2
\end{aligned} \tag{2.46}$$

which by cross-substitution yields

$$\begin{aligned}
\|y_{1T}\| &\leq \gamma_1(\gamma_2(\|y_{1T}\| + \|e_{2T}\|) + \|e_{1T}\| + b_2) + b_1 \\
\|y_{2T}\| &\leq \gamma_2(\gamma_1(\|y_{2T}\| + \|e_{1T}\|) + \|e_{2T}\| + b_1) + b_2.
\end{aligned} \tag{2.47}$$

One may wonder under what conditions on γ_1 and γ_2 the inequalities (2.47) imply that

$$\begin{aligned}
\|y_{1T}\| &\leq \delta_1(\|e_{1T}\|, \|e_{2T}\|) + d_1 \\
\|y_{2T}\| &\leq \delta_2(\|e_{1T}\|, \|e_{2T}\|) + d_2
\end{aligned} \tag{2.48}$$

for certain constants d_1, d_2 and functions $\delta_i : \mathbb{R}^+ \times \mathbb{R}^+ \to \mathbb{R}^+$, $i = 1, 2$, which are of class $\mathcal{K}$ in *both* their arguments. Indeed, this would imply that the closed-loop system $G_1 \|_f G_2$ has finite nonlinear gain from e_1, e_2 to y_1, y_2. As shown in Mareels & Hill [194] this is the case if there exist functions $g, h \in \mathcal{K}$ and a constant $c \geq 0$, such that

$$\gamma_1 \circ (\mathrm{id} + g) \circ \gamma_2(z) \leq z - h(z) + c, \quad \text{for all } z, \tag{2.49}$$

with id denoting the identity mapping. Condition (2.49) can be interpreted as a direct generalization of the small-gain condition $\gamma_1 \cdot \gamma_2 < 1$. See also [149] for another formulation.

5. There is an extensive literature related to the theory presented in Sects. 2.1 and 2.2. Among the many contributions we mention the work of Safonov [262] & Teel [337] on conic relations, the work on nonlinear small-gain theorems in Mareels & Hill [194], Jiang, Teel & Praly [149], Teel [336] briefly discussed in the pre-

---

[1] A function $\gamma : \mathbb{R}^+ \to \mathbb{R}^+$ is of class $\mathcal{K}$ (denoted $\gamma \in \mathcal{K}$) if it is zero at zero, strictly increasing and continuous.

vious Note 4, and work on robust stability, see e.g., Georgiou [111], Georgiou & Smith [112], as well as the important contributions on stability theory within the "Petersburg school", see e.g., the classical paper Yakubovich [359], and developments inspired by this, see e.g., Megretski & Rantzer [215]. The developments stemming from dissipative systems theory will be treated in Chaps. 3, 4, and 8.

6. For further ramifications and implications of the loop transformations sketched in Sect. 2.3 we refer to Vidyasagar [343], Scherer, Gahinet & Chilali [306], Scherer [307], and the references quoted therein.

7. The scattering relation between L_2-gain and passivity is classical, and can be found in Desoer & Vidyasagar [83], see also Anderson [6]. The geometric, coordinate-free, treatment given in Sect. 2.4 is developed in Maschke & van der Schaft [208], Stramigioli, van der Schaft, Maschke & Melchiorri [190, 331], Cervera, van der Schaft & Banos [63].

Chapter 3
Dissipative Systems Theory

In this chapter the general theory of *dissipative systems* is treated, laying much of the foundation for subsequent chapters. The theory will be shown to provide a state space interpretation of the notions of finite L_2-gain and passivity for input–output maps as discussed in Chaps. 1 and 2, and to generalize the concept of *Lyapunov functions* for autonomous dynamical systems to systems with inputs and outputs.

3.1 Dissipative Systems

Throughout we consider state space systems with inputs and outputs of the general form

$$\Sigma : \quad \begin{aligned} \dot{x} &= f(x, u), \quad u \in U \\ y &= h(x, u), \quad y \in Y \end{aligned} \tag{3.1}$$

where $x = (x_1, \ldots, x_n)$ are local coordinates for an n-dimensional state space manifold $\mathcal{X}$, and U and Y are linear spaces, of dimension m, respectively p. Throughout this chapter, as well as in the subsequent chapters, we will make the following assumption; see also the discussion in Sect. 1.3.

Assumption 3.1.1 There exists a unique solution trajectory $x(\cdot)$ on the infinite time interval $[0, \infty)$ of the differential equation $\dot{x} = f(x, u)$, for all initial conditions $x(0)$ and all input functions $u(\cdot) \in L_{2e}(U)$. Furthermore it will be assumed that the thus generated output functions $y(\cdot) = h(x(\cdot), u(\cdot))$ are in $L_{2e}(Y)$.

On the combined space $U \times Y$ of inputs and outputs consider a function

$$s : U \times Y \to \mathbb{R}, \tag{3.2}$$

called the *supply rate*. Denote as before $\mathbb{R}^+ = [0, \infty)$.

© Springer International Publishing AG 2017
A. van der Schaft, L_2-Gain and Passivity Techniques in Nonlinear Control,
Communications and Control Engineering, DOI 10.1007/978-3-319-49992-5_3

Definition 3.1.2 A state space system Σ is said to be *dissipative* with respect to the supply rate s if there exists a function $S : \mathcal{X} \rightarrow \mathbb{R}^+$, called the *storage function*, such that for all initial conditions $x(t_0) = x_0 \in \mathcal{X}$ at any time t_0, and for all allowed input functions $u(\cdot)$ and all $t_1 \geq t_0$ the following inequality holds[1]

$$S(x(t_1)) \leq S(x(t_0)) + \int_{t_0}^{t_1} s(u(t), y(t))dt \tag{3.3}$$

If (3.3) holds with *equality* for all x_0, $t_1 \geq t_0$, and all $u(\cdot)$, then Σ is *conservative* with respect to s. Finally, Σ is called *cyclo-dissipative* with respect to s if there exists a function $S : \mathcal{X} \rightarrow \mathbb{R}$ (not necessarily nonnegative) such that (3.3) holds.

The inequality (3.3) is called the *dissipation inequality*. It expresses the fact that the "stored energy" $S(x(t_1))$ of Σ at any *future* time t_1 is at most equal to the stored energy $S(x(t_0))$ at *present* time t_0, *plus* the total *externally supplied energy* $\int_{t_0}^{t_1} s(u(t), y(t))dt$ during the time interval $[t_0, t_1]$. Hence, there can be no internal "creation of energy"; only internal *dissipation* of energy is possible.

Remark 3.1.3 Note that cyclo-dissipativity implies

$$\int_{t_0}^{t_1} s(u(t), y(t))dt \geq 0 \tag{3.4}$$

for all trajectories $u(\cdot), x(\cdot), y(\cdot)$ of Σ on the time interval $[t_0, t_1]$ which are such that $x(t_1) = x(t_0)$ (whence the terminology *cyclo*-dissipative).

The two most important choices of supply rates will be seen to correspond to the notions of passivity, respectively finite L_2-gain, as treated for input–output maps in the preceding chapters.

For simplicity of exposition we will identify throughout this chapter the linear input and output spaces U and Y with $\mathbb{R}^m$, respectively $\mathbb{R}^p$, equipped with the standard Euclidean inner product and norm. Throughout the Euclidean inner product of two vectors $v, z \in \mathbb{R}^m$ will be denoted by $v^T z$, and the Euclidean norm of a vector $v \in \mathbb{R}^m$ by $\|v\|$.

Definition 3.1.4 A *state space system* Σ with $U = Y = \mathbb{R}^m$ is *passive* if it is dissipative with respect to the supply rate $s(u, y) = u^T y$. Σ is *input strictly passive* if there exists $\delta > 0$ such that Σ is dissipative with respect to $s(u, y) = u^T y - \delta \|u\|^2$. Σ is *output strictly passive* if there exists $\varepsilon > 0$ such that Σ is dissipative with respect to $s(u, y) = u^T y - \varepsilon \|y\|^2$. Finally, Σ is *lossless* if it is conservative with respect to $s(u, y) = u^T y$.

Definition 3.1.4 is directly seen to extend the definitions of (input/output) strict passivity for input–output maps G as given in the previous Chap. 2. Based on Assumption 3.1.1 we consider for every $x_0 \in \mathcal{X}$ the input–output map $G_{x_0} : L_{2e}(U) \rightarrow$

---

[1] Here it is additionally assumed that for allowed input functions $u(\cdot)$ and generated output functions $y(\cdot)$ the integral $\int_{t_0}^{t_1} s(u(t), y(t))dt$ is well defined.

$L_{2e}(Y)$, given as the map from allowed input functions $u(\cdot)$ on $[0, \infty)$ to output functions $y(\cdot)$ on $[0, \infty)$ specified as $y(t) = h(x(t), u(t))$, where $x(t)$ is the state at time $t \geq 0$ resulting from initial condition $x(0) = x_0$ and input function $u(\cdot)$. Note that all these input–output maps G_{x_0}, $x_0 \in \mathcal{X}$, are *causal* (Definition 1.1.3), and *time invariant* (Definition 1.1.5).

Proposition 3.1.5 *Consider the state space system Σ with $U = Y = \mathbb{R}^m$, and consider for any x_0 the input–output map G_{x_0}. If Σ is passive, input strictly passive, respectively output strictly passive, then so are the input–output maps G_{x_0} for every x_0.*

Proof Suppose Σ is dissipative with respect to the supply rate $s(u, y) = u^T y$. Then for some function $S \geq 0$

$$\int_0^T u^T(t)y(t)dt \ \geq \ S(x(T)) - S(x(0)) \ \geq \ -S(x(0)) \tag{3.5}$$

for all $x(0) = x_0$, and all $T \geq 0$ and all input functions $u(\cdot)$. This means precisely that the input–output maps G_{x_0} of Σ, for every $x_0 \in \mathcal{X}$, are passive in the sense of Definition 2.2.1 (with bias β given as $S(x_0)$). The (input or output) strict passivity case follows similarly.

□

A second important class of supply rates is

$$s(u, y) = \frac{1}{2}\gamma^2||u||^2 - \frac{1}{2}||y||^2 \ , \qquad \gamma \geq 0 \ , \tag{3.6}$$

where $||u||$ and $||y||$ denote the Euclidian norms on $U = \mathbb{R}^m$, respectively $Y = \mathbb{R}^p$.

Definition 3.1.6 A *state space system* Σ with $U = \mathbb{R}^m$, $Y = \mathbb{R}^p$ has L_2-*gain* $\leq \gamma$ if it is dissipative with respect to the supply rate $s(u, y) = \frac{1}{2}\gamma^2||u||^2 - \frac{1}{2}||y||^2$. The L_2-gain of Σ is defined as $\gamma(\Sigma) := \inf\{\gamma \mid \Sigma$ has L_2-gain $\leq \gamma\}$. Σ is said to have L_2-gain $< \gamma$ if there exists $\tilde{\gamma} < \gamma$ such that Σ has L_2-gain $\leq \tilde{\gamma}$. Finally Σ is called *inner* if it is conservative with respect to $s(u, y) = \frac{1}{2}||u||^2 - \frac{1}{2}||y||^2$.

Definition 3.1.6 is immediately seen to extend the definition of finite L_2-gain from Chaps. 1 and 2.

Proposition 3.1.7 *Suppose Σ is dissipative with respect to $s(u, y) = \frac{1}{2}\gamma^2||u||^2 - \frac{1}{2}||y||^2$ for some $\gamma > 0$. Then all input–output maps $G_{x_0} : L_{2e}(U) \to L_{2e}(Y)$ have L_2-gain $\leq \gamma$. Furthermore, the infimum of the L_2-gains of G_{x_0} over all x_0 is equal to the L_2-gain of Σ.*

Proof If Σ is dissipative with respect to $s(u, y) = \frac{1}{2}\gamma^2||u||^2 - \frac{1}{2}||y||^2$ then there exists $S \geq 0$ such that for all $T \geq 0$, $x(0)$, and $u(\cdot)$

$$\frac{1}{2}\int_0^T (\gamma^2 ||u(t)||^2 - ||y(t)||^2)dt \geq S(x(T)) - S(x(0))|! \geq S(x(0)) \qquad (3.7)$$

and thus

$$\int_0^T ||y(t)||^2 dt \leq \gamma^2 \int_0^T ||u(t)||^2 dt + 2S(x(0)) \qquad (3.8)$$

This implies by Proposition 1.2.7 that the input–output maps G_{x_0} for every initial condition $x(0) = x_0$ have L_2-gain $\leq \gamma$. The rest of the statements follows directly.
□

Remark 3.1.8 Note that by considering supply rates $s(u, y) = \tilde{\gamma}||u||^q - ||y||^q$ we may also treat L_q-gain for $q \neq 2$; this will not be further discussed.

In the subsequent chapters we will elaborate on the special cases $s(u, y) = u^T y$ and $s(u, y) = \frac{1}{2}\gamma^2||u||^2 - \frac{1}{2}||y||^2$ corresponding to, respectively, passivity (Chap. 4) and finite L_2-gain (Chap. 8), in much more detail. Instead, in the current chapter we will focus on the *general* theory of dissipative systems.

Before doing so we mention one immediate generalization of the definition of dissipativity. In Chaps. 1 and 2 we already noticed that the notions of finite L_2-gain and passivity can be extended from input–output *maps* to *relations*. In the same vein, the definition of dissipativity for input–state–output systems Σ as in (3.1) can be extended to state space systems described by a mixture of differential *and* algebraic equations, where we do not distinguish a priori between input and output variables. That is, we may consider systems of the general form

$$F(x, \dot{x}, w) = 0, \qquad (3.9)$$

where $x = (x_1, \ldots, x_n)$ are local coordinates for an n-dimensional state space manifold $\mathcal{X}$, and $w \in W = \mathbb{R}^s$ denotes the total vector of *external variables*. Note that this entails two generalizations of (3.1): (i) we replace the combined vector $(u, y) \in Y \times U$ by a vector $w \in W$ (where we do not make an a priori splitting into input and output variables), and (ii) we replace the explicit differential and algebraic equations in (3.1) by a general mixture, called a set of *differential–algebraic equations* (DAEs). Note that systems (3.9) include *implicit* and *constrained* state space systems. In this more general context the supply rate s is now simply defined as a function

$$s : W \to \mathbb{R},$$

while the DAE system (3.9) is called dissipative with respect to s if there exists a function $S : \mathcal{X} \to \mathbb{R}^+$ such that

$$S(x(t_1)) \leq S(x(t_0)) + \int_{t_0}^{t_1} s(w(t))dt, \qquad (3.10)$$

for all[2] solutions $x(\cdot)$, $w(\cdot)$ of (3.9).

---

[2]Here we naturally restrict to continuous solutions.

Let us now return to the definition of dissipativity given in Definition 3.1.2. First, we notice that in general the storage function of a dissipative system is far from unique. Nonuniqueness already arises from the fact that we may always add a non-negative constant to a storage function, and so obtain another storage function. Indeed, the dissipation inequality (3.3) is invariant under addition of a constant to the storage function S. However, apart from this rather trivial nonuniqueness, more often than not dissipative systems will admit really *different* storage functions. Furthermore, if S_1 and S_2 are storage functions then any convex combination $\alpha S_1 + (1 - \alpha)S_2, \alpha \in [0, 1]$, is also a storage function, as immediately follows from substitution into the dissipation inequality. Hence the set of storage functions is always a *convex* set.

The storage function *is* guaranteed to be unique (up to a constant) in case the system is *conservative* and a controllability condition is met, as formulated in the following proposition.

Proposition 3.1.9 *Consider a system Σ that is conservative with respect to some supply rate s. Assume that the system is "connected" in the sense that for every two states x_a, x_b, there exists a number of intermediate states $x_1, x_2, \ldots, x_m$ with $x_1 = x_a, x_m = x_b$, such that for every pair x_i, x_{i+1} either x_i can be steered (by the application of a suitable input function) to x_{i+1}, or, conversely, x_{i+1} can be steered to $x_i, i = 1, 2 \ldots, m$. Then the storage function is unique up to a constant.*

Proof Let S^1, S^2 be two storage functions. By the dissipation equality, the difference $S^2 - S^1$ is constant along any state trajectory of the system. By the above property of "connectedness" this constant is the same for every state trajectory. Hence $S^2 = S^1$ up to this constant. □

Remark 3.1.10 A simple physical example of a dissipative system that is not conservative, but still has unique (up to a constant) storage function will be provided in Example 4.1.7 in Chap. 4.

A fundamental question is how we may *decide* if Σ is dissipative with respect to a given supply rate s. The following theorem gives an intrinsic *variational* characterization of dissipativity.

Theorem 3.1.11 *Consider the system Σ and supply rate $s(u, y)$. Then Σ is dissipative with respect to s if and only if*

$$S_a(x) := \sup_{\substack{u(\cdot) \\ T \geq 0}} - \int_0^T s(u(t), y(t))dt \,, \qquad x(0) = x, \qquad (3.11)$$

is finite ($<\infty$) for all $x \in \mathcal{X}$. Furthermore, if S_a is finite for all $x \in \mathcal{X}$ then S_a is a storage function, called the available storage, and all other possible storage functions S satisfy

$$S_a(x) \leq S(x) - \inf_x S(x) \,, \qquad \forall x \in \mathcal{X} \qquad (3.12)$$

Moreover,

$$\inf_x S_a(x) = 0 \tag{3.13}$$

Proof Suppose S_a is finite. Clearly $S_a \geq 0$ (take $T = 0$ in (3.11)). Compare now $S_a(x(t_0))$ with $S_a(x(t_1)) - \int_{t_0}^{t_1} s(u(t), y(t))dt$, for a given $u : [t_0, t_1] \to \mathbb{R}^m$ and resulting state $x(t_1)$. Since S_a is given as the *supremum* over all $u(\cdot)$ in (3.11) it immediately follows that

$$S_a(x(t_0)) \geq S_a(x(t_1)) - \int_{t_0}^{t_1} s(u(t), y(t))dt , \tag{3.14}$$

and thus S_a is a storage function, proving that Σ is dissipative with respect to the supply rate s.

Suppose conversely that Σ is dissipative. Then there exists $S \geq 0$ such that for all $u(\cdot)$

$$S(x(0)) + \int_0^T s(u(t), y(t))dt \geq S(x(T)) \geq 0, \tag{3.15}$$

which shows that

$$S(x(0)) \geq \sup - \int_0^T s(u(t), y(t))dt = S_a(x(0)), \tag{3.16}$$

proving finiteness of S_a. On the other hand, $S' := S - \inf_x S(x)$ is a storage function as well (since we have just subtracted the constant $\inf_x S(x)$ from S and thus the dissipation inequality remains to hold, while clearly $S' \geq 0$). Hence also $S'(x_0) \geq S_a(x_0)$ for all x_0, proving (3.12). Moreover, since $\inf_x S'(x) = 0$, also $\inf_x S_a(x) = 0$. □

The quantity $S_a(x_0)$ can be interpreted as the maximal "energy" which can be extracted from the system Σ starting at initial condition x_0. Theorem 3.1.11 thus states that Σ is dissipative if and only if this "extractable energy" is finite for every initial condition.

Under additional conditions the following equivalent characterizations of the available storage S_a can be obtained.

Proposition 3.1.12 (*i*) *Assume the system Σ and supply rate $s(u, y)$ are such that for any x there exists $u(x)$ such that*

$$s(u(x), h(x, u(x))) \leq 0, \quad x \in \mathcal{X} \tag{3.17}$$

Then

$$S_a(x) = \sup_{u(\cdot)} - \int_0^\infty s(u(t), y(t))dt, \quad x(0) = x \tag{3.18}$$

(ii) *Assume* Σ *and* $s(u, y)$ *are such that there exists a state feedback* $u(x)$ *such that* (3.17) *holds, while furthermore* x^* *is a globally asymptotically stable equilibrium of the closed-loop system*[3] $\dot{x} = f(x, u(x))$. *Then*

$$S_a(x) = \sup_{u(\cdot),\, x \to x^*} - \int_0^\infty s(u(t), y(t))dt, \qquad x(0) = x \qquad (3.19)$$

Proof (i) By letting $T \to \infty$ in (3.11) we have $\sup_{u(\cdot)} - \int_0^\infty s(u(t), y(t))dt \le S_a(x)$. Conversely, note that

$$\begin{aligned} - \int_0^\infty s(u(t), y(t))dt &= - \int_0^T s(u(t), y(t))dt - \int_T^\infty s(u(t), y(t))dt \\ &\ge - \int_0^T s(u(t), y(t))dt\,, \end{aligned} \qquad (3.20)$$

whenever $u(\cdot)$ is such that $s(u(t), y(t)) \le 0$ for all $t \in [T, \infty)$. Hence for any $\bar{u} : [0, T] \to U$ there exists $u : [0, \infty) \to U$ with $u_T = \bar{u}$ such that $- \int_0^\infty s(u(t), y(t))dt \ge - \int_0^T s(\bar{u}(t), y(t))dt$. Therefore, by taking the supremum at both sides of this inequality we obtain the inequality $\sup_{u(\cdot)} - \int_0^\infty s(u(t), y(t))dt \ge S_a(x)$.

(ii) As in the proof of part (i) we have $\sup_{u(\cdot),\, x \to x^*} - \int_0^\infty s(u(t), y(t))dt \le S_a(x)$. For the reverse inequality we apply the same reasoning as in the proof of part (i), by considering extensions of $\bar{u} : [0, T] \to U$ to $u : [0, \infty) \to U$ which are such that $x(t) \to x^*$ for $t \to \infty$. $\qquad \square$

Remark 3.1.13 Note that part (i) of Proposition 3.1.12 applies to the (input strict or output strict) passivity supply rate and to the L_2-gain supply rate by taking $u = 0$. Furthermore, part (ii) applies whenever Σ has a globally asymptotically stable equilibrium x^* for $u = 0$.

The next proposition shows that if the system is *reachable* from some state, then the finiteness of extractable energy needs only to be checked for this initial condition.

Proposition 3.1.14 *Assume that* Σ *is reachable from* $x^* \in \mathcal{X}$. *Then* Σ *is dissipative if and only if* $S_a(x^*) < \infty$.

Proof (*Only if*) Trivial. (If) Suppose there exists $x \in \mathcal{X}$ such that $S_a(x) = \infty$. Since by reachability we can steer x^* to x in finite time, this would imply (using time invariance) that also $S_a(x^*) = \infty$. $\qquad \square$

Corollary 3.1.15 *Assume that* Σ *is reachable from* $x^* \in \mathcal{X}$. *Then* Σ *is passive if and only if the input–output map* G_{x^*} *is passive, and* Σ *has* L_2-gain $\le \gamma$ *if and only if* G_{x^*} *has* L_2-gain $\le \gamma$, *while* $\gamma(\Sigma) = \gamma(G_{x^*})$. *Furthermore, if* G_{x^*} *is passive with zero bias or has* L_2-gain $\le \gamma$ *with zero bias, then* $S_a(x^*) = 0$.

---

[3]Here it is assumed that $\dot{x} = f(x, u(x))$ has unique solutions on $[0, \infty)$ for all initial conditions.

Proof Suppose the input–output map G_{x^*} is passive, then $\exists\ \beta < \infty$ such that (cf. Definition 2.2.1)

$$\int_0^T u^T(t)y(t)dt \ \geq\ -\beta \tag{3.21}$$

for all $u(\cdot)$, $T \geq 0$. Therefore

$$S_a(x^*) = \sup_{u(\cdot),\,T\geq 0}\ -\int_0^T u^T(t)y(t)dt \ \leq\ \beta < \infty, \qquad x(0) = x^* \tag{3.22}$$

and by Proposition 3.1.14 Σ is passive. If $\beta = 0$ then $S_a(x^*) = 0$.

Similarly, let G_{x^*} have L_2-gain $\leq \gamma$, then (cf. Proposition 1.2.7) for all $\tilde\gamma > \gamma$ there exists a constant c such that

$$\int_0^T \|y(t)\|^2 dt \ \leq\ \tilde\gamma^2 \int_0^T \|u(t)\|^2 dt + c \tag{3.23}$$

yielding (with $x(0) = x^*$)

$$S_a(x^*) \ = \ \sup_{u(\cdot),\,T\geq 0}\ -\int_0^T \left(\frac{1}{2}\tilde\gamma^2\|u(t)\|^2 - \frac{1}{2}\|y(t)\|^2\right) dt \ \leq\ \frac{c}{2}, \tag{3.24}$$

implying that Σ has L_2-gain $\leq \tilde\gamma$ for all $\tilde\gamma > \gamma$. If $c = 0$, then clearly $S_a(x^*) = 0$. It also follows that $\gamma(\Sigma) = \gamma(G_{x^*})$. $\qquad\square$

If Σ is reachable from a state x^* then, in addition to the available storage S_a, there exists another canonically defined storage function. Contrary to the available storage, which is the *minimal* storage function (see (3.12)), this storage function has a *maximality* property, in the following sense.

Theorem 3.1.16 *Assume that Σ is reachable from $x^* \in \mathcal{X}$. Define the required supply (from x^*) $S_r : \mathcal{X} \to \mathbb{R} \cup \{-\infty\}$ as*

$$S_r(x) := \inf_{u(\cdot),\,T\geq 0} \int_{-T}^0 s(u(t), y(t))dt\ , \quad x(-T) = x^*,\ x(0) = x \tag{3.25}$$

Then S_r satisfies the dissipation inequality (3.3). Furthermore, Σ is dissipative if and only if there exists $K > -\infty$ such that $S_r(x) \geq K$ for all $x \in \mathcal{X}$. Moreover, if S is a storage function for Σ, then

$$S(x) \ \leq\ S_r(x) + S(x^*), \qquad \forall x \in \mathcal{X}, \tag{3.26}$$

and $S_r(x) + S(x^*)$ is itself a storage function. In particular, $S_r(x) + S_a(x^*)$ is a storage function.

Proof The fact that S_r satisfies the dissipation inequality (3.3) follows from the variational definition of S_r in (3.25). Indeed, in taking the system from x^* at $t = -T$ to $x(t_1)$ at time t_1 we can restrict to those input functions $u(\cdot) : [-T, t_1] \to U$ which first take x^* to $x(t_0)$ at time $t_0 \le t_1$, and then are equal to a given input $u(\cdot) : [t_0, t_1] \to U$ transferring $x(t_0)$ to $x(t_1)$. This will a be suboptimal control policy, whence

$$S_r(x(t_0)) + \int_{t_0}^{t_1} s(u(t), y(t))dt \ \ge \ S_r(x(t_1)) \tag{3.27}$$

For the second claim, note that by definition of S_a and S_r

$$S_a(x^*) = \sup_x -S_r(x), \tag{3.28}$$

from which by Proposition 3.1.14 it follows that Σ is dissipative if and only if $\exists K > -\infty$ such that $S_r(x) \ge -K$ for all x.

Finally, let S satisfy the dissipation inequality (3.3). Then for any $u(\cdot) : [-T, 0] \to \mathbb{R}^m$ transferring $x(-T) = x^*$ to $x(0) = x$ we have by the dissipation inequality

$$S(x) - S(x^*) \ \le \ \int_{-T}^{0} s(u(t), y(t))dt \tag{3.29}$$

Taking the infimum on the right-hand side over all those $u(\cdot)$ yields (3.26). Furthermore if $S \ge 0$, then by (3.26) $S_r + S(x^*) \ge 0$, and by adding $S(x^*)$ to both sides of (3.27) it follows that also $S_r + S(x^*)$ satisfies the dissipation inequality. □

Remark 3.1.17 Let Σ be reachable from x^*. Then under the additional assumption of existence of u^* such that $f(x^*, u^*) = 0$, $h(x^*, u^*) = 0$ it can be verified that the required supply is equivalently given as

$$S_r(x) = \lim_{t_1 \to -\infty} \inf_{u(\cdot),\, x(t_1)=x^*,\, x(0)=x} \int_{t_1}^{0} s(u(t), y(t))dt \tag{3.30}$$

Furthermore, we note that in case Σ is dissipative with a storage function S which attains its global *minimum* at some point $x^* \in \mathcal{X}$, then also $S - S(x^*)$ will be a storage function, which is *zero* at x^*. Hence in this case any motion starting from x^* at time 0 satisfies by the dissipation inequality

$$\int_{0}^{T} s(u(t), y(t))dt \ \ge \ 0, \qquad x(0) = x^*, \ \forall T \ge 0 \tag{3.31}$$

Thus if we start from the state of "minimal energy" x^* then the net supply flow is always directed *into* the system.[4] This leads to the following *alternative* definition of dissipativity.

Definition 3.1.18 Consider a system Σ and supply rate s. The system is called *dissipative from* x^* if (3.31) holds.

Proposition 3.1.19 *Let Σ be dissipative with storage function S satisfying $S(x^*) = 0$. Then the system is also dissipative from x^*. Conversely, if the system is dissipative from x^* then $S_a(x^*) = 0$. If additionally the system is reachable from x^* then the system is dissipative, while its required supply satisfies $S_r(x^*) = 0$.*

Proof The fact that dissipativity with storage function S satisfying $S(x^*) = 0$ implies (3.31) was already observed in (3.31). Conversely, assume that the system is dissipative from x^*. Then by definition of S_a in (3.11) it directly follows that $S_a(x^*) = 0$. Furthermore by Proposition 3.1.14 it follows that the system is dissipative, while clearly $S_r(x^*) = 0$. $\square$

Hence, if Σ is dissipative as well dissipative from x^*, then both S_a and S_r attain their minimum 0 at x^*. Under an additional assumption it can be shown that all other storage functions attain their minimum at x^* as well, as formulated in the following proposition.

Proposition 3.1.20 *Let Σ be dissipative and dissipative from x^*. Suppose furthermore the supply rate s is such that there exists a feedback $u(x)$ satisfying (3.17) for which x^* is a globally asymptotically equilibrium for the closed-loop system $\dot{x} = f(x, u(x))$. Then any storage function S attains its minimum at x^*, implying that $S(x) - S(x^*)$ is a storage function that is zero at x^*. Furthermore*

$$S_a(x) \leq S(x) - S(x^*), \qquad x \in \mathcal{X} \tag{3.32}$$

Proof Consider the dissipation inequality for any storage function S, rewritten as

$$-\int_0^T s(u(t), y(t))dt \leq S(x) - S(x(T)) \tag{3.33}$$

with $x(0) = x$. Extend $u(\cdot) : [0, T] \to U$ to the infinite time interval $[0, \infty)$ by considering on (T, ∞) a feedback $u(x)$ as in (3.17) such that x^* is a globally asymptotically equilibrium $\dot{x} = f(x, u(x))$. It follows from (3.20) and convergence of $x(t)$ to x^* for $t \to \infty$ that

$$-\int_0^T s(u(t), y(t))dt \leq S(x) - S(x^*) \tag{3.34}$$

---

[4]Note however that there does not always exist such a state of minimal internal energy. In particular $\inf_x S_a(x) = 0$ but not necessarily the minimum is attained.

Hence by taking the supremum at the left-hand side over all $u(\cdot) : [0, T] \to U$ and $T \geq 0$ we obtain (3.32), also implying that S attains its minimum at x^*. ☐

The above developments can be summarized as follows.

Corollary 3.1.21 *Consider a system (3.1) that is dissipative from x^*, reachable from x^*, and for which there exists a feedback $u(x)$ satisfying (3.17) such that x^* is a globally asymptotically equilibrium of $\dot{x} = f(x, u(x))$. Then any storage function S attains its minimum at x^* and the storage function $S'(x) := S(x) - S(x^*)$ satisfies*

$$S_a(x) \leq S'(x) \leq S_r(x), \quad \text{for all } x \in \mathcal{X}, \tag{3.35}$$

where $S_a(x^*) = S_r(x^*) = 0$.

Remark 3.1.22 For a linear system $\dot{x} = Ax + Bu$, $y = Cx + Du$ with $x^* = 0$ satisfying the assumptions of Corollary 3.1.21 it can be proved by standard optimal control arguments [351] that S_a and S_r are given by quadratic functions $\frac{1}{2}x^T Q_a x$, respectively $\frac{1}{2}x^T Q_r x$, with Q_a, Q_r symmetric matrices satisfying $Q_a \leq Q_r$.

3.2 Stability of Dissipative Systems

In this section we will elaborate on the close connection between dissipative systems theory and the theory of *Lyapunov functions* for autonomous dynamical systems $\dot{x} = f(x)$.

Consider the dissipation inequality (3.3), where we assume throughout this section that the storage functions S are C^1 (continuously differentiable); see the discussion in the Notes for this chapter for generalizations. By dividing the dissipation inequality by $t_1 - t_0$, and letting $t_1 \to t_0$ we see that (3.3) is equivalent to

$$S_x(x)f(x, u) \leq s(u, h(x, u)), \quad \text{for all } x, u, \tag{3.36}$$

with $S_x(x)$ denoting the *row vector* of partial derivatives

$$S_x(x) = \left(\frac{\partial S}{\partial x_1}(x), \ldots, \frac{\partial S}{\partial x_n}(x) \right) \tag{3.37}$$

The inequality (3.36) is called the *differential dissipation inequality*, and is much easier to check than (3.3) since we do not have to compute the system trajectories (which for most nonlinear systems is even not possible).

In order to make the connection with the theory of Lyapunov functions we recall some basic notions and results from Lyapunov stability theory. Consider the set of differential equations

$$\dot{x} = f(x) \tag{3.38}$$

Here x are local coordinates for an n-dimensional manifold $\mathcal{X}$, and thus (3.38) is the local coordinate expression of a *vector field* on $\mathcal{X}$. Throughout we assume that f is locally Lipschitz continuous; implying existence and uniqueness of solutions of (3.38), at least for small time. The solution of (3.38) for initial condition $x(0) = x_0$ will be denoted as $x(t; x_0)$, with $t \in [0, T(x_0))$ and $T(x_0) > 0$ maximal.

Definition 3.2.1 Let x^* be an equilibrium of (3.38), that is $f(x^*) = 0$, and thus $x(t; x^*) = x^*$, for all t. The equilibrium x^* is

(a) *stable*, if for each $\varepsilon > 0$ there exists $\delta(\varepsilon)$ such that

$$\| x_0 - x^* \| < \delta(\varepsilon) \Rightarrow \| x(t; x_0) - x^* \| < \varepsilon, \quad \forall t \geq 0 \qquad (3.39)$$

(b) *asymptotically stable*, if it is stable and additionally there exists $\bar{\delta} > 0$ such that

$$\| x_0 - x^* \| < \bar{\delta} \Rightarrow \lim_{t \to \infty} x(t, x_0) = x^* \qquad (3.40)$$

(c) *globally asymptotically stable*, if it is stable and $\lim_{t \to \infty} x(t; x_0) = x^*$ for all $x_0 \in \mathcal{X}$.
(d) *unstable*, if it is not stable.

Remark 3.2.2 If x^* is a globally asymptotically stable equilibrium then necessarily $\mathcal{X}$ is diffeomorphic to $\mathbb{R}^n$.

An important tool in the stability analysis of equilibria are *Lyapunov functions*.

Definition 3.2.3 Let x^* be an equilibrium of (3.38). A C^1 function $V : \mathcal{X} \to \mathbb{R}^+$ satisfying
$$V(x^*) = 0, \quad V(x) > 0, \quad x \neq x^* \qquad (3.41)$$

(that is, V is *positive definite* at x^*), as well as

$$\dot{V}(x) := V_x(x)f(x) \leq 0, \quad x \in \mathcal{X}, \qquad (3.42)$$

is called a *Lyapunov function* for the equilibrium x^*.

Theorem 3.2.4 *Let x^* be an equilibrium of (3.38). If there exists a Lyapunov function for the equilibrium x^*, then x^* is a stable equilibrium. If moreover*

$$\dot{V}(x) < 0, \quad \forall x \in \mathcal{X}, \ x \neq x^*, \qquad (3.43)$$

then x^* is an asymptotically stable equilibrium, which is globally asymptotically stable if V is proper (that is, the sets $\{x \in \mathcal{X} \mid 0 \leq V(x) \leq c\}$ are compact for every $c \in \mathbb{R}^+$).

Remark 3.2.5 Theorem 3.2.4 can be also applied to any neighborhood $\tilde{\mathcal{X}}$ of x^*. In particular, if (3.41) and (3.42), or (3.41) and (3.43) hold on a neighborhood of x^*, then x^* is still a stable, respectively, asymptotically stable, equilibrium.

Remark 3.2.6 For $\mathcal{X} = \mathbb{R}^n$ the requirement of properness amounts to V being *radially unbounded*; that is, $V(x) \to \infty$ whenever $\|x\| \to \infty$.

With the aid of Theorem 3.2.4 the following stability result for dissipative systems is readily established.

Proposition 3.2.7 *Let $s(u, y)$ be a supply rate, and $S : \mathcal{X} \to \mathbb{R}^+$ be a C^1 storage function for Σ. Assume that s satisfies*

$$s(0, y) \leq 0, \qquad \forall y \in Y \tag{3.44}$$

Assume furthermore that $x^* \in \mathcal{X}$ is a strict local minimum for S. Then x^* is a stable equilibrium of the unforced system $\dot{x} = f(x, 0)$ with Lyapunov function $V(x) := S(x) - S(x^*)$ for x around x^*, while $s(0, h(x^*, 0)) = 0$. If additionally, $\dot{S}(x) < 0$, for all $x \neq x^*$, then x^* is an asymptotically stable equilibrium.

Proof By (3.36) and (3.44) $S_x(x)f(x, 0) \leq s(0, h(x, 0)) \leq 0$, and thus S is nonincreasing along solutions of $\dot{x} = f(x, 0)$. Since S has a strict minimum at x^* this implies $f(x^*, 0) = 0$, and thus $s(0, h(x^*, 0)) = 0$. The rest follows directly from Theorem 3.2.4. $\square$

An important weakness in the asymptotic stability statement of Proposition 3.2.7 concerns the condition $\dot{S}(x) < 0$ for all $x \neq x^*$. In general, this condition can*not* be inferred from the dissipation inequality (unless e.g., $y = x$). An important generalization of Theorem 3.2.4 to remedy this weakness is based on *LaSalle's Invariance principle*. Recall that a set $\mathcal{N} \subset \mathcal{X}$ is *invariant* for $\dot{x} = f(x)$ if $x(t; x_0) \in \mathcal{N}$ for all $x_0 \in \mathcal{N}$ and for all $t \in \mathbb{R}$, and is *positively invariant* if this holds for all $t \geq 0$, where $x(t; x_0)$, $t \geq 0$, denotes the solution of $\dot{x} = f(x)$ for $x(0) = x_0$.

Theorem 3.2.8 *Let $V : \mathcal{X} \to \mathbb{R}$ be a C^1 function for which $\dot{V}(x) := V_x(x) f(x) \leq 0$, for all $x \in \mathcal{X}$. Suppose there exists a compact set C which is positively invariant for $\dot{x} = f(x)$. Then for any $x_0 \in C$ the solution $x(t; x_0)$ converges for $t \to \infty$ to the largest subset of $\{x \in \mathcal{X} \mid \dot{V}(x) = 0\} \cap C$ that is invariant for $\dot{x} = f(x)$.*

The usual way of applying Theorem 3.2.8 is as follows. Since $\dot{V}(x) \leq 0$, the connected component of $\{x \in \mathcal{X} \mid V(x) \leq V(x_0)\}$ containing x_0 is positively invariant. If additionally V is assumed to be *positive definite* at x^* then the connected component of $\{x \in \mathcal{X} \mid V(x) \leq V(x_0)\}$ containing x_0 will be compact for x_0 close enough to x^*, and hence may serve as the compact set C in the above theorem.

Using this reasoning Theorem 3.2.8 yields the following connection between dissipativity and asymptotic stability.

Proposition 3.2.9 *Let $S : \mathcal{X} \to \mathbb{R}^+$ be a C^1 storage function for Σ. Assume that the supply rate s satisfies*

$$s(0, y) \leq 0, \qquad for\ all\ y \tag{3.45}$$

Assume that $x^* \in \mathcal{X}$ is a strict local minimum for S. Furthermore, assume that no solution of $\dot{x} = f(x, 0)$ other than $x(t) \equiv x^*$ remains in $\{x \in \mathcal{X} \mid s(0, h(x, 0)) = 0\}$ for all t. Then x^* is an asymptotically stable equilibrium of $\dot{x} = f(x, 0)$, which is globally asymptotically stable if $V \geq 0$ is proper.

Proof Note that $\dot{S}(x) = 0$ implies $s(0, h(x, 0)) = 0$, which by assumption implies $h(x, 0) = 0$. The statement now directly follows from LaSalle's Invariance principle.   □

Remark 3.2.10 The requirement $s(0, y) \leq 0$ for all y is satisfied by the (output and input strict) passivity and L_2-gain supply rates.

A main condition on the storage function S in the previous statements is the requirement that S has a *strict* (local) minimum at the equilibrium x^*. This is *not* part of the standard definition of a storage function. On the other hand, in case $S(x^*) = 0$ (see also Proposition 3.1.20), then the property of a strict (local) minimum may be sometimes *derived* making use of an additional *observability* condition.

In the rest of this section we assume that Σ has no *feedthrough terms*, i.e., $y = h(x)$. Without loss of generality take $x^* = 0$. Moreover, assume $h(0) = 0$.

Definition 3.2.11 Σ with $y = h(x)$ is *zero-state observable* if $u(t) = 0$, $y(t) = 0, \forall t \geq 0$, implies $x(t) = 0, \forall t \geq 0$.

Proposition 3.2.12 *Let $S \geq 0$ be a C^1 storage function with $S(0) = 0$ for a supply rate s satisfying $s(0, y) \leq 0$ for all y, and such that $s(0, y) = 0$ implies $y = 0$. Suppose Σ_a is zero-state observable, then $S(x) > 0$ for all $x \neq 0$.*

Proof By substituting $u = 0$ in (3.3) we obtain

$$S(x(T)) - S(x(0)) \leq \int_0^T s(0, y(t))dt$$

implying, since $S(x(T)) \geq 0$,

$$S(x(0)) \geq \int_0^T s(0, y(t))dt$$

which is > 0 for $x(0) \neq 0$.   □

Remark 3.2.13 The same result follows for any supply rate $s(u, y)$ for which there exists an output feedback $u = \alpha(y)$ such that $s(\alpha(y), y) \leq 0$ for all y, and $s(\alpha(y), y) = 0$ implies $y = 0$. Just consider in the above proof $u = \alpha(y)$ instead of $u = 0$.

Remark 3.2.14 Note that the L_2-gain and output strict passivity supply rate satisfy the conditions in the above Proposition 3.2.12, while the passivity supply rate satisfies the conditions of Remark 3.2.13 (take $u = -y$).

A weaker property of observability, called *zero-state detectability*, is instrumental for proving asymptotic stability based on LaSalle's Invariance principle.

Definition 3.2.15 Σ_a is *zero-state detectable* if $u(t) = 0$, $y(t) = 0$, $\forall t \geq 0$, implies $\lim_{t \to \infty} x(t) = 0$.

Proposition 3.2.16 *Let S be a C^1 storage function with $S(0) = 0$ and $S(x) > 0$, $x \neq 0$, for a supply rate s satisfying $s(0, y) \leq 0$ and such that $s(0, y) = 0$ implies $y = 0$, where $h(0) = 0$. Suppose that Σ_a is zero-state detectable. Then $x = 0$ is an asymptotically stable equilibrium of $\dot{x} = f(x, 0)$. If additionally S is proper then 0 is globally asymptotically stable.*

Proof By Proposition 3.2.9 $x = 0$ is a stable equilibrium of $\dot{x} = f(x, 0)$. Furthermore

$$\dot{S}(x) = S_x(x) f(x) \leq s(0, h(x, 0)),$$

and asymptotic stability follows by LaSalle's Invariance principle, since $\dot{S}(x) = 0$ implies $h(x, 0) = 0$. □

Finally, let us investigate the case that the storage function S has a local minimum at x^*, which is however *not* a strict minimum. In this case, $S(x) - S(x^*)$ is *not* a standard Lyapunov function, and thus stability, let alone asymptotic stability, of x^* is not guaranteed. Nevertheless, even in this case one can still obtain (asymptotic) stability, provided additional conditions are satisfied. The tool for doing this is formulated in the following theorem; see the references in the Notes for this chapter.

Theorem 3.2.17 *Let x^* be an equilibrium of $\dot{x} = f(x)$, and let $V : \mathcal{X} \to \mathbb{R}^1$ be a C^1 function which is positive semi-definite at x^*, that is,*

$$V(x^*) = 0, \quad V(x) \geq 0 \tag{3.46}$$

Furthermore, suppose that $\dot{V}(x) := V_x(x) f(x) \leq 0$, for all $x \in \mathcal{X}$.

(i) *Define $\mathcal{V}_0 := \{x \in \mathcal{X} \mid V(x) = 0\}$. If x^* is asymptotically stable conditionally to $\mathcal{V}_0$, that is (3.39) and (3.40) hold for $x_0 \in \mathcal{V}_0$, then x^* is a stable equilibrium of $\dot{x} = f(x)$.*

(ii) *Define $\mathcal{V} := \{x \in \mathcal{X} \mid \dot{V}(x) = 0\}$, and let $\mathcal{V}^*$ be the largest positively invariant (with respect to $\dot{x} = f(x)$) set contained in $\mathcal{V}$. Then x^* is an asymptotically stable equilibrium of $\dot{x} = f(x)$ if and only if x^* is an asymptotically stable equilibrium conditionally to $\mathcal{V}^*$, that is, (3.39) and (3.40) hold for $x_0 \in \mathcal{V}^*$.*

Remark 3.2.18 By replacing the condition $\dot{V}(x) = V_x(x) f(x) \leq 0$ by the condition that the function V is nonincreasing along solution trajectories, the above theorem also holds for functions V which are not C^1.

With the aid of Theorem 3.2.17 we obtain the following stability result extending Proposition 3.2.9.

Proposition 3.2.19 *Let* $S \geq 0$ *with* $S(x^*) = 0$ *be a solution to the dissipation inequality, where the supply rate* $s(u, y)$ *is such that*

$$s(0, y) \leq 0 \text{ for all } y, \ s(0, y) = 0 \text{ if and only if } y = 0$$

Let $\mathcal{H}^*$ *be the largest positively invariant set contained in the set* $\mathcal{H}:=\{x \mid h(x, 0)=0\}$. *If* x^* *is asymptotically stable conditionally to* $\mathcal{H}^*$, *then* x^* *is an asymptotically stable equilibrium.*

Proof In view of the dissipation inequality $\dot{S}(x) \leq s(0, h(x, 0)) \leq 0$. Since $s(0, y) = 0$ if and only if $y = 0$, it follows that the largest positively invariant set where $\dot{S}(x) = 0$ is contained in $\mathcal{H}^*$. Application of Theorem 3.2.17 yields the claim. $\qquad\qquad\square$

Remark 3.2.20 Note that the L_2-gain and output strict passivity supply rates satisfy the conditions of Proposition 3.2.19.

Remark 3.2.21 The property of $x^* = 0$ being asymptotically stable conditionally to the largest positively invariant set contained in the set $\{x \mid h(x, 0) = 0\}$ is very close to zero-state detectability. In fact, this latter property implies that $\lim_{t \to \infty} x(t) = 0$ whenever $y(t) = 0$, $t \geq 0$, for all initial conditions x_0 close to 0.

For later use we state the following closely related result.

Proposition 3.2.22 *Consider the* C^1 *system*

$$\dot{x} = f(x) + g(x)k(x), \quad f(x^*) = 0, \quad k(x^*) = 0, \qquad (3.47)$$

and assume that x^* *is an asymptotically stable equilibrium of* $\dot{x} = f(x)$, *and that there exists a* C^1 *function* $S \geq 0$ *which is positive semi-definite at* x^* *and satisfies*

$$S_x(x)\,[f(x) + g(x)k(x)] \leq -\varepsilon||k(x)||^2, \qquad (3.48)$$

for some $\varepsilon > 0$. *Then* x^* *is an asymptotically stable equilibrium of (3.47).*

Proof Similarly to the proof of Proposition 3.2.19, let $\mathcal{K}^*$ be the largest positively invariant set contained in $\mathcal{K} := \{x \mid k(x) = 0\}$. Since x^* is an asymptotically stable equilibrium of $\dot{x} = f(x)$ it follows that x^* is asymptotically stable conditionally to $\mathcal{K}^*$. Since $S_x(x)[f(x) + g(x)k(x)] = 0$ implies $k(x) = 0$, the rest of the proof is the same as that of Proposition 3.2.19. $\qquad\qquad\square$

Remark 3.2.23 Note that the condition of x^* being an asymptotically stable equilibrium of $\dot{x} = f(x)$ can be regarded as a *zero detectability* assumption on $\dot{x} = f(x) + g(x)k(x)$, $y = k(x)$.

3.3 Interconnections of Dissipative Systems

Dissipative systems theory can be viewed as an extension of Lyapunov function theory to systems with external variables (inputs and outputs). Furthermore, it provides a systematic way to construct Lyapunov functions for large-scale interconnected systems by starting from the storage functions of the component systems, and requiring a compatibility between the interconnection equations and the supply rates of the component systems. In fact, this will be a leading theme in the state space versions of the passivity theorems in Chap. 4 and the small-gain theorems in Chap. 8. Also this is a continuing thread in the theory of port-Hamiltonian systems in Chap. 6. While in these subsequent chapters the attention will be confined to the passivity supply rate (in the case of passive and port-Hamiltonian systems) and the L_2-gain supply rate (in the case of the small-gain theorems) the current section will be devoted to a *general* theory of interconnections of dissipative systems.

Consider k systems Σ_i of the form (3.1) with input, state, and output spaces $U_i, \mathcal{X}_i, Y_i, i = 1, \ldots, k$. Suppose Σ_i are dissipative with respect to the supply rates

$$s_i(u_i, y_i), \quad u_i \in U_i, \ y_i \in Y_i, \quad i = 1, \ldots, k, \tag{3.49}$$

and storage functions $S_i(x_i), i = 1, \ldots, k$.

Now consider an interconnection of $\Sigma_i, i = 1, \ldots, k$, defined through an *interconnection subset*

$$I \subset U_1 \times Y_1 \times \cdots \times U_k \times Y_k \times U^e \times Y^e \tag{3.50}$$

where U^e, Y^e are spaces of external input and output variables u^e, y^e. This defines an interconnected system Σ_I with state space $\mathcal{X}_1 \times \cdots \times \mathcal{X}_k$ and inputs and outputs u^e, y^e, by imposing the interconnection equations

$$\big((u_1, y_1), \ldots, (u_k, y_k), (u^e, y^e)\big) \in I \tag{3.51}$$

Note that in general the interconnected system Σ_I is of the DAE form (3.9). The following result is immediate.

Proposition 3.3.1 *Suppose the supply rates $s_1, \ldots, s_k$ and the interconnection subset I are such that there exists a supply rate $s^e : U^e \times Y^e \to \mathbb{R}$ for which*

$$
\begin{aligned}
s_1(u_1, y_1) + \cdots + s_k(u_k, y_k) \le s^e(u^e, y^e), \\
\text{for all } ((u_1, y_1), \ldots, (u_k, y_k), (u^e, y^e)) \in I
\end{aligned}
\tag{3.52}
$$

Then the interconnected system Σ_I is dissipative with respect to the supply rate s^e, with storage function

$$S(x_1, \ldots, x_k) := S_1(x_1) + \cdots + S_k(x_k) \tag{3.53}$$

Proof Just add the k dissipation inequalities

$$S_i(x_i(t_1)) \leq S_i(x_i(t_0)) + \int_{t_0}^{t_1} s_i(u_i(t), y_i(t))dt, \quad i = 1, \ldots, k$$

and invoke the inequality (3.52). □

Note that for the purpose of stability analysis of the interconnected system Σ_I the external inputs and outputs u^e, y^e and the supply rate s^e can be left out, in which case (3.52) reduces to

$$s_1(u_1, y_1) + \cdots + s_k(u_k, y_k) \leq 0,$$
$$\text{for all } ((u_1, y_1), \ldots, (u_k, y_k)) \in I \tag{3.54}$$

Example 3.3.2 Consider a system having inputs and outputs (u_c, y_c) accessible to control interaction, and another set of inputs and outputs (u_e, y_e) via which the system interacts with its environment. Suppose the system is passive, with respect to the combined set of variables (u_c, y_c) and (u_e, y_e); that is, there exists a storage function S such that

$$\frac{dS}{dt} \leq u_c^T y_c + (u^e)^T y^e$$

An example is a robotic mechanism interacting with its environment via generalized forces u_e and generalized velocities y_e, and controlled by collocated sensors (generalized velocities y_c) and actuators (generalized forces u_c). Closing the loop with a passive controller with storage function S_c, that is,

$$\frac{dS_c}{dt} \leq -y_c^T u_c,$$

results in a system which is passive with respect to (u^e, y^e), since

$$\frac{d}{dt}(S + S_c) \leq (u^e)^T y^e$$

Note that the storage function of the interconnected system Σ_I in Proposition 3.3.1 is simply the *sum* of the storage functions of the component systems Σ_i. A useful extension of Proposition 3.3.1 is obtained by allowing instead for *weighted* combinations of the storage functions of the component systems. For simplicity we will only consider the case without external inputs and outputs u^e, y^e.

Proposition 3.3.3 *Suppose the supply rates $s_1, \ldots, s_k$ and the interconnection subset I are such that there exist positive constants $\alpha_1, \ldots, \alpha_k$ for which*

$$\alpha_1 s_1(u_1, y_1) + \cdots + \alpha_k s_k(u_k, y_k) \leq 0,$$
$$\text{for all } ((u_1, y_1), \ldots, (u_k, y_k)) \in I \tag{3.55}$$

Then the nonnegative function

$$S_\alpha(x_1, \ldots, x_k) := \alpha_1 S_1(x_1) + \cdots + \alpha_k S_k(x_k) \tag{3.56}$$

satisfies $\frac{d}{dt} S_\alpha \leq 0$ along all solutions of the interconnected system Σ_I.

Proof Multiply each i-th dissipation inequality

$$S_i(x_i(t_1)) \leq S_i(x_i(t_0)) + \int_{t_0}^{t_1} s_i(u_i(t), y_i(t))dt$$

by α_i, $i = 1, \ldots, k$, add them, and use the inequality (3.55). $\square$

In Sect. 8.2 we will see how this proposition underlies the small-gain theorem and extensions of it. Furthermore, it will appear naturally in the network interconnection of passive systems in Sect. 4.4.

3.4 Scattering of State Space Systems

The generalization (3.10) of the definition of dissipativity to DAE systems without a priori splitting of the external variables into inputs u and y is also useful in discussing the extension of the notion of *scattering*, as treated in Sect. 2.4 for input–output maps, to the state space system context.

Consider a state space system Σ given in standard input-state-output form (3.1), which is assumed to be passive, i.e.,

$$S(x(t_1)) - S(x(t_0)) \leq \int_{t_0}^{t_1} u^T(t)y(t)dt \tag{3.57}$$

for some storage function $S \geq 0$. Consider the scattering representation (v, z) of (u, y) defined as, see (2.41),

$$v = \frac{1}{\sqrt{2}}(u + y), \qquad z = \frac{1}{\sqrt{2}}(-u + y) \tag{3.58}$$

The inverse transformation of (3.58) is $u = \frac{1}{\sqrt{2}}(v - z)$, $y = \frac{1}{\sqrt{2}}(v + z)$, and substitution of these expressions in (3.57) yields

$$S(x(t_1)) - S(x(t_0)) \leq \frac{1}{2} \int_{t_0}^{t_1} \left(||v(t)||^2 - ||z(t)||^2 \right) dt \tag{3.59}$$

This shows that Σ is passive with respect to u and y if and only if Σ with transformed external variables v and z has L_2-gain ≤ 1 from v to z, while the storage function remains the same. Similarly, it follows that Σ is lossless with respect to u and y if and only if it is inner with respect to v and z.

Note that in general the transformed system Σ with external variables $w = (v, z)$ is in the format (3.9). In case the original Σ is an affine system without feedthrough term, i.e., of the form

$$\Sigma_a : \quad \begin{aligned} \dot{x} &= f(x) + g(x)u \\ y &= h(x) \end{aligned} \tag{3.60}$$

the substitution $u = \frac{1}{\sqrt{2}}(v - z)$, $y = \frac{1}{\sqrt{2}}(v + z)$, leads to an input–state–output representation in the wave vectors v, z, namely

$$\Sigma_s : \quad \begin{aligned} \dot{x} &= f(x) - g(x)h(x) + \sqrt{2}g(x)v \\ z &= \sqrt{2}h(x) - v \end{aligned} \tag{3.61}$$

Summarizing

Proposition 3.4.1 Σ_a *is passive (lossless) with storage function S if and only if Σ_s has L_2-gain ≤ 1 (is inner) with storage function S.*

3.5 Dissipativity and the Return Difference Inequality

Dissipative systems theory turns out to provide an insightful framework for the study of the *Inverse problem of optimal control*, as originally introduced in [155] for the linear quadratic optimal control problem.

Consider the nonlinear optimal control problem (see Sects. 9.4 and 11.2 for further information)

$$\min_u \int_0^\infty (||u(t)||^2 + \ell(x(t)))dt , \tag{3.62}$$

for the system

$$\dot{x} = f(x) + g(x)u, \quad f(0) = 0, \tag{3.63}$$

where $\ell \geq 0$ is a cost function with $\ell(0) = 0$.

Denote the minimal cost (*value*) defined by (3.62) for initial condition $x(0) = x_0$ by $V(x_0)$. The function $V : \mathcal{X} \to \mathbb{R}^+$ is called the *value function*. Suppose that the value function V is well defined for all initial conditions and is C^1. Then it is known from optimal control theory that V is a nonnegative solution to the Hamilton–Jacobi–Bellman equation

$$V_x(x)f(x) - \frac{1}{2}V_x(x)g(x)g^T(x)V_x^T(x) + \frac{1}{2}\ell(x) = 0, \; V(0) = 0 \tag{3.64}$$

Fig. 3.1 Feedback system

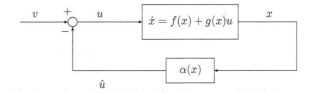

Furthermore the optimal control is given in feedback form as

$$u = -\alpha(x) := -g^T(x)V_x^T(x) \tag{3.65}$$

If additionally $V(x) > 0$ for $x \neq 0$ and the system $\dot{x} = f(x)$, $y = \ell(x)$ is *zero-state detectable* (cf. Definition 3.2.15), it follows from LaSalle's Invariance principle that this optimal feedback is actually stabilizing, since (3.64) can be rewritten as

$$V_x(x)[f(x) - g(x)\alpha(x)] = -\frac{1}{2}\alpha^T(x)\alpha(x) - \ell(x), \tag{3.66}$$

and thus asymptotic stability of $x = 0$ follows as in Proposition 3.2.16.

As we will now show, the optimal control feedback $u = -\alpha(x) := -g^T(x)V_x^T(x)$ has a direct *dissipativity interpretation*. Indeed, (3.64) and (3.65) can be rewritten as

$$V_x(x)f(x) - \tfrac{1}{2}\alpha^T(x)\alpha(x) = -\ell(x) \leq 0 \quad V(0) = 0$$
$$V_x(x)g(x) = \alpha^T(x) \tag{3.67}$$

implying that the following system (the "loop transfer" from u to minus the optimal feedback $-\alpha(x)$)

$$\dot{x} = f(x) + g(x)u$$
$$\hat{u} = \alpha(x) \tag{3.68}$$

is dissipative with respect to the supply rate

$$s(u, \hat{u}) := \frac{1}{2}||\hat{u}||^2 + \hat{u}^T u \tag{3.69}$$

This leads to the following interesting consequence. By further rewriting the supply rate $s(u, \hat{u}) = \frac{1}{2}||\hat{u}||^2 + \hat{u}^T u$ as

$$\frac{1}{2}||\hat{u}||^2 + \hat{u}^T u = \frac{1}{2}||u + \hat{u}||^2 - \frac{1}{2}||u||^2 = \frac{1}{2}||v||^2 - \frac{1}{2}||u||^2, \tag{3.70}$$

it means that the feedback system in Fig. 3.1 with external inputs v satisfies the property

$$\frac{1}{2} \int_0^T ||u(t)||^2 dt \leq \frac{1}{2} \int_0^T ||v(t)||^2 dt + V(x(0)) - V(x(T))$$

$$\leq \frac{1}{2} \int_0^T ||v(t)||^2 dt + V(x(0)) , \qquad (3.71)$$

for all initial conditions $x(0)$, and for all external input signals v. Thus the L_2-gain *from the external inputs v to the internal inputs u is less than or equal to one.*

In the linear case the frequency domain version of the inequality (3.71) is called the *return difference inequality*. The inequality expresses the favorable property that in the closed-loop system of Fig. 3.1 the L_2-norm of the optimal feedback $u = -\alpha(x)$ is *attenuated* with regard to the L_2-norm of any *external* control signal $v(\cdot)$.

Conversely, it can be shown that any stabilizing feedback $u = -\alpha(x)$ for which (3.71) holds is actually *optimal* with respect to *some* cost function $\ell \geq 0$ with $\ell(0) = 0$. Indeed, consider $u = -\alpha(x)$ such that (3.71) is satisfied for some function $V \geq 0$, with $V(0) = 0$. Equivalently, the system (3.68) is dissipative with respect to the supply rate $s(u, \hat{u}) = \frac{1}{2}||\hat{u}||^2 + \hat{u}^T u$, with storage function $V \geq 0$, $V(0) = 0$. Then it follows that (assuming V is C^1)

$$V_x(x)f(x) - \frac{1}{2}\alpha^T(x)\alpha(x) \leq 0$$
$$V_x(x)g(x) = \alpha^T(x) \qquad (3.72)$$

Hence we may *define* the cost function $\ell \geq 0$ as

$$\ell(x) := -V_x(x)f(x) + \frac{1}{2}\alpha^T(x)\alpha(x), \qquad (3.73)$$

satisfying $\ell(0) = 0$. It follows that V is actually a nonnegative solution of the Hamilton–Jacobi–Bellman equation (3.64) of the optimal control problem (3.62) for this cost function ℓ. As will be shown in Sect. 11.2, it follows that V is the value function of this optimal control problem and that $u = -\alpha(x)$ is the optimal feedback control.

Summarizing, we have obtained the following theorem.

Theorem 3.5.1 *Consider the system (3.63). Let $\ell \geq 0$ be a cost function with $\ell(0) = 0$ such that $\dot{x} = f(x)$, $y = \ell(x)$, is zero-state detectable, and that the value function V of the optimal control problem is well defined, C^1, and satisfies $V(x) > 0$, $x \neq 0$. Then the feedback $u = -\alpha(x) := -g^T(x)V_x^T(x)$ is stabilizing, and the resulting feedback system in Fig. 3.1 satisfies property (3.71).*

Conversely, $u = -\alpha(x)$ is a stabilizing feedback such that the feedback system in Fig. 3.1 satisfies property (3.71) for a C^1 function $V \geq 0$ with $V(0) = 0$, then $u = -\alpha(x)$ is the optimal control for (3.62) with the cost function $\ell(x) := -V_x(x)f(x) + \frac{1}{2}\alpha^T(x)\alpha(x)$.

Thus, loosely speaking, a feedback $u = -\alpha(x)$ is optimal with regard to *some* optimal control problem of the form (3.62) *if and only if* the return difference inequality (3.71) holds.

Remark 3.5.2 In view of Proposition 3.1.14, we additionally note that if the system $\dot{x} = f(x) + g(x)u$ is *reachable* from $x = 0$, then property (3.74) holds for a feedback $u = -\alpha(x)$ and a $V \geq 0$ with $V(0) = 0$, if and only if

$$\int_0^T ||u(t)||^2 dt \leq \int_0^T ||v(t)||^2 dt , \qquad x(0) = 0, \tag{3.74}$$

for all external inputs v, and all $T \geq 0$.

Remark 3.5.3 Notice that the optimal regulator in Fig. 3.1, that is the system

$$\begin{aligned} \dot{x} &= [f(x) - g(x)\alpha(x)] + g(x)v \\ \hat{u} &= \alpha(x) = g^T(x)V_x^T(x) \end{aligned} \tag{3.75}$$

is *output strictly passive*, as follows directly from (3.66).

3.6 Notes for Chapter 3

1. The main part of the theory exposed in Sect. 3.1 is based on Willems' seminal and groundbreaking paper [350]. The developments around Propositions 3.1.19, 3.1.20 and Corollary 3.1.21 are relatively new, although inspired by similar arguments in Willems [349, 351].

2. Other expositions of (parts of) the theory of dissipative systems can be found in Hill & Moylan [126], Moylan [225], Brogliato, Lozano, Maschke & Egeland [52], as well as Isidori [139], Arcak, Meissen & Packard [11].

3. We refer to e.g., Weiland & Willems [346], and Trentelman & Willems [338, 339], Willems & Trentelman [353] for developments on dissipative systems theory within a (linear) behavioral systems theory framework; generalizing the concept of quadratic supply rates to quadratic differential forms that may also involve derivatives of the inputs and outputs. Note that in these papers "dissipativity" is often used in the meaning of "cyclo-dissipativity".

4. The first part of Sect. 3.2 is also mainly based on Willems' paper [350], together with important contributions due to Hill & Moylan [123–126]. The exposition on stability in Sect. 3.2 was influenced by Sepulchre, Jankovich & Kokotovic [312].

5. The theory of dissipative systems is closely related to the theory of *Integral Quadratic Constraints* (IQCs); see e.g., Megretski & Rantzer [215], Jönsson [152] and the references quoted therein.
   Basically, in the theory of Integral Quadratic Constraints (IQCs) the system Σ_1 denotes the given *linear* "nominal" part of the system to be studied, specified by a transfer matrix G, while $\Delta := \Sigma_2$ denotes the "troublemaking" (nonlinear,

time-delay, time-varying, uncertain) components. In order to assess stability of
the overall system one searches for IQCs for Δ. These are given by a Hermitian
matrix valued function $\Pi(j\omega)$, $\omega \in \mathbb{R}$, such that

$$\int_{-\infty}^{\infty} \begin{bmatrix} \widehat{u}_2(j\omega) \\ \widehat{y}_2(j\omega) \end{bmatrix}^* \Pi(j\omega) \begin{bmatrix} \widehat{u}_2(j\omega) \\ \widehat{y}_2(j\omega) \end{bmatrix} d\omega \geq 0$$

for all L_2 signals u_2, y_2 compatible with Δ. Here $\widehat{}$ denotes Fourier transform,
and $*$ is complex conjugate and transpose. For rational Π that are bounded on
the imaginary axis, the time domain version of the IQC is

$$\int_0^{\infty} \sigma(x_\pi(t), y_2(t), u_2(t))dt \geq 0$$

for a certain quadratic form σ, where x_π is solution of an auxiliary system

$$\dot{x}_\pi = A_\pi x_\pi + B_{y_2} y_2 + B_{u_2} u_2, \quad x_\pi(0) = 0.$$

The main theorem (Theorem 1 in Megretski & Rantzer [215]) states that if we
can find a Π such that for every $\tau \in [0, 1]$ the interconnection of G and $\tau\Delta$ is
well posed and Π is an IQC for $\tau\Delta$, while there exists a $\varepsilon > 0$ such that

$$\begin{bmatrix} G(j\omega) \\ I \end{bmatrix}^* \Pi(j\omega) \begin{bmatrix} G(j\omega) \\ I \end{bmatrix} \leq -\varepsilon I, \quad \forall \omega \in \mathbb{R}$$

then the closed-loop system is stable. Compared with the setup of dissipative
systems theory there are two major differences. One is that Π is not necessarily a
constant matrix, and that therefore in the time domain formulation the function σ,
which replaces the supply rate s of dissipative systems theory, also depends on an
auxiliary dynamical system (acting as an additional filter for the signals u_2, y_2).
Secondly, the IQC should only hold for all L_2 signals u_2, y_2, and therefore in the
time domain version the integral is from 0 to ∞, instead of from 0 to any $T \geq 0$,
as in dissipative systems theory formulation. The first aspect constitutes a major
extension with respect to dissipative systems theory. The second difference is
more of a technical nature, closely related to an extension of dissipativity to
cyclo-dissipativity. The L_2-stability problems caused by the second difference
are taken care of by an ingenious homotopy argument based on the variation of τ
from 0 to 1 (nominal value). In Veenman & Scherer [341] it has been shown how
in most situations IOC stability analysis can be proved by dissipative systems
theory.
On a methodological level, the philosophy of the theory of IQCs is somewhat
different from dissipative systems theory in the sense that in IQC theory the
emphasis is on stability analysis by splitting between nominal linear dynamics
and "troublemaking" nonlinearities or time delays, whose disturbing properties
are sought to be bounded by a suitable IQC. Dissipative systems theory, on the

other hand, is primarily a compositional theory of complex systems (rooted in network dynamics), where nonlinear dynamical components are not necessarily considered to be detrimental. The theory of IQCs is especially useful for stability analysis of systems with "small-scale" nonlinearities or time delays. It yields sharp results on the classical cases of "noncausal multipliers" and the Popov criterion.

6. In Sect. 3.2 we have assumed throughout that there exist storage functions which are continuously differentiable (C^1), in order to make the link with Lyapunov stability theory, and, very importantly, in order to be able to rewrite the dissipation inequality (3.3) as the differential dissipation inequality (3.36). Now, for Lyapunov stability theory the Lyapunov functions do not necessarily have to be C^1, see e.g., Sontag [317]. Moreover, often storage functions for nonlinear systems are *not* everywhere differentiable (in particular this may happen for the available storage S_a and the required supply S_r, being solutions to an optimal control problem). Since it is much easier to work with differential dissipation inequalities than with dissipation inequalities in integral form, it would thus be desirable to have a *generalized solution* concept for differential dissipation inequalities (3.36), admitting solutions S that are *not* everywhere differentiable. In fact, this is possible using the concept of a *viscosity solution* (see e.g., Fleming & Soner [99], for a clear exposé), as shown in James [144] (see also James & Baras [145], Ball & Helton [22]). We also like to refer to Clarke, Ledyaev, Stern & Wolenski [67, 68] for a broader discussion of generalized solution concepts for Hamilton–Jacobi inequalities or equalities, showing equivalence between apparently different solution concepts.

7. See e.g., Khalil [159] for a coverage of Lyapunov stability theory and LaSalle's Invariance principle.

8. Theorem 3.2.17 is due to Iggidr, Kalitine & Outbib [134]. I thank Laurent Praly for pointing out an error in the presentation of the consequences of this theorem in the second edition of this book.

9. Proposition 3.2.22 is due to Imura, Sugie & Yoshikawa [132], Imura, Maeda, Sugie & Yoshikawa [131], where an alternative proof is given.

10. Section 3.3 is largely based on Willems [350], Moylan & Hill [227], see also Moylan [225].

11. The stability analysis of an interconnected system the approach taken in Sect. 3.3 can be turned around as well. Given the component systems $\Sigma_1, \ldots, \Sigma_k$ and the interconnection subset I, one may *search* for (suitably defined) supply rates $s_1, \ldots, s_k$, for which the systems are dissipative and (3.54) holds. This point of view is already (implicitly) present in classical papers on dissipative systems such as Moylan & Hill [227], see also Moylan [225], and was recently emphasized and explored in Meissen, Lessard, Arcak & Packard [217], Arcak, Meissen, Packard [11].

Furthermore, in Jokic & Nakic [151] the *converse* result is obtained stating that if an interconnected linear system has an additive quadratic Lyapunov function (a sum of terms only depending on the state variables of the subsystems) then there exist interconnection neutral supply rates with respect to which the subsystems are dissipative.

12. Section 3.5 is an extended and simplified exposition of basic ideas developed in Moylan & Anderson [226]. For related work on the inverse optimal control problem and its applications to robust control design we refer to Sepulchre, Jankovic & Kokotovic [312], and Freeman & Kokotovic [106].

13. The differential dissipation inequality (3.36) admits the following *factorization* perspective. For concreteness, assume that $f(0,0) = 0$, $h(0,0) = 0$ as well as $s(0,0) = 0$. Then, under technical assumptions (see Chap. 9), satisfaction of (3.36) will imply that there exists a map $\bar{h} : \mathcal{X} \times \mathbb{R}^m \to \mathbb{R}^{\bar{p}}$ such that

$$S_x(x) f(x, u) - s(u, h(x, u)) = -\|\bar{h}(x, u)\|^2 \qquad (3.76)$$

Equivalently, the system is *conservative* with respect to the new supply rate

$$\bar{s}(u, y, \bar{y}) := s(u, y) - \|\bar{y}\|^2, \qquad (3.77)$$

involving, next to y, the new output $\bar{y} = \bar{h}(x, u)$. This factorization perspective will be further discussed in the context of the passivity supply rate in Sect. 4.1, and will be key in the developments on the L_2-gain supply rate in Sect. 9.4.

Chapter 4
Passive State Space Systems

In this chapter we focus on passive systems as an outstanding subclass of dissipative systems, firmly rooted in the mathematical modeling of physical systems.

4.1 Characterization of Passive State Space Systems

Recall from Chap. 3 the definitions of (input and/or output strict) passivity of a state space system, cf. Definition 3.1.4.

Definition 4.1.1 A *state space system* Σ with equal number of inputs and outputs

$$\begin{aligned}
\dot{x} &= f(x, u), \quad x \in \mathcal{X}, \; u \in U = \mathbb{R}^m \\
y &= h(x, u), \quad y \in Y = \mathbb{R}^m
\end{aligned} \tag{4.1}$$

is *passive* if it is dissipative with respect to the supply rate $s(u, y) = u^T y$. Furthermore, Σ is called *cyclo-passive* if the storage function is *not* necessarily satisfying the nonnegativity condition. Σ is called *lossless* if it is conservative with respect to $s(u, y) = u^T y$. The system Σ is *input strictly passive* if there exists $\delta > 0$ such that Σ is dissipative with respect to $s(u, y) = u^T y - \delta||u||^2$ (also called δ-input strictly passive). Σ is *output strictly passive* if there exists $\varepsilon > 0$ such that Σ is dissipative with respect to $s(u, y) = u^T y - \varepsilon||y||^2$ (ϵ-output strictly passive).

Also recall from Chap. 3 that there is a minimal storage function S_a (the *available storage*), and under a reachability condition, a storage function S_r (the *required supply* from x^*), which is maximal in the sense of (3.26); see also Corollary 3.1.21. The storage function in the case of the passivity supply rate often has the interpretation of a (generalized) energy function, and $S_a(x)$ equals the *maximal* energy that can be extracted from the system being in state x, while $S_r(x)$ is the *minimal* energy that is needed to bring the system toward state x, while starting from a ground state x^*.

© Springer International Publishing AG 2017
A. van der Schaft, *L₂-Gain and Passivity Techniques in Nonlinear Control*,
Communications and Control Engineering, DOI 10.1007/978-3-319-49992-5_4

In physical examples, the true physical energy usually will be "somewhere in the middle" between S_a and S_r.

Assuming differentiability of the storage function (as will be done throughout this section), passivity, respectively input or output strict passivity, can be characterized through the differential dissipation inequalities (3.36). These take a particularly explicit form for systems which are *affine* in the input u (as often encountered in applications), and given as

$$\Sigma_a^{ft} : \begin{aligned} \dot{x} &= f(x) + g(x)u \\ y &= h(x) + j(x)u, \end{aligned} \qquad (4.2)$$

with $g(x)$ an $n \times m$ matrix, and $j(x)$ an $m \times m$ matrix. In case of the passivity supply rate $s(u, y) = u^T y$ the differential dissipation inequality then takes the form

$$\frac{d}{dt} S = S_x(x)[f(x) + g(x)u] \le u^T[h(x) + j(x)u], \quad \forall x, u, \qquad (4.3)$$

where, as before, the notation $S_x(x)$ stands for the row vector of partial derivatives of the function $S : \mathcal{X} \to \mathbb{R}$. Note that

$$S_x(x)[f(x) + g(x)u] - u^T[h(x) + j(x)u] =$$
$$\tfrac{1}{2} [1 \ u^T] \begin{bmatrix} 2S_x(x)f(x) & S_x(x)g(x) - h^T(x) \\ g^T(x)S_x^T(x) - h(x) & -(j(x) + j^T(x)) \end{bmatrix} \begin{bmatrix} 1 \\ u \end{bmatrix} \qquad (4.4)$$

while similar expressions are obtained in the case of the output and input strict passivity supply rates.

This leads to the following characterizations.

Proposition 4.1.2 *Consider the system Σ_a^{ft} given by (4.2). Then:*
(i) Σ_a^{ft} is passive with C^1 storage function S if and only if for all x

$$\begin{bmatrix} 2S_x(x)f(x) & S_x(x)g(x) - h^T(x) \\ g^T(x)S_x^T(x) - h(x) & -(j(x) + j^T(x)) \end{bmatrix} \le 0 \qquad (4.5)$$

(ii) Σ_a^{ft} is ε-output strictly passive with C^1 storage function S if and only if for all x

$$\begin{bmatrix} 2S_x(x)f(x) + 2\varepsilon h^T(x)h(x) & S_x(x)g(x) - h^T(x) + k^T(x) \\ g^T(x)S_x^T(x) - h(x) + k(x) & \ell(x) - (j(x) + j^T(x)) \end{bmatrix} \le 0, \qquad (4.6)$$

where $k(x) := 4\varepsilon h^T(x)j(x)$, $\ell(x) := 2\varepsilon j(x)j^T(x)$.
(iii) Σ_a^{ft} is δ-input strictly passive with C^1 storage function S if and only if for all x

$$\begin{bmatrix} 2S_x(x)f(x) & S_x(x)g(x) - h^T(x) \\ g^T(x)S_x^T(x) - h(x) & 2\delta I_m - (j(x) + j^T(x)) \end{bmatrix} \le 0 \qquad (4.7)$$

The proof of this proposition is based on the following basic lemma.

Lemma 4.1.3 *Let $R = R^T$ be an $m \times m$ matrix, q an m-vector, and p a scalar. Then*

$$u^T R u + 2u^T q + p \leq 0, \text{ for all } u \in \mathbb{R}^m, \tag{4.8}$$

if and only if

$$\begin{bmatrix} p & q^T \\ q & R \end{bmatrix} \leq 0 \tag{4.9}$$

Proof (of Lemma 4.1.3) Obviously, the inequality (4.9) implies

$$u^T R u + 2u^T q + p = \begin{bmatrix} 1 & u^T \end{bmatrix} \begin{bmatrix} p & q^T \\ q & R \end{bmatrix} \begin{bmatrix} 1 \\ u \end{bmatrix} \leq 0, \quad \forall u \in \mathbb{R}^m \tag{4.10}$$

In order to prove[1] the converse implication assume that

$$v^T \begin{bmatrix} p & q^T \\ q & R \end{bmatrix} v > 0 \tag{4.11}$$

for some $(m + 1)$-dimensional vector v. If the first component of v is different from zero we can directly scale the vector v to a vector of the form $\begin{bmatrix} 1 \\ u \end{bmatrix}$ while still (4.11) holds, leading to a contradiction. If the first component of v equals zero then we can consider a small perturbation of v for which the first component of v is nonzero while still (4.11) holds, and we use the previous argument. $\square$

Proof (of Proposition 4.1.2) Write out the dissipation inequalities in the form $u^T R(x)u + 2u^T q(x) + p(x) \leq 0$, and apply Lemma 4.1.3 with R, q, p additionally depending on x. $\square$

Example 4.1.4 It follows from (4.7) that an input strictly passive system necessarily has a nonzero feedthrough term $j(x)u$. An example is provided by a proportional–integral (PI) controller

$$\begin{aligned} \dot{x}_c &= u_c \\ y_c &= k_I x_c + k_P u_c \end{aligned} \tag{4.12}$$

with $k_P, k_I \geq 0$ the proportional, respectively integral, control coefficients. This is a k_P-input strictly system with storage function is $\frac{1}{2} k_I x_c^2$, since

$$\frac{d}{dt} \frac{1}{2} k_I x_c^2 = u_c y_c - k_P u_c^2 \tag{4.13}$$

A drastic simplification of the conditions for (output strict) passivity occurs for systems *without feedthrough term* $(j(x) = 0)$ given as

---

[1] With thanks to Anders Rantzer for a useful conversation.

$$\Sigma_a : \begin{array}{l} \dot{x} = f(x) + g(x)u \\ y = h(x) \end{array} \qquad (4.14)$$

Corollary 4.1.5 *Consider the system Σ_a given by (4.14). Then:*
(i) Σ_a is passive with C^1 storage function S if and only if for all x

$$\begin{array}{l} S_x(x)f(x) \le 0 \\ S_x(x)g(x) = h^T(x) \end{array} \qquad (4.15)$$

(ii) Σ_a is ε-output strictly passive with C^1 storage function S if and only if for all x

$$\begin{array}{l} S_x(x)f(x) \le -\varepsilon h^T(x)h(x) \\ S_x(x)g(x) = h^T(x) \end{array} \qquad (4.16)$$

(iii) Σ_a is not δ-input strictly passive for any $\delta > 0$.

Proof Use the well-known fact that $\begin{bmatrix} k & q^T \\ q & 0_m \end{bmatrix} \le 0$ (with 0_m denoting the $m \times m$ zero matrix) if and only if $q = 0$ and $k \le 0$. $\qquad\qquad\square$

Remark 4.1.6 For a *linear* system

$$\begin{array}{l} \dot{x} = Ax + Bu \\ y = Cx + Du \end{array} \qquad (4.17)$$

with quadratic storage function $S(x) = \frac{1}{2}x^T Q x$, $Q = Q^T \ge 0$, the passivity condition (4.5) amounts to the linear matrix inequality (LMI)

$$\begin{bmatrix} A^T Q + QA & QB - C^T \\ B^T Q - C & -D - D^T \end{bmatrix} \le 0 \qquad (4.18)$$

Obvious extensions to input/output strict passivity are left to the reader. In case $D = 0$ (no feedthrough) the conditions (4.18) simplify to the LMI

$$A^T Q + QA \le 0, \quad B^T Q = C \qquad (4.19)$$

The relation of these LMIs to frequency-domain conditions is known as the *Kalman–Yakubovich–Popov Lemma*; see the Notes at the end of this chapter for references.

The inequalities in Proposition 4.1.2 and Corollary 4.1.5, as well as the resulting LMIs (4.18) and (4.19) in the linear system case, admit the following *factorization* perspective. Given a matrix inequality $P(x) \le 0$, where $P(x)$ is an $k \times k$ symmetric matrix depending smoothly on x, we may always, by standard linear-algebraic factorization for every constant x, construct an $\ell \times k$ matrix $F(x)$ such that $P(x) = -F^T(x)F(x)$, where ℓ is equal to the maximal rank of $P(x)$ (over x).

Furthermore, by an application of the implicit function theorem, locally on a neighborhood where the rank of $P(x)$ is constant, this can be done in such a way that $F(x)$ is depending smoothly on x. Applied to (minus) the matrices appearing in Proposition 4.1.2 and Corollary 4.1.5 this leads to the following result. For concreteness, focus on the inequality (4.5); similar statements hold for the other cases. Inequality 4.5 holds if and only if

$$
\begin{bmatrix} 2S_x(x)f(x) & S_x(x)g(x) - h^T(x) \\ g^T(x)S_x^T(x) - h(x) & -(j(x) + j^T(x)) \end{bmatrix} = -F^T(x)F(x) \leq 0 \tag{4.20}
$$

for a certain matrix

$$
F(x) = \begin{bmatrix} \phi(x) & \Psi(x) \end{bmatrix} \tag{4.21}
$$

with $\phi(x)$ an ℓ-dimensional column vector, and $\psi(x)$ an $\ell \times m$ matrix, with ℓ the (local) rank of the matrix in (4.5). Writing out (4.20) yields

$$
\begin{aligned}
2S_x(x)f(x) &= -\phi^T(x)\phi(x) \\
S_x(x)g(x) - h^T(x) &= -\phi^T(x)\Psi(x) \\
j(x) + j^T(x) &= \Psi^T(x)\Psi(x)
\end{aligned} \tag{4.22}
$$

It follows that by defining the new, artificial, output equation

$$
\bar{y} = \phi(x) + \Psi(x)u \tag{4.23}
$$

one obtains

$$
S_x(x)[f(x) + g(x)u] - u^T[h(x) + j(x)u] = -\frac{1}{2}\|\bar{y}\|^2, \tag{4.24}
$$

and therefore

$$
\frac{d}{dt}S = u^T y - \frac{1}{2}\|\bar{y}\|^2. \tag{4.25}
$$

Hence, by factorization we have turned the dissipativity of the system Σ_a^{ft} with respect to the passivity supply rate $s(u, y) = u^T y$ into the fact that Σ_a^{ft} is *conservative* with respect to the *new* supply rate

$$
s_{\text{new}}(u, y) = u^T y - \frac{1}{2}\|\bar{y}\|^2, \tag{4.26}
$$

defined in terms of the inputs u, outputs y, *as well as* the new outputs $\bar{y}$ defined by (4.23). The same can be done for the output and input strict passivity supply rates; in fact, for any supply rate which is quadratic in u, y. Within the context of the L_2-gain supply rate this[2] will be exploited in Chap. 9; see especially Sect. 9.4.

---

[2]In fact, in Sect. 9.4 we will see how this can be extended to *general* systems Σ.

Let us briefly focus on the *linear* passive system case, corresponding to the LMIs (4.18). As was already mentioned in Remark 3.1.22 for general supply rates, the available storage S_a of a linear passive system (4.17) with $D = 0$ is given as $\frac{1}{2}x^T Q_a x$ where Q_a is the *minimal* solution to the LMI (4.18), while the required supply is $\frac{1}{2}x^T Q_r x$ where Q_r is the *maximal* solution to this same LMI.

Although in general (4.18) has a convex *set* of solutions $Q \geq 0$, this set may sometimes reduce to a *unique* solution; even for systems with nonzero internal energy dissipation. This is illustrated by the following simple physical example.

Example 4.1.7 Consider the ubiquitous mass–spring–damper system

$$\begin{bmatrix} \dot{q} \\ \dot{p} \end{bmatrix} = \begin{bmatrix} 0 & \frac{1}{m} \\ -k & -\frac{d}{m} \end{bmatrix} \begin{bmatrix} q \\ p \end{bmatrix} + \begin{bmatrix} 0 \\ 1 \end{bmatrix} u, \quad u = \text{force}$$

$$y = \begin{bmatrix} 0 & \frac{1}{m} \end{bmatrix} \begin{bmatrix} q \\ p \end{bmatrix} = \text{velocity}$$

(4.27)

with physical energy $H(q, p) = \frac{1}{2m}p^2 + \frac{1}{2}kq^2$ (q extension of the linear spring with spring constant k, p momentum of mass m), and internal energy dissipation corresponding to a linear damper with damping coefficient $d > 0$. The LMI (4.19) takes the form

$$\begin{bmatrix} 0 & -k \\ \frac{1}{m} & -\frac{d}{m} \end{bmatrix} \begin{bmatrix} q_{11} & q_{12} \\ q_{12} & q_{22} \end{bmatrix} + \begin{bmatrix} q_{11} & q_{12} \\ q_{12} & q_{22} \end{bmatrix} \begin{bmatrix} 0 & \frac{1}{m} \\ -k & -\frac{d}{m} \end{bmatrix} \leq 0$$

$$\begin{bmatrix} 0 & 1 \end{bmatrix} \begin{bmatrix} q_{11} & q_{12} \\ q_{12} & q_{22} \end{bmatrix} = \begin{bmatrix} 0 & \frac{1}{m} \end{bmatrix}$$

(4.28)

The last equation yields $q_{12} = 0$ as well as $q_{22} = \frac{1}{m}$. Substituted in the inequality this yields the unique solution $q_{11} = k$, corresponding to the *unique* quadratic storage function $H(q, p)$, which is equal to $S_a = S_r$. The explanation for the perhaps surprising equality of S_a and S_r in this case is the fact that the definitions of S_a and S_r involve sup and inf (instead of max and min).

We note for later use that passivity of a *static* nonlinear map $y = F(u)$, with $F : \mathbb{R}^m \to \mathbb{R}^m$, amounts to requiring that

$$u^T F(u) \geq 0, \quad \text{for all } u \in \mathbb{R}^m,$$

(4.29)

which for $m = 1$ reduces to the condition that the graph of the function F is in the first and third quadrant. This definition immediately extends to *relations* instead of mappings.

Furthermore, passivity of the dynamical system Σ implies the following *static* passivity property of the steady-state values of its inputs and outputs. Let Σ be an input-state-output system in the general form (4.1). For any constant input $\bar{u}$ consider the existence of a steady-state $\bar{x}$, and corresponding steady-state output value $\bar{y}$, satisfying

$$0 = f(\bar{x}, \bar{u}), \quad \bar{y} = h(\bar{x}, \bar{u}) \tag{4.30}$$

This defines the following relation between $\bar{u}$ and $\bar{y}$, called the *steady-state input–output relation* Σ_{ss} corresponding to Σ:

$$\Sigma_{ss} := \{(\bar{u}, \bar{y}) \mid \exists \bar{x} \text{ s.t. (4.30) holds}\} \tag{4.31}$$

In case of a cyclo-passive system (4.1) with storage function S satisfying $\frac{d}{dt} S \leq u^T y$ it follows that

$$0 = \frac{d}{dt} S(\bar{x}) \leq \bar{u}^T \bar{y}, \quad \text{for any } (\bar{u}, \bar{y}) \in \Sigma_{ss}, \tag{4.32}$$

with the obvious interpretation that at its steady states every cyclo-passive system necessarily dissipates energy.

Note that in general Σ_{ss} need not be the graph of a *mapping* from $\bar{u}$ to $\bar{y}$. For example, Σ_{ss} corresponding to the (multi-dimensional) nonlinear integrator

$$\dot{x} = u, \quad y = \frac{\partial H}{\partial x}(x), \quad x, u, y \in \mathbb{R}^m \tag{4.33}$$

(which is a cyclo-passive system with, possibly indefinite, storage function H), is given as

$$\Sigma_{ss} = \left\{ (\bar{u} = 0, \bar{y}) \mid \exists \bar{x} \text{ s.t. } \bar{y} = \frac{\partial H}{\partial x}(\bar{x}) \right\} \tag{4.34}$$

This will be further explored within the context of port-Hamiltonian systems in Chap. 6, Sect. 6.5.

4.2 Stability and Stabilization of Passive Systems

Many of the stability results as established in Chap. 3 for dissipative systems involving additional conditions on the supply rate directly apply to the passivity supply rate. In particular Propositions 3.2.7, 3.2.9 (see Remark 3.2.10) and Proposition 3.2.12 (see Remark 3.2.14) hold for passive systems. Moreover, Propositions 3.2.15 and 3.2.19 apply to output strictly passive systems; see Remark 3.2.20.

Loosely speaking, equilibria of passive systems are typically *stable*, but not necessarily *asymptotically stable*. On the other hand, there is no obvious relation between passivity and stability of the *input–output maps*. This is already illustrated by the simplest example of a passive (in fact, lossless) system; namely the *integrator*

$$\dot{x} = u, \quad y = x, \quad x, u, y \in \mathbb{R}$$

Obviously, 0 is a stable equilibrium with Lyapunov function $\frac{1}{2}x^2$, while the input–output mappings of this system map $L_{2e}(\mathbb{R})$ into $L_{2e}(\mathbb{R})$, but *not* $L_2(\mathbb{R})$ into $L_2(\mathbb{R})$. The same applies to a *nonlinear* integrator, with output equation $y = x$ replaced by

$y = S_x(x)$ for some nonnegative function S having its minimum at 0. The situation becomes different by changing $\dot{x} = u$, $y = x$ into $\dot{x} = -x + u$, $y = x$, leading to a system with asymptotically stable equilibrium 0 and finite L_2-gain input–output map. On the other hand, the minor modification $\dot{x} = -x^3 + u$ displays 0 as an asymptotically stable equilibrium, but does *not* define a mapping from $L_2(\mathbb{R})$ to $L_2(\mathbb{R})$. To explain the differences, notice that of the three preceding examples only $\dot{x} = -x + u$, $y = x$ is *output strictly* passive. Indeed, output strict passivity implies *finite L_2-gain*, as formulated in the following state space version of Theorem 2.2.13.

Proposition 4.2.1 *If Σ is ε-output strictly passive, then it has L_2-gain $\leq \frac{1}{\varepsilon}$.*

Proof If Σ is ε-output strictly passive there exists $S \geq 0$ such that for all $t_1 \geq t_0$ and all u

$$S(x(t_1)) - S(x(t_0)) \leq \int_{t_0}^{t_1} (u^T(t)y(t) - \varepsilon\|y(t)\|^2)dt \qquad (4.35)$$

Therefore

$$\varepsilon \int_{t_0}^{t_1} \|y(t)\|^2)dt \;\leq\; \int_{t_0}^{t_1} u^T(t)y(t)dt - S(x(t_1)) + S(x(t_0)) \;\leq$$

$$\int_{t_0}^{t_1} (u^T(t)y(t) + \tfrac{1}{2}\|\tfrac{1}{\sqrt{\varepsilon}}u(t) - \sqrt{\varepsilon}y(t)\|^2)dt - S(x(t_1)) + S(x(t_0)) \;=$$

$$\int_{t_0}^{t_1} (\tfrac{1}{2\varepsilon}\|u(t)\|^2 + \tfrac{\varepsilon}{2}\|y(t)\|^2)dt - S(x(t_1)) + S(x(t_0)) \;,$$

whence

$$S(x(t_1)) - S(x(t_0)) \;\leq\; \int_{t_0}^{t_1} \left(\frac{1}{2\varepsilon}\|u(t\|^2 - \frac{\varepsilon}{2}\|y(t)\|^2\right) dt \;, \qquad (4.36)$$

implying that Σ has L_2-gain $\leq \frac{1}{\varepsilon}$ (with storage function $\frac{1}{\varepsilon}S$). $\qquad\qquad\square$

Further implications of output strict passivity for the input–output stability will be discussed in the context of L_2-gain analysis of state space systems in Chap. 8.

The importance of output strict passivity for asymptotic and input–output stability directly motivates the consideration of the following simple class of feedbacks which *render* a passive system output strictly passive. Indeed, consider a passive system Σ as given in (4.1) with C^1 storage function S, that is

$$\frac{d}{dt}S \leq u^T y \qquad (4.37)$$

If the system is not output strictly passive, then an obvious way to *render* the system output strictly passive is to apply a *static output feedback*

$$u = -dy + v, \quad d > 0, \qquad (4.38)$$

with $v \in \mathbb{R}^m$ the new input, and d a positive scalar.[3] Then the closed-loop system satisfies

$$\frac{d}{dt} S \leq v^T y - d\|y\|^2, \tag{4.39}$$

and thus is d-output strictly passive, and has L_2-gain $\leq \frac{1}{d}$ (from v to y). Hence, we obtain the following corollary of Propositions 3.2.16 and 3.2.19.

Corollary 4.2.2 *Consider the passive system Σ with storage function S satisfying $S(0) = 0$. Assume that S is positive definite at 0 and that the system $\dot{x} = f(x, 0)$, $y = h(x, 0)$, is zero-state detectable. Alternatively, assume 0 is an asymptotically stable equilibrium of $\dot{x} = f(x, 0)$ conditionally to $\{x \mid h(x, 0) = 0\}$. In both cases the feedback $u = -dy$, $d > 0$, asymptotically stabilizes the system around the equilibrium 0.*

Finally, we remark that in certain cases the verification of the property of zero-state detectability or asymptotic stability conditionally to $y = h(x, 0) = 0$ can be reduced to the verification of the same property for a *lower-dimensional* system. Consider as a typical case the feedback interconnection of Σ_1 and Σ_2 as in Fig. 1.1 with $e_2 = 0$ (see Fig. 4.1 later on). Suppose that Σ_1 satisfies the property

$$y_1(t) = 0, \quad t \geq 0 \Rightarrow x_1(t) = 0, \quad t \geq 0 \text{ and } u_1(t) = 0, \quad t \geq 0 \tag{4.40}$$

(This is a strong zero-state observability property.) Now, let $y_1(t) = 0$, $t \geq 0$, and $e_1(t) = 0$, $t \geq 0$. Then $u_2(t) = 0$, $t \geq 0$, and by (4.40), $y_2(t) = 0$, $t \geq 0$. Hence, checking zero-state detectability or asymptotic stability conditionally to $y_1 = h_1(x_1) = 0$ for the closed-loop system is the same as checking the same property for Σ_2. Summarizing, we have obtained the following.

Proposition 4.2.3 *Consider the closed-loop system $\Sigma_1 \|_f \Sigma_2$ with $e_2 = 0$, having input e_1 and output y_1. Suppose that Σ_1 satisfies property (4.40). Then the closed-loop system is zero-state detectable, respectively asymptotically stable conditionally to $y_1 = 0$, if and only if Σ_2 is zero-state detectable, respectively, asymptotically stable conditionally to $y_2 = 0$.*

Example 4.2.4 (Euler's equations) Euler's equations of the dynamics of the angular velocities of a fully actuated rigid body, spinning around its center of mass (in the absence of gravity), are given by

$$I\dot{\omega} = -S(\omega)I\omega + u \tag{4.41}$$

Here I is the positive diagonal inertia matrix, $\omega = (\omega_1, \omega_2, \omega_3)^T$ is the vector of angular velocities in body coordinates, $u = (u_1, u_2, u_3)^T$ is the vector of inputs, while the skew-symmetric matrix $S(\omega)$ is given as

---

[3]This can be extended to $u = -Dy + v$, with D a matrix satisfying $D + D^T > 0$.

$$S(\omega) = \begin{bmatrix} 0 & -\omega_3 & \omega_2 \\ \omega_3 & 0 & -\omega_1 \\ -\omega_2 & \omega_1 & 0 \end{bmatrix} \quad (4.42)$$

Since $\frac{d}{dt}\frac{1}{2}\omega^T I \omega = u^T \omega$ it follows that the system (4.41) with output $y = \omega$ is passive (in fact, lossless). Stabilization to $\omega = 0$ is achieved by output feedback $u = -Dy$ for any positive matrix D. In Sect. 7.1 we will see how this can be extended to the underactuated case by making use of the underlying Hamiltonian structure of (4.41).

Example 4.2.5 (*Rigid body kinematics*) The dynamics of the orientation of a rigid body around its center of mass is described as

$$\dot{R} = RS(\omega) \quad (4.43)$$

where $R \in SO(3)$ is an orthonormal rotation matrix describing the orientation of the body with respect to an inertial frame, $\omega = (\omega_1, \omega_2, \omega_3)^T$ is the vector of angular velocities as in the previous example, and $S(\omega)$ is given by (4.42). The rotation matrix $R \in SO(3)$ can be parameterized by a rotation φ around a unit vector k as follows:

$$R = I_3 + \sin\varphi \, S(k) + (1 - \cos\varphi)S^2(k) \quad (4.44)$$

The *Euler parameters* (ε, η) corresponding to R are now defined as

$$\varepsilon = \sin\left(\frac{\varphi}{2}\right) k, \qquad \eta = \cos\left(\frac{\varphi}{2}\right), \quad (4.45)$$

and satisfy

$$\varepsilon^T \varepsilon + \eta^2 = 1 \quad (4.46)$$

It follows that

$$R = (\eta^2 - \varepsilon^T \varepsilon)I_3 + 2\varepsilon\varepsilon^T + 2\eta S(\varepsilon), \quad (4.47)$$

and thus R can be represented as an element (ε, η) of the three-dimensional unit sphere S^3 in $\mathbb{R}^4$. Note that (ε, η) and $(-\varepsilon, -\eta)$ correspond to the same matrix R. In particular, $(0, 1)$ and $(0, -1)$ both correspond to $R = I_3$. Thus the unit sphere S^3 defines a double covering of the matrix group $SO(3)$. In this representation the dynamics (4.43) is given as

$$\begin{bmatrix} \dot{\varepsilon} \\ \dot{\eta} \end{bmatrix} = \frac{1}{2} \begin{bmatrix} \eta I_3 + S(\varepsilon) \\ -\varepsilon^T \end{bmatrix} \omega, \quad (4.48)$$

evolving on S^3 in $\mathbb{R}^4$. Define the function $V : S^3 \to \mathbb{R}$ as

$$V(\varepsilon, \eta) = \varepsilon^T \varepsilon + (1 - \eta)^2, \quad (4.49)$$

which by (4.46) is equal to $V(\varepsilon, \eta) = 2(1 - \eta)$. Differentiating V along (4.48) yields

Fig. 4.1 Standard feedback configuration $\Sigma_1 \|_f \Sigma_2$

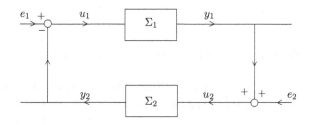

$$\frac{d}{dt} V = \omega^T \varepsilon \tag{4.50}$$

Hence the dynamics (4.48), with inputs ω and outputs ε, is passive (in fact, lossless) with storage function[4] V. As a consequence, the feedback control $\omega = -\varepsilon$ will asymptotically stabilize the system (4.48) toward $(0, \pm 1)$, that is, $R = I_3$. In Chap. 7 we will see how Examples 4.2.4 and 4.2.5 can be *combined* for the control of the total dynamics of the rigid body described by (4.43), (4.41) with inputs u.

4.3   The Passivity Theorems Revisited

The state space version of the passivity theorems as derived for passive input–output maps in Chap. 2, see in particular Theorem 2.2.11, follows the lines of the general theory of interconnection of dissipative systems as treated in Chap. 3, Sect. 3.3. Let us consider the standard feedback closed-loop system $\Sigma_1 \|_f \Sigma_2$ of Fig. 4.1, which is the same as Fig. 1.1 with the input–output maps G_1 and G_2 replaced by the state space systems

$$\Sigma_i : \begin{array}{ll} \dot{x}_i = f_i(x_i, u_i), & x_i \in \mathcal{X}_i, \quad u_i \in U_i \\ y_i = h_i(x_i, u_i), & y_i \in Y_i \end{array} \quad i = 1, 2, \tag{4.51}$$

with $U_1 = Y_2$, $U_2 = Y_1$. Suppose that both Σ_1 and Σ_2 in (4.51) (with $U_1 = U_2 = Y_1 = Y_2$) are *passive* or *output strictly passive*, with storage functions $S_1(x_1)$, respectively $S_2(x_2)$, i.e.,

$$\begin{aligned} S_1(x_1(t_1)) &\leq S_1(x_1(t_0)) + \int_{t_0}^{t_1} (u_1^T(t) y_1(t) - \varepsilon_1 \|y_1(t)\|^2) dt \\ S_2(x_2(t_1)) &\leq S_2(x_2(t_0)) + \int_{t_0}^{t_1} (u_2^T(t) y_2(t) - \varepsilon_2 \|y_2(t)\|^2) dt, \end{aligned} \tag{4.52}$$

with $\varepsilon_1 > 0$, $\varepsilon_2 > 0$ in case of output strict passivity, and $\varepsilon_1 = \varepsilon_2 = 0$ in case of mere passivity. Substituting the standard feedback interconnection equations (see (1.30))

---

[4]Note that this storage function does *not* have an interpretation in terms of physical energy. It is instead a function that is directly related to the *geometry* of the dynamics (4.48) on S^3, integrating ω.

$$u_1 = e_1 - y_2,$$
$$u_2 = e_2 + y_1, \tag{4.53}$$

the addition of the two inequalities (4.52) results in

$$\begin{aligned}
S_1(x_1(t_1)) + S_2(x_2(t_1)) &\le S_1(x_1(t_0)) + S_2(x_2(t_0)) + \\
\int_{t_0}^{t_1} (e_1^T(t)y_1(t) + e_2^T(t)y_2(t) &- \varepsilon_1 ||y_1(t)||^2 - \varepsilon_2 ||y_2(t)||^2) \, dt \\
&\le S_1(x_1(t_0)) + S_2(x_2(t_0)) + \\
\int_{t_0}^{t_1} (e_1^T(t)y_1(t) + e_2^T(t)y_2(t) &- \varepsilon[||y_1(t)||^2 + ||y_2(t)||^2]) \, dt
\end{aligned} \tag{4.54}$$

with $\varepsilon = \min(\varepsilon_1, \varepsilon_2)$. Hence the closed-loop system with inputs (e_1, e_2) and outputs (y_1, y_2) is output strictly passive if $\varepsilon > 0$, respectively, passive if $\varepsilon = 0$, with storage function

$$S(x_1, x_2) = S_1(x_1) + S_2(x_2), \quad (x_1, x_2) \in \mathcal{X}_1 \times \mathcal{X}_2 \tag{4.55}$$

Using Lemmas 3.2.9 and 3.2.16 we arrive at the following proposition, which can be regarded as the state space version of Theorems 2.2.6 and 2.2.11.

Proposition 4.3.1 (Passivity theorem) *Assume that for every pair of allowed external input functions $e_1(\cdot)$, $e_2(\cdot)$ there exist allowed input functions $u_1(\cdot)$, $u_2(\cdot)$ of the closed-loop system $\Sigma_1 \|_f \Sigma_2$.*
(i) Suppose Σ_1 and Σ_2 are passive or output strictly passive. Then $\Sigma_1 \|_f \Sigma_2$ with inputs (e_1, e_2) and outputs (y_1, y_2) is passive, and output strictly passive if both Σ_1 and Σ_2 are output strictly passive.
(ii) Suppose Σ_1 is passive and Σ_2 is input strictly passive, or Σ_1 is output strictly passive and Σ_2 is passive, then $\Sigma_1 \|_f \Sigma_2$ with $e_2 = 0$ and input e_1 and output y_1 is output strictly passive.
(iii) Suppose that S_1, S_2 satisfying (4.52) are C^1 and have strict local minima at x_1^*, respectively x_2^*. Then (x_1^*, x_2^*) is a stable equilibrium of $\Sigma_1 \|_f \Sigma_2$ with $e_1 = e_2 = 0$.
(iv) Suppose that Σ_1 and Σ_2 are output strictly passive and zero-state detectable, and that S_1, S_2 satisfying (4.52) are C^1 and have strict local minima at $x_1^* = 0$, respectively $x_2^* = 0$. Then $(0, 0)$ is an asymptotically stable equilibrium of $\Sigma_{\Sigma_1, \Sigma_2}^f$ with $e_1 = e_2 = 0$. If additionally S_1, S_2 have global minima at $x_1^* = 0$, respectively $x_2^* = 0$, and are proper, then $(0, 0)$ is a globally asymptotically stable equilibrium.

Proof (i) has been proved above, cf. (4.54), while (ii) follows similarly. (iii) results from application of Lemma 3.2.9 to $\Sigma_1 \|_f \Sigma_2$ with inputs (e_1, e_2) and outputs (y_1, y_2). (iv) follows from Proposition 3.2.16 applied to $\Sigma_1 \|_f \Sigma_2$. □

Remark 4.3.2 The standard negative feedback interconnection $u_1 = -y_2 + e_1$, $u_2 = y_1$ for $e_2 = 0$ has the following alternative interpretation. It can be also regarded as the *series interconnection* $u_2 = y_1$ of Σ_1 and Σ_2, together with the additional negative unit feedback loop $u_1 = -y_2 + e_1$. This interpretation will be used in Chap. 5, Theorem 5.2.1.

Remark 4.3.3 Note the inherent *robustness* property expressed in Proposition 4.3.1: the statements continue to hold for perturbed systems Σ_1 and Σ_2, as long as they remain (output strictly) passive and their storage functions satisfy the required properties.

Remark 4.3.4 As in Lemma 3.2.12 the strict positivity of S_1 and S_2 outside $x_1^* = 0$, $x_2^* = 0$ can be ensured by zero-state observability of Σ_1 and Σ_2.

In case $S_1(x_1) - S_1(x_1^*)$ and/or $S_2(x_2) - S_2(x_2^*)$ are not positive definite but only positive *semi*definite at x_1^*, respectively x_2^*, then Proposition 4.3.1 can be refined as in Theorem 3.2.19. We leave the details to the reader; see also [312].

In Theorem 2.2.18, see also Remark 2.2.19, we have seen how "lack of passivity" of one of the output maps G_1, G_2 can be compensated by "surplus of passivity" of the other. The argument generalizes to the state space setting as follows.

Corollary 4.3.5 *Suppose the systems $\Sigma_i, i = 1, 2$, are dissipative with respect to the supply rates*

$$s_i(u_i, y_i) = u_i^T y_i - \varepsilon_i \|y_i\|^2 - \delta_i \|u_i\|^2, \quad i = 1, 2, \tag{4.56}$$

where the constants $\varepsilon_i, \delta_i, i = 1, 2$, satisfy

$$\varepsilon_1 + \delta_2 > 0, \quad \varepsilon_2 + \delta_1 > 0 \tag{4.57}$$

Then the standard feedback interconnection $\Sigma_1 \|_f \Sigma_2$ has finite L_2-gain from inputs e_1, e_2 to outputs y_1, y_2.

Proof Since Σ_i is dissipative with respect to the supply rates s_i we have

$$\dot{S}_i \leq u_i^T y_i - \varepsilon_i \|y_i\|^2 - \delta_i \|u_i\|^2, \quad i = 1, 2 \tag{4.58}$$

for certain storage functions $S_i, i = 1, 2$ (assumed to be differentiable; otherwise use the integral version of the dissipation inequalities). Substitution of $u_1 = e_1 - y_2$, $u_2 = e_2 + y_1$ into the sum of these two inequalities yields

$$\begin{aligned} \dot{S}_1 + \dot{S}_2 &\leq e_1^T y_1 + e_2^T y_2 \\ &- \varepsilon_1 \|y_1\|^2 - \delta_1 \|e_1 - y_2\|^2 - \varepsilon_2 \|y_2\|^2 - \delta_2 \|e_2 + y_1\|^2 \end{aligned} \tag{4.59}$$

which, multiplying both sides by -1, can be rearranged as

$$\begin{aligned} -\delta_1 \|e_1\|^2 - \delta_2 \|e_2\|^2 - \dot{S}_1 - \dot{S}_2 &\geq \\ (\varepsilon_1 + \delta_2) \|y_1\|^2 + (\varepsilon_2 + \delta_1) \|y_2\|^2 - 2\delta_1 e_1^T y_2 - 2\delta_2 e_2^T y_1 - e_1^T y_1 - e_2^T y_2 \end{aligned} \tag{4.60}$$

Then, completely similar to the proof of Theorem 2.2.18, by the positivity assumption on $\alpha_1^2 := \varepsilon_1 + \delta_2$, $\alpha_2^2 := \varepsilon_2 + \delta_1$ we can perform "completion of the squares" on the right-hand side of the inequality (4.60), to obtain an expression of the form

$$\left\| \begin{bmatrix} \alpha_1 y_1 \\ \alpha_2 y_2 \end{bmatrix} - A \begin{bmatrix} e_1 \\ e_2 \end{bmatrix} \right\|^2 \le c^2 \left\| \begin{bmatrix} e_1 \\ e_2 \end{bmatrix} \right\|^2 - \dot{S}_1 - \dot{S}_2, \tag{4.61}$$

for a certain 2×2 matrix A and constant c. In combination with the triangle inequality (2.29) this gives the desired result.   □

This corollary is illustrated by the following example, which contains a further interesting extension.

Example 4.3.6 (Lur'e functions) Consider an input-state-output system

$$\Sigma_1 : \begin{array}{l} \dot{x}_1 = f(x_1, u_1) \\ y_1 = h(x_1) \end{array} \quad u_1, y_1 \in \mathbb{R}, \tag{4.62}$$

and a system Σ_2 given by a *static nonlinearity*

$$\Sigma_2 : y_2 = F(u_2), \qquad u_2, y_2 \in \mathbb{R}, \tag{4.63}$$

interconnected by negative feedback $u_1 = -y_2$, $u_2 = y_1$.

Suppose the static nonlinearity F is passive in the sense of (4.29), that is, $uF(u) \ge 0$ for all $u \in \mathbb{R}$ (its graph is in the first and third quadrant). Obviously, if Σ_1 is passive with C^1 storage function $S_1(x_1)$, then by a direct application of the passivity theorem (Proposition 4.3.1) the closed-loop system satisfies $\dot{S}_1 \le 0$.

Now suppose that Σ_1 is *not* passive, but only dissipative with respect to the supply rate

$$s_1(u_1, y_1) = u_1 y_1 + \frac{u_1^2}{k}, \tag{4.64}$$

for some $k > 0$, having C^1 storage function S_1. On the other hand, suppose that F is $\frac{1}{k}$-*output strictly* passive; that is, dissipative with respect to the supply rate

$$s_2(u_2, y_2) = u_2 y_2 - \frac{y_2^2}{k} \tag{4.65}$$

for the same k as above. Then by application of Corollary 4.3.5 the closed-loop system satisfies $\dot{S}_1 \le 0$. Note that dissipativity of F with respect to s_2 can be equivalently expressed by the *sector condition*

$$0 \le \frac{F(u_2)}{u_2} \le k \tag{4.66}$$

The story can be continued as follows. Suppose that Σ_1 is *not* dissipative with respect to s_1, but that instead $\Sigma_{1\alpha}$, defined as

$$\Sigma_{1\alpha} : \begin{array}{l} \dot{x}_1 = f(x_1, u_1) \\ \widehat{y}_1 := y_1 + \alpha \dot{y}_1 = h(x_1) + \alpha \frac{dh}{dx_1}(x_1) f_1(x_1, u_1) \end{array} \tag{4.67}$$

is dissipative with respect to s_1 for some $\alpha > 0$. Suppose as above that the static nonlinearity F satisfies (4.66) (and thus is output strictly passive). Then consider instead of the static nonlinearity Σ_2 defined by F the *dynamical* system

$$\Sigma_{2\alpha} : \quad \begin{aligned} \alpha \dot{x}_2 &= -x_2 + u_2, \quad x_2 \in \mathbb{R} \\ y_2 &= F(x_2) \end{aligned} \tag{4.68}$$

It readily follows that $\Sigma_{2\alpha}$ is dissipative with respect to s_2, with storage function

$$S_2(x_2) := \alpha \int_0^{x_2} F(\sigma) d\sigma \geq 0 \tag{4.69}$$

Indeed, by (4.66)

$$\dot{S}_2 = \alpha F(x_2) \dot{x}_2 = F(x_2)(-x_2 + u_2) \leq u_2 F(x_2) - \frac{F^2(x_2)}{k}$$

Hence, (again by Corollary 4.3.5) the closed-loop system of $\Sigma_{1\alpha}$ and $\Sigma_{2\alpha}$ satisfies $\dot{S}_1 + \dot{S}_2 \leq 0$. Finally note that

$$\alpha \dot{x}_2 + x_2 = u_2 = y_1 + \alpha \dot{y}_1,$$

and thus $\alpha(\dot{x}_2 - \dot{y}_1) = -(x_2 - y_1)$, implying that the level set $x_2 = h(x_1)$ is an (attractive) invariant set. Hence, we can *restrict* the closed-loop system to the level set $x_2 = h(x_1)$, where the system has total storage function

$$S(x_1) := S_1(x_1) + \alpha \int_0^{h_1(x_1)} F(\sigma) d\sigma$$

satisfying $\dot{S} \leq 0$. In case of a *linear* system Σ_1 with quadratic storage function S_1 the obtained function S is called a *Lur'e function*. Depending on the properties of S, we may derive stability, and under strengthened conditions, (global) asymptotic stability, for Σ_1 with the static nonlinearity F in the negative feedback loop. This yields the *Popov criterion*; see the references in the Notes at the end of Chap. 2.

Example 4.3.7 Consider the system

$$\begin{aligned} \dot{x} &= f(x) + g_1(x)u_1 + g_2(x)u_2, & u_1 \in \mathbb{R}^{m_1}, u_2 \in \mathbb{R}^{m_2} \\ y_1 &= h_1(x), & y_1 \in \mathbb{R}^{m_1} \\ y_2 &= h_2(x), & y_2 \in \mathbb{R}^{m_2} \end{aligned} \tag{4.70}$$

which is passive with respect to the inputs u_1, u_2 and outputs y_1, y_2, with storage function $S(x)$. Consider the static nonlinearity

$$\perp := \{(v, z) \in \mathbb{R}^{m_2} \times \mathbb{R}^{m_2} \mid v \geq 0, \ z \geq 0, \ v^T z = 0\}, \tag{4.71}$$

where $v \geq 0, z \geq 0$ means that all elements of v, z are nonnegative. Clearly this is a passive static system. Interconnect $\perp$ to the system by setting $u_2 = -z$, $y_2 = v$. The resulting system satisfies

$$\frac{d}{dt} S \leq u_1^T y_1, \tag{4.72}$$

and thus defines a passive system (although not of a standard input-state-output type). This type of systems occurs, e.g., in electrical circuits with ideal diodes; see the Notes at the end of this chapter.

The passivity theorems given so far are one-way: the feedback interconnection of two passive systems is again passive. As we will now see, the *converse* also holds: if the feedback interconnection of two systems is passive then necessarily these systems are passive. This will be shown to have immediate consequences for the set of storage functions of the interconnected system, which always contains an *additive* one.

Proposition 4.3.8 (Converse passivity theorem) *Consider Σ_i with state spaces $\mathcal{X}_1, i = 1, 2$, and with allowed input functions $u_1(\cdot), u_2(\cdot)$, in standard feedback configuration $u_1 = e_1 - y_2, u_2 = e_2 + y_2$. Assume that for every pair of allowed external input functions $e_1(\cdot), e_2(\cdot)$ there exist allowed input functions $u_1(\cdot), u_2(\cdot)$ of the closed-loop system $\Sigma_1 \|_f \Sigma_2$. Conversely, assume that for all allowed input functions $u_1(\cdot), u_2(\cdot)$ there exist allowed external input functions $e_1(\cdot), e_2(\cdot)$ satisfying at any time-instant $u_1 = e_1 - y_2, u_2 = e_2 + y_2$. Then $\Sigma_1 \|_f \Sigma_2$ with inputs e_1, e_2 and outputs y_1, y_2 is passive if and only if both Σ_1 and Σ_2 are passive. Furthermore, the available storage S_a and required supply S_r of $\Sigma_1 \|_f \Sigma_2$ (assuming Σ_i is reachable from some $x_i^*, i = 1, 2$) are additive, that is*

$$\begin{aligned} S_a(x_1, x_2) &= S_{a1}(x_1) + S_{a2}(x_2) \\ S_r(x_1, x_2) &= S_{r1}(x_1) + S_{r2}(x_2) \end{aligned} \tag{4.73}$$

with S_{ai}, S_{ri} denoting the available storage, respectively required supply, of Σ_i, $i = 1, 2$.

Proof The "if" part is Proposition 4.3.1. For the converse statement we note that $\Sigma_1 \|_f \Sigma_2$ is passive if and only

$$S_a(x_1, x_2) := \sup_{e_1(\cdot), e_2(\cdot), T \geq 0} - \int_0^T \left(e_1^T(t) y_1(t) + e_2^T(t) y_2(t) \right) dt < \infty \tag{4.74}$$

for all $(x_1, x_2) \in \mathcal{X}$. Substituting the "inverse" interconnection equations $e_1 = u_1 + y_2$ and $e_2 = u_2 - y_1$ this is equivalent to

$$\sup_{e_1(\cdot), e_2(\cdot), T \geq 0} - \int_0^T \left(u_1^T(t) y_1(t) + u_2^T(t) y_2(t) \right) dt < \infty \tag{4.75}$$

for all (x_1, x_2). Using the assumption that for all allowed $u_1(\cdot), u_2(\cdot)$ there exist allowed external input functions $e_1(\cdot), e_2(\cdot)$ this is equal to

$$\sup_{u_1(\cdot), u_2(\cdot), T \geq 0} - \int_0^T \left(u_1^T(t) y_1(t) + u_2^T(t) y_2(t) \right) =$$

$$\sup_{u_1(\cdot), T \geq 0} - \int_0^T u_1^T(t) y_1(t) dt + \sup_{u_2(\cdot), T \geq 0} - \int_0^T u_2^T(t) y_2(t) dt < \infty$$

for all (x_1, x_2). Hence $\Sigma_1 \|_f \Sigma_2$ is passive iff Σ_1 and Σ_2 are passive, in which case $S_a(x_1, x_2) = S_{a1}(x_1) + S_{a2}(x_2)$. The same reasoning leads to the second equality of (4.73). $\qquad \square$

A similar statement, for any storage function of $\Sigma_1 \|_f \Sigma_2$, can be obtained from the differential dissipation inequality as follows.

Proposition 4.3.9 *Consider $\Sigma_i, i = 1, 2$, of the form (4.14) with equilibria $x_i^* \in \mathcal{X}_i$ satisfying $f_i(x_i^*) = 0, i = 1, 2$. Assume that $\Sigma_1 \|_f \Sigma_2$ is passive (lossless) with C^1 storage function $S(x_1, x_2)$. Then also $\Sigma_i, i = 1, 2$, are passive (lossless) with storage functions $S_1(x_1) := S(x_1, x_2^*), S_2(x_2) := S(x_1^*, x_2)$.*

Proof We will only prove the passive case; the same arguments apply to the lossless case. $\Sigma_1 \|_f \Sigma_2$ being passive is equivalent to the existence of $S : \mathcal{X}_1 \times \mathcal{X}_2 \to \mathbb{R}^+$ satisfying

$$\begin{aligned}
&S_{x_1}(x_1, x_2) [f_1(x_1) - g_1(x_1) h_2(x_2)] \\
&+ S_{x_2}(x_1, x_2) [f_2(x_2) + g_2(x_2) h_1(x_1)] \leq 0 \\
&S_{x_1}(x_1, x_2) g_1(x_1) = h_1^T(x_1) \\
&S_{x_2}(x_1, x_2) g_2(x_2) = h_2^T(x_2)
\end{aligned} \tag{4.76}$$

This results in

$$S_{x_1}(x_1, x_2) f_1(x_1) - \underbrace{S_{x_1}(x_1, x_2) g_1(x_1)}_{=h_1^T(x_1)} h_2(x_2)$$

$$+ S_{x_2}(x_1, x_2) f_2(x_2) + \underbrace{S_{x_2}(x_1, x_2) g_2(x_2)}_{=h_2^T(x_2)} h_1(x_1) \tag{4.77}$$

$$= S_{x_1}(x_1, x_2) f_1(x_1) + S_{x_2}(x_1, x_2) f_2(x_2) \leq 0$$

For $x_2 = x_2^*$, (4.77) amounts to

$$\begin{aligned}
&S_{x_1}(x_1, x_2^*) f_1(x_1) + S_{x_2}(x_1, x_2^*) f_2(x_2^*) \\
&= S_{x_1}(x_1, x_2^*) f_1(x_1) = S_{1x_1}(x_1) f_1(x_1) \leq 0
\end{aligned} \tag{4.78}$$

since $f_2(x_2^*) = 0$. Furthermore, the second line of (4.76) becomes

$$S_{1x_1}(x_1) g_1(x_1) = S_{x_1}(x_1, x_2^*) g_1(x_1) = h_1^T(x_1) \tag{4.79}$$

Hence, $S_1(x_1)$ is a storage function for Σ_1. In the same way $S_2(x_2)$ is a storage function for Σ_2. $\qquad \square$

An important consequence of Propositions 4.3.8 and 4.3.9 is the fact that among the storage functions of the passive system $\Sigma_1 \|_f \Sigma_2$ there always exist *additive* storage functions $S(x_1, x_2) = S_1(x_1) + S_2(x_2)$. In fact, the available storage and required supply functions are additive by Proposition 4.3.8, while by Proposition 4.3.9 an arbitrary storage function $S(x_1, x_2)$ for $\Sigma_1 \|_f \Sigma_2$ can be replaced by the additive storage function $S(x_1, x_2^*) + S(x_1^*, x_2)$.

4.4   Network Interconnection of Passive Systems

In many complex network systems—from mass–spring–damper systems, electrical circuits, communication networks, chemical reaction networks, and transportation networks to power networks—the passivity of the overall network system naturally arises from the properties of the network interconnection structure and the passivity of the subsystems. In this section this will be illustrated by three different scenarios of network interconnection of passive systems.

The interconnection structure of a network system can be advantageously encoded by a directed *graph*. Recall the following standard notions and facts from (algebraic) graph theory; see [48, 114] and the Notes at the end of the chapter for further information. A *graph* $\mathcal{G}$, is defined by a set $\mathcal{V}$ of N *vertices* (nodes) and a set $\mathcal{E}$ of M *edges* (links, branches), where $\mathcal{E}$ is identified with a set of unordered pairs $\{i, j\}$ of vertices $i, j \in \mathcal{V}$. We allow for multiple edges between vertices, but not for self-loops $\{i, i\}$. By endowing the edges with an orientation, turning the unordered pairs $\{i, j\}$ into ordered pairs (i, j), we obtain a *directed graph*. In the following "graph" will throughout mean "directed graph." A directed graph with N vertices and M edges is specified by its $N \times M$ *incidence matrix*, denoted by D. Every column of D corresponds to an edge of the graph, and contains one -1 at the row corresponding to its tail vertex and one $+1$ at the row corresponding to its head vertex, while all other elements are 0. In particular, $\mathbb{1}^T D = 0$ where $\mathbb{1}$ is the vector of all ones. Furthermore, $\ker D^T = \text{span } \mathbb{1}$ if and only if the graph is *connected* (any vertex can be reached from any other vertex by a sequence of—undirected—edges). In general, the dimension of $\ker D^T$ is equal to the number of connected components of the graph. A directed graph is *strongly connected* if any vertex can be reached from any other vertex by a sequence of directed edges.

The *first* case of network interconnection of passive systems concerns the interconnection of passive systems which are partly associated to the *vertices*, and partly to the *edges* of an underlying graph. As illustrated later on, this is a common case in many physical networks. Thus to each i-th vertex there corresponds a passive system with scalar inputs and outputs (see Remark 4.4.2 for generalizations)

$$
\begin{aligned}
\dot{x}_i^v &= f_i^v(x_i^v, u_i^v), \quad x_i^v \in \mathcal{X}_i^v, \ u_i^v \in \mathbb{R} \\
y_i^v &= h_i^v(x_i^v, u_i^v), \quad y_i^v \in \mathbb{R}
\end{aligned}
\tag{4.80}
$$

with storage function S_i^v, $i = 1, \ldots, N$, and to each j-th edge (branch) there corresponds a passive single-input single-output system

$$
\begin{aligned}
\dot{x}_i^b &= f_i^b(x_i^b, u_i^b), \quad x_i^b \in \mathcal{X}_i^b, \; u_i^b \in \mathbb{R} \\
y_i^b &= h_i^b(x_i^b, u_i^b), \quad y_i^b \in \mathbb{R}
\end{aligned}
\tag{4.81}
$$

with storage function S_i^b, $i = 1, \ldots, M$. Collecting the scalar inputs and outputs into vectors

$$
\begin{aligned}
u^v &= \left[u_1^v, \ldots, u_N^v\right]^T, \quad y^v = \left[y_1^v, \ldots, y_N^v\right]^T \\
u^b &= \left[u_1^b, \ldots, u_M^b\right]^T, \quad y^b = \left[y_1^b, \ldots, y_M^b\right]^T
\end{aligned}
\tag{4.82}
$$

these passive systems are interconnected to each other by the interconnection equations

$$
\begin{aligned}
u^v &= -Dy^b + e^v \\
u^b &= D^T y^v + e^b
\end{aligned}
\tag{4.83}
$$

where $e^v \in \mathbb{R}^N$ and $e^b \in \mathbb{R}^M$ are external inputs. Since the interconnection (4.83) satisfies

$$
(u^v)^T y^v + (u^b)^T y^b = (e^v)^T y^v + (e^b)^T y^b
$$

the following result directly follows.

Proposition 4.4.1 *Consider a graph with incidence matrix D, with passive systems (4.80) with storage functions S_i^v associated to the vertices and passive systems (4.81) with storage functions S_i^b associated to the edges, interconnected by (4.83). Then the interconnected system is again passive with inputs e^v, e^b and outputs y^v, y^b, with total storage function*

$$
S_1^v(x_1^v) + \cdots + S_N^v(x_N^v) + S_1^b(x_1^b) + \cdots + S_1^b(x_M^b)
\tag{4.84}
$$

Remark 4.4.2 The setup can be generalized to multi-input multi-output systems with u_i^v, y_i^v, u_j^b, y_j^b all in $\mathbb{R}^m$ by replacing the incidence matrix D in the above by the Kronecker product $D \otimes I_m$ and D^T by $D^T \otimes I_m$, with I_m denoting the $m \times m$ identity matrix.

Remark 4.4.3 Proposition 4.4.1 continues to hold in cases where some of the edges or vertices correspond to *static* passive systems. Simply define the total storage function as the sum of the storage functions of the *dynamic* passive systems.

Example 4.4.4 (Power networks) Consider a power system of synchronous machines, interconnected by a network of purely inductive transmission lines. Modeling the synchronous machines by swing equations, and assuming that all voltage and current signals are sinusoidal of the same frequency and all voltages have constant amplitude one arrives at the following model. Associated to the N vertices each i-th synchronous machine is described by the passive system

$$\dot{p}_i = -A_i \omega_i + u_i^v$$
$$y_i^v = \omega_i \qquad\qquad\qquad\qquad (4.85)$$

where ω_i is the frequency deviation from nominal frequency (e.g., 50 Hz), $p_i = J_i \omega_i$ is the momentum deviation (with J_i related to the inertia of the synchronous machine), A_i the damping constant, and u_i^v is the incoming power, $i = 1, \ldots, N$. Furthermore, denoting the phase differences across the j-th line by q_j, the dynamics of the j-th line (associated to the j-th edge of the graph) is given by the passive system

$$\dot{q}_j = u_j^b$$
$$y_j^b = \gamma_j \sin q_j \qquad\qquad\qquad (4.86)$$

with the constant γ_j determined by the susceptance of the line and the voltage amplitude at the adjacent vertices, $j = 1, \ldots, M$. Here y_j^b equals the (average or active) power through the line. Denoting $p = (p_1, \ldots, p_N)^T$, $\omega = (\omega_1, \ldots, \omega_N)^T$, and $q = (q_1, \ldots, q_M)^T$, the final system resulting from the interconnection (4.83) is given as

$$\begin{bmatrix} \dot{q} \\ \dot{p} \end{bmatrix} = \begin{bmatrix} 0 & D^T \\ -D & -A \end{bmatrix} \begin{bmatrix} \Gamma \mathrm{Sin}\, q \\ \omega \end{bmatrix} + \begin{bmatrix} 0 \\ u \end{bmatrix}, \quad p = J\omega$$
$$y = \omega, \qquad\qquad\qquad\qquad\qquad\qquad\qquad (4.87)$$

with A and J denoting diagonal matrices with elements A_i, J_i, $i = 1, \ldots, N$, and Γ the diagonal matrix with elements γ_j, $j = 1, \ldots, M$. Furthermore $\mathrm{Sin} : \mathbb{R}^M \to \mathbb{R}^M$ denotes the element-wise sinus function, i.e., $\mathrm{Sin}\, q = (\sin q_1, \ldots, \sin q_M)$. Finally, the input u denotes the vector of generated/consumed power and the output y the vector of frequency deviations, both associated to the vertices. The final system (4.87) is a passive system with additive storage function

$$H(q, p) := \frac{1}{2} p^T J^{-1} p - \sum_{j=1}^{M} \gamma_j \cos q_j \qquad\qquad (4.88)$$

Example 4.4.5 (Mass-spring systems) Consider N masses moving in one-dimensional space interconnected by M springs. Associate the masses to the vertices of a graph with incidence matrix D, and the springs to the edges. Furthermore, let $p_1, \ldots, p_N$ be the momenta of the masses, and $q_1, \ldots, q_M$ the extensions of the springs. Then the equations of motion of the total system are given as

$$\begin{bmatrix} \dot{p} \\ \dot{q} \end{bmatrix} = \begin{bmatrix} 0 & -D \\ D^T & 0 \end{bmatrix} \begin{bmatrix} \frac{\partial K}{\partial p}(p) \\ \frac{\partial P}{\partial q}(q) \end{bmatrix} + \begin{bmatrix} e^v \\ e^b \end{bmatrix}, \qquad (4.89)$$

where $p = (p_1, \ldots, p_N)^T$ and $q = (q_1, \ldots, q_M)^T$, and where $K(p) = \sum \frac{1}{2m_i} p_i^2$ is the total kinetic energy of the masses, and $P(q)$ the total potential energy of the springs. This defines a passive system with inputs e^v, e^b (external forces, respectively, external velocity flows) and outputs $\frac{\partial K}{\partial p}(p)$, $\frac{\partial P}{\partial q}(q)$ (velocities, respectively, spring forces), and additive storage function $K(p) + P(q)$.

Similar to Remark 4.4.2 this can be generalized to a mass-spring system in $\mathbb{R}^3$, by considering $p_i, q_j \in \mathbb{R}^3$, and replacing the incidence matrix D by the Kronecker product $D \otimes I_3$ and D^T by $D^T \otimes I_3$. Furthermore, by Remark 4.4.3 the setup can be extended to *mass–spring–damper systems*, in which case part of the edges correspond to *dampers*.

In Chap. 6 we will see how Examples 4.4.4 and 4.4.5 actually define passive *port-Hamiltonian* systems.

A *second* case of network interconnection of passive systems is that of a multi-agent system, where the input of each passive agent system depends on the outputs of the other systems and of itself. Thus consider N passive systems Σ_i associated to the vertices of a graph, given by

$$
\begin{aligned}
\dot{x}_i &= f_i(x_i, u_i), \quad x_i \in \mathcal{X}_i, \ u_i \in \mathbb{R} \\
y_i &= h_i(x_i, u_i), \quad y_i \in \mathbb{R}
\end{aligned}
\tag{4.90}
$$

with storage functions $S_i, i = 1, \ldots, N$. Collecting the inputs into the vector $u = (u_1, \ldots, u_N)^T$ and the outputs into $y = (y_1, \ldots, y_N)^T$ we consider interconnection equations

$$
u = -Ly + e
\tag{4.91}
$$

where e is a vector of external inputs, and L is a *Laplacian matrix*, defined as follows.

Definition 4.4.6 A Laplacian matrix of a graph with N vertices is defined as an $N \times N$ matrix L with positive diagonal elements, and non-positive off-diagonal elements, with either the row sums of L equal to zero (a *communication Laplacian matrix*) or the column sums equal to zero (*flow Laplacian* matrix). If both the row and sums are zero then L is called a *balanced Laplacian* matrix.

This means that any communication Laplacian L_c satisfies $L_c \mathbb{1} = 0$, and can be written as $L_c = -K_c D^T$ for an incidence matrix D of the communication graph, and a matrix K_c of nonnegative elements. In fact, the nonzero elements of the i-th row of K_c are the weights of the edges incoming to vertex i. Dually, any flow Laplacian L_f satisfies $\mathbb{1}^T L_f = 0$, and can be written as $L_f = -DK_f$ for a certain incidence matrix, and a matrix K_f of nonnegative elements. The nonzero elements of the i-th column of K_f are the weights of the edges originating from vertex i.

A communication Laplacian matrix L_c, respectively flow Laplacian matrix L_f is balanced if and only [70]

$$
L_c + L_c^T \geq 0, \text{ respectively, } L_f + L_f^T \geq 0
\tag{4.92}
$$

Remark 4.4.7 A special case of a balanced Laplacian matrix is a *symmetric* balanced Laplacian matrix L, which can be written as $L = DKD^T$, where D is the incidence matrix and K is an $M \times M$ diagonal matrix of positive weights corresponding to the M edges of the graph.

Remark 4.4.8 The interconnection (4.91) with L a communication Laplacian matrix corresponds to feeding back the *differences* of the output values

$$u_i = -\sum_k a_{ik}(y_i - y_k), \quad i = 1, \dots, N, \tag{4.93}$$

where the summation index k is running over all vertices that are connected to the i-th vertex by an edge directed toward i, and a_{ik} is the positive weight of this edge. On the other hand, the interconnection (4.91) with L a flow Laplacian matrix corresponds to an output feedback satisfying $\mathbb{1}^T u = 0$, corresponding to a distribution of the material flow through the network. This occurs for transportation and distribution networks, including chemical reaction networks.

Proposition 4.4.9 *Consider the passive systems (4.90) interconnected by (4.91), where L is a balanced Laplacian matrix. Then the interconnected system is passive with additive storage function $S_1(x_1) + \cdots + S_N(x_N)$.*

Proof Follows from the fact that by (4.92)

$$u^T y = -(Ly + e)^T y = -\frac{1}{2} y^T (L + L^T) y + e^T y \le e^T y \qquad \square$$

Proposition 4.4.9 can be generalized to flow and communication Laplacian matrices that are *not balanced* by additionally assuming that the connected components of the underlying graph are *strongly connected*[5] In fact, under this assumption, any flow or communication Laplacian matrix can be *transformed* into a balanced one. Furthermore, this can be done in a constructive way by employing a general form of *Kirchhoff's Matrix Tree theorem*, which for our purposes can be described as follows (see the Notes at the end of this chapter).

Let L be a *flow* Laplacian matrix, and assume for simplicity that the graph is connected, implying that $\dim \ker L = 1$. Denote the (i, j)-th cofactor of L by $C_{ij} = (-1)^{i+j} M_{i,j}$, where $M_{i,j}$ is the determinant of the (i, j)-th minor of L, which is the matrix obtained from L by deleting its i-th row and j-th column. Define the adjoint matrix $\text{adj}(L)$ as the matrix with (i, j)-th element given by C_{ji}. It is well known that

$$L \cdot \text{adj}(L) = (\det L)I_N = 0 \tag{4.94}$$

---

[5]In fact, balancedness of a communication or flow Laplacian matrix *implies* that all connected components are strongly connected; cf. [70].

Furthermore, since $\mathbb{1}^T L = 0$ the sum of the rows of L is zero, and hence by the properties of the determinant function the quantities C_{ij} do not depend on i, implying that $C_{ij} = \gamma_j$, $i = 1, \ldots, N$. Hence, by defining $\gamma := (\gamma_1, \ldots, \gamma_N)^T$, it follows from (4.94) that $L\gamma = 0$. Moreover, γ_i is equal to *the sum of the products of weights of all the spanning trees of $\mathcal{G}$ directed toward* vertex i. In particular, it follows that $\gamma_j \geq 0$, $j = 1, \ldots, N$. In fact, $\gamma \neq 0$ if and only if $\mathcal{G}$ has a spanning tree. Since for every vertex i there exists at least one spanning tree directed toward i if and only if the graph is strongly connected, we conclude that $\gamma \in \mathbb{R}_+^N$ if and only if the graph is strongly connected.

In case the graph $\mathcal{G}$ is not connected the same analysis can be performed on each of its connected components. Hence, if all connected components of $\mathcal{G}$ are strongly connected, Kirchhoff's matrix tree theorem provides us with a vector $\gamma \in \mathbb{R}_+^N$ such that $L\gamma = 0$. It immediately follows that the transformed matrix $L\Gamma$, where Γ is the positive $N \times N$-dimensional diagonal matrix with diagonal elements $\gamma_1, \ldots, \gamma_N$, is a balanced Laplacian matrix.

Dually, if L is a communication Laplacian matrix and the connected components of the graph are strongly connected, then there exist a positive $N \times N$ diagonal matrix Γ such that ΓL is balanced. Summarizing, we obtain the following.

Proposition 4.4.10 *Consider a flow Laplacian matrix L_f (communication Laplacian matrix L_c). Then there exists a positive diagonal matrix Γ_f (Γ_c) such that $L_f \Gamma_f$ ($\Gamma_c L_c$) is balanced if and only if the connected components of the graph are all strongly connected.*

This has the following consequence for the passivity of the interconnection of passive systems Σ_i, $i = 1, \ldots, N$, under the interconnection (4.91).

Proposition 4.4.11 *Consider passive systems $\Sigma_1, \ldots, \Sigma_N$ with storage functions $S_1, \ldots, S_N$, interconnected by $u = -Ly + e$, where L is either a flow Laplacian L_f or a communication Laplacian L_c, and assume that the connected components of the interconnection graph are strongly connected. Let L_f be a flow Laplacian, and consider a positive diagonal matrix $\Gamma_f = \mathrm{diag}(\gamma_1^f, \ldots, \gamma_N^f)$ such that $L_f \Gamma_f$ is balanced. Then the interconnected system with inputs e and scaled outputs $\frac{1}{\gamma_1^f} y_1, \ldots, \frac{1}{\gamma_N^f} y_N$ is passive with storage function*

$$S^f(x_1, \ldots, x_N) := \frac{1}{\gamma_1^f} S_1(x_1) + \cdots + \frac{1}{\gamma_N^f} S_N(x_N) \tag{4.95}$$

Alternatively, let L_c be a communication Laplacian, and consider a positive diagonal matrix $\Gamma_c = \mathrm{diag}(\gamma_1^c, \ldots, \gamma_N^c)$ such that $\Gamma_c L_c$ is balanced. Then the interconnected system with inputs e and scaled outputs $\gamma_1^c y_1, \ldots, \gamma_N^c y_N$, is passive with storage function

$$S^c(x_1, \ldots, x_N) := \gamma_1^c S_1(x_1) + \cdots + \gamma_N^c S_N(x_N) \tag{4.96}$$

Proof The first statement follows by passivity from

$$\frac{d}{dt}S^f \le y^T \Gamma_f^{-1} u = -y^T \Gamma_f^{-1} L_f y + y^T \Gamma_f^{-1} e$$
$$= -(\Gamma_f^{-1} y)^T L_f \Gamma_f (\Gamma_f^{-1} y) + (\Gamma_f^{-1} y)^T e \qquad (4.97)$$

and balancedness of $L_f \Gamma_f$. Similarly, the second statement follows from

$$\frac{d}{dt}S^c \le y^T \Gamma_c u = -y^T \Gamma_c L_c y + y^T \Gamma_c e \qquad (4.98)$$

and balancedness of $\Gamma_c L_c$. □

Remark 4.4.12 The result continues to hold in case some of the systems Σ_i are *static* passive nonlinearities. Indeed, since for each j-th static passive nonlinearity $u_j y_j \ge 0$, the same inequalities continue to hold, with the storage functions S^f or S^c now being the weighted sum of the storage functions of the *dynamical* passive systems Σ_i.

Remark 4.4.13 The notion of a balanced Laplacian matrix is also instrumental in defining the *effective resistance* from one vertex of the connected network to another. In fact, let L be a balanced Laplacian matrix. For any vertex i and j note that $e_i - e_j \in \text{im } L$, where e_i and e_j are the standard basis vectors with 1 at the i-th or j-th element, and 0 everywhere else. Thus there exists a vector v satisfying

$$Lv = e_i - e_j, \qquad (4.99)$$

which is moreover unique up to addition of a multiple of the vector $\mathbb{1}$ of all ones. This means that the quantity

$$R_{ji} := v_i - v_j, \qquad (4.100)$$

is independent of the choice of v satisfying (4.99). It is called the *effective resistance* of the network from vertex j to vertex i.

The same idea of taking *weighted combinations* of storage functions is used in the following *third* case of interconnection of passive systems. Consider again a multi-agent system, composed of N passive agent systems Σ_i with scalar inputs and outputs u_i, y_i, and storage functions $S_i(x_i)$, $i = 1, \ldots, N$. These are interconnected by

$$u = Ky + e \qquad (4.101)$$

where $u = (u_1, \ldots, u_N)^T$, $y = (y_1, \ldots, y_N)^T$, and the $N \times N$ matrix K has the following special structure:

$$K = \begin{bmatrix} -\alpha_1 & 0 & \cdot & \cdot & 0 & -\beta_N \\ \beta_1 & -\alpha_2 & \cdot & \cdot & 0 & 0 \\ 0 & \beta_2 & -\alpha_3 & \cdot & 0 & 0 \\ \cdot & \cdot & \cdot & \cdot & \cdot & \cdot \\ 0 & 0 & \cdot & \beta_{N-2} & -\alpha_{N-1} & 0 \\ 0 & 0 & \cdot & 0 & \beta_{N-1} & -\alpha_N \end{bmatrix} \qquad (4.102)$$

for positive constants $\alpha_i, \beta_i, i = 1, \ldots, N$. This represents a circular graph, where the first $N - 1$ gains $\beta_1, \ldots, \beta_{N-1}$ are *positive*, but the last interconnection gain $-\beta_N$ (from vertex N to vertex 1) is *negative*.

The main differences with the case $u = -Ly + e$ considered before, where L is either a flow or communication Laplacian matrix, are the special structure of the graph (a circular graph instead of a general graph), the fact that the right-upper element of K, given by $-\beta_N$, is *negative*, and the fact neither the row or column sums of K are zero. Nevertheless, also the matrix K can be transformed by a diagonal matrix into a matrix satisfying a property similar to (4.92), provided the constants $\alpha_i, \beta_i, i = 1, \ldots, N$, satisfy the following condition.

Theorem 4.4.14 ([12]) *Consider the $N \times N$ matrix K given in (4.102). There exists a positive $N \times N$ diagonal matrix Γ such that $\Gamma K + K^T \Gamma < 0$ if and only the positive constants $\alpha_i, \beta_i, i = 1, \ldots, N$, satisfy*[6]

$$\frac{\beta_1 \cdots \beta_N}{\alpha_1 \cdots \alpha_N} < \sec \left(\frac{\pi}{N}\right)^N \tag{4.103}$$

The condition (4.103) is referred to as the *secant condition*. Proceeding in the same way as for the Laplacian matrix interconnection case we obtain the following interconnection result.

Proposition 4.4.15 *Consider passive systems $\Sigma_1, \ldots, \Sigma_N$ with storage functions $S_1, \ldots, S_N$, interconnected by $u = -Ky + e$, where K is given by (4.102) with $\alpha_i, \beta_i, i = 1, \ldots, N$, satisfying (4.103). Take any positive diagonal matrix $\Gamma = \mathrm{diag}(\gamma_1, \ldots, \gamma_N)$ such that $\Gamma K + K^T \Gamma < 0$. Then the interconnected system with inputs e and scaled outputs $\gamma_1 y_1, \ldots, \gamma_N y_N$ is output strictly passive with storage function*

$$S^K(x_1, \ldots, x_N) := \gamma_1 S_1(x_1) + \cdots + \gamma_N S_N(x_N) \tag{4.104}$$

Proof This follows from

$$\frac{d}{dt} S^K \leq y^T \Gamma u = y^T \Gamma K y + y^T \Gamma e = y^T \Gamma K y + y^T \Gamma e \tag{4.105}$$

and $\Gamma K + K^T \Gamma < 0$. □

Remark 4.4.16 The stability of the interconnected system can be alternatively considered from the *small-gain* point of view; cf. Chaps. 2 and 8. Indeed, the interconnected system can be also formulated as the circular interconnection, with gains $+1$ for the first $N - 1$ interconnections and gain -1 for the interconnection from vertex N to vertex 1, of the *modified* systems $\widehat{\Sigma}_i$ with inputs v_i and outputs $\widehat{y}_i$ obtained from Σ_i by substituting $u_i = -\alpha_i y_i + v_i, \widehat{y}_i = \beta_i y_i, i = 1, \ldots, N$. Then by output strict

---

[6]Note that the secant function is given as $\sec \phi = \frac{1}{\cos \phi}$.

passivity of $\widehat{\Sigma}_i$ the L_2-gain of $\widehat{\Sigma}_i$ is $\leq \frac{\alpha_i}{\beta_i}$. Application of the small-gain condition, cf. Chap. 8, then yields stability for all $\alpha_i, \beta_i, i = 1, \ldots, N$, satisfying (4.103) with the right-hand side replaced by 1. This latter condition is however (much) *stronger* than (4.103). For instance, sec $(\frac{\pi}{N})^N = 8$ for $N = 3$.

4.5 Passivity of Euler–Lagrange Equations

A standard method for deriving the equations of motion for physical systems is via the *Euler–Lagrange equations*

$$\frac{d}{dt}\left(\frac{\partial L}{\partial \dot{q}}(q,\dot{q})\right) - \frac{\partial L}{\partial q}(q,\dot{q}) = \tau, \qquad (4.106)$$

where $q = (q_1, \ldots, q_n)^T$ are generalized configuration coordinates for the system with n degrees of freedom, L is the Lagrangian function,[7] and $\tau = (\tau_1 \ldots, \tau_n)^T$ is the vector of generalized forces acting on the system. Furthermore, $\frac{\partial L}{\partial \dot{q}}(q,\dot{q})$ denotes the column vector of partial derivatives of $L(q,\dot{q})$ with respect to the generalized velocities $\dot{q}_1, \ldots, \dot{q}_n$, and similarly for $\frac{\partial L}{\partial q}(q,\dot{q})$.

By defining the vector of generalized *momenta* $p = (p_1, \ldots, p_n)^T$ as

$$p := \frac{\partial L}{\partial \dot{q}}(q,\dot{q}), \qquad (4.107)$$

and assuming that the map $\dot{q} \mapsto p$ is invertible for every q, this defines the $2n$-dimensional state vector $(q_1, \ldots, q_n, p_1, \ldots, p_n)^T$, in which case the n *second-order* equations (4.106) transform into $2n$ *first-order* equations

$$\dot{q} = \frac{\partial H}{\partial p}(q,p)$$
$$\dot{p} = -\frac{\partial H}{\partial q}(q,p) + \tau, \qquad (4.108)$$

where the *Hamiltonian function* H is the Legendre transform of L, defined implicitly as

$$H(q,p) = p^T\dot{q} - L(q,\dot{q}), \quad p = \frac{\partial L}{\partial \dot{q}}(q,\dot{q}) \qquad (4.109)$$

Equation (4.108) are called the *Hamiltonian equations* of motion. In physical systems the Hamiltonian H usually can be identified with the total energy of the system. It immediately follows from (4.108) that

---

[7]Not to be confused with the Laplacian matrix of the previous section; too many mathematicians with a name starting with "L."

$$\frac{d}{dt}H = \frac{\partial^T H}{\partial q}(q, p)\dot{q} + \frac{\partial^T H}{\partial p}(q, p)\dot{p}$$

$$= \frac{\partial^T H}{\partial p}(q, p)\tau = \dot{q}^T \tau, \tag{4.110}$$

expressing that the increase in energy of the system is equal to the supplied work (*conservation of energy*). This directly translates into the following statement regarding passivity (in fact, losslessness) of the Hamiltonian and Euler–Lagrange equations.

Proposition 4.5.1 *Assume the Hamiltonian H is bounded from below, i.e., $\exists\, C > -\infty$ such that $H(q, p) \geq C$. Then (4.106) with state vector $(q, \dot{q})$, and (4.108) with state vector (q, p), are lossless systems with respect to the supply rate $y^T \tau$, with output $y = \dot{q}$ and storage function $E(q, \dot{q}) := H(q, \frac{\partial L}{\partial \dot{q}}(q, \dot{q})) - C$, respectively $H(q, p) - C$.*

Proof Clearly $H(q, p) - C \geq 0$. The property of being lossless directly follows from (4.110).   □

Remark 4.5.2 If the map from $\dot{q}$ to p is *not* invertible this means that there are algebraic constraints $\phi_i(q, p) = 0, i = 1, \ldots, k$, relating the momenta p, and that the Hamiltonian $H(q, p)$ is only defined up to addition with an arbitrary combination of the constraint functions $\phi_i(q, p), i = 1, \ldots, k$. This leads to a *constrained* Hamiltonian representation; see the Notes at the end of this chapter for further information.

The Euler–Lagrange equations (4.106) describe dynamics without internal energy dissipation, resulting in losslessness. The equations can be extended to

$$\frac{d}{dt}\left(\frac{\partial L}{\partial \dot{q}}(q, \dot{q})\right) - \frac{\partial L}{\partial q}(q, \dot{q}) + \frac{\partial R}{\partial \dot{q}}(\dot{q}) = \tau, \tag{4.111}$$

where $R(\dot{q})$ is a *Rayleigh dissipation* function, satisfying

$$\dot{q}^T \frac{\partial R}{\partial \dot{q}}(\dot{q}) \geq 0, \quad \text{for all } \dot{q} \tag{4.112}$$

Then the time evolution of $H(q, \frac{\partial L}{\partial \dot{q}}(q, \dot{q}))$ satisfies

$$\frac{d}{dt}H = -\dot{q}^T \frac{\partial R}{\partial \dot{q}}(\dot{q}) + \dot{q}^T \tau \tag{4.113}$$

Hence if H is bounded from below, then, similar to Proposition 4.5.1, the systems (4.111) and (4.112) with inputs τ and outputs $\dot{q}$ are passive.

We may interpret (4.111) as the closed-loop system depicted in Fig. 4.2. Equation (4.111) thus can be seen as the feedback interconnection of the lossless system Σ_1 given by the Euler–Lagrange equations (4.106) with input τ', and the static passive system Σ_2 given by the map $\dot{q} \mapsto \frac{\partial R}{\partial \dot{q}}(\dot{q})$. If (4.112) is strengthened to

Fig. 4.2 Feedback
representation of (4.111)

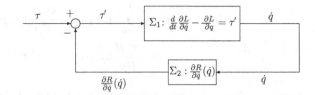

$$\dot{q}^T \frac{\partial R}{\partial \dot{q}}(\dot{q}) \geq \delta \|\dot{q}\|^2 \qquad (4.114)$$

(assuming an inner product structure on the output space of generalized velocities)
for some $\delta > 0$, then the nonlinearity (4.114) defines an δ-input strictly passive map
from $\dot{q}$ to $\frac{\partial R}{\partial \dot{q}}(\dot{q})$, and (4.111) with output $\dot{q}$ becomes output strictly passive; as also
follows from Proposition 4.3.1(ii).

Furthermore, we can apply Theorem 2.2.15 as follows. Consider any initial condition $(q(0), \dot{q}(0))$, and the corresponding input–output map of the system Σ_1. Assume
that for any $\tau \in L_{2e}(\mathbb{R}^n)$ there are solutions $\tau' = \frac{\partial R}{\partial \dot{q}}(\dot{q}), \dot{q} \in L_{2e}(\mathbb{R}^n)$. Then the
map $\tau \mapsto \dot{q}$ has L_2-gain $\leq \frac{1}{\delta}$. In particular, if $\tau \in L_2(\mathbb{R}^n)$ then $\dot{q} \in L_2(\mathbb{R}^n)$. Note
that not necessarily the signal $\frac{\partial R}{\partial \dot{q}}(\dot{q})$ will be in $L_2(\mathbb{R}^n)$; in fact this will depend on
the properties of the Rayleigh function R.

Finally, (4.113) for $\tau = 0$ yields

$$\frac{d}{dt} H = -\dot{q}^T \frac{\partial R}{\partial \dot{q}}(\dot{q}) \qquad (4.115)$$

Hence, if we assume that H has a *strict minimum* at some some point $(q_0, 0)$, and by
(4.114) and La Salle's invariance principle, $(q_0, 0)$ will be an asymptotically stable
equilibrium of the system whenever R is such that $\dot{q}^T \frac{\partial R}{\partial \dot{q}}(\dot{q}) =$ if and only if $\dot{q} = 0$
(in particular, if (4.114) holds).

4.6   Passivity of Second-Order Systems and Riemannian Geometry

In standard mechanical systems the Lagrangian function $L(q, \dot{q})$ is given by the
difference

$$L(q, \dot{q}) = \frac{1}{2} \dot{q}^T M(q) \dot{q} - P(q) \qquad (4.116)$$

of the *kinetic energy* $\frac{1}{2} \dot{q}^T M(q) \dot{q}$ and the *potential energy* $P(q)$. Here $M(q)$ is an
$n \times n$ inertia (generalized mass) matrix, which is symmetric and positive definite
for all q. It follows that the vector of generalized momenta is given as $p = M(q) \dot{q}$,
and thus that the map from $\dot{q}$ to $p = M(q) \dot{q}$ is invertible. Furthermore, the resulting

Hamiltonian H is given as

$$H(q, p) = \frac{1}{2}p^T M^{-1}(q)p + P(q), \tag{4.117}$$

which equals the *total energy* (kinetic energy plus potential energy).

It turns out to be of interest to work out the Euler–Lagrange equations (4.106) and the property of conservation of total energy in more detail for this important case. This will lead to a direct connection to the passivity of a "*virtual system*" that can be associated to the Euler–Lagrange equations, and which has a clear geometric interpretation.

Let $m_{ij}(q)$ be the (i, j)-th element of $M(q)$. Writing out

$$\frac{\partial L}{\partial \dot{q}_k}(q, \dot{q}) = \sum_j m_{kj}(q)\dot{q}_j$$

and

$$\frac{d}{dt}\left(\frac{\partial L}{\partial \dot{q}_k}(q, \dot{q})\right) = \sum_j m_{kj}(q)\ddot{q}_j + \sum_j \frac{d}{dt}m_{kj}(q)\dot{q}_j$$

$$= \sum_j m_{kj}(q)\ddot{q}_j + \sum_{i,j} \frac{\partial m_{kj}}{\partial q_i}\dot{q}_i\dot{q}_j,$$

as well as

$$\frac{\partial L}{\partial q_k}(q, \dot{q}) = \frac{1}{2}\sum_{i,j} \frac{\partial m_{ij}}{\partial q_k}(q)\dot{q}_i\dot{q}_j - \frac{\partial P}{\partial q_k}(q),$$

the Euler–Lagrange equations (4.106) for $L(q, \dot{q}) = \frac{1}{2}\dot{q}^T M(q)\dot{q} - P(q)$ take the form

$$\sum_j m_{kj}(q)\ddot{q}_j + \sum_{i,j}\left\{\frac{\partial m_{kj}}{\partial q_i}(q) - \frac{1}{2}\frac{\partial m_{ij}}{\partial q_k}\right\}(q)\dot{q}_i\dot{q}_j - \frac{\partial P}{\partial q_k}(q) = \tau_k,$$

for $k = 1, \ldots, n$. Furthermore, since

$$\sum_{i,j} \frac{\partial m_{kj}}{\partial q_i}(q)\dot{q}_i\dot{q}_j = \sum_{i,j} \frac{1}{2}\left\{\frac{\partial m_{kj}}{\partial q_i}(q) + \frac{\partial m_{ki}}{\partial q_j}\right\}(q)\dot{q}_i\dot{q}_j,$$

by defining the *Christoffel symbols* of the first kind

$$c_{ijk}(q) := \frac{1}{2}\left\{\frac{\partial m_{kj}}{\partial q_i} + \frac{\partial m_{ki}}{\partial q_j} - \frac{\partial m_{ij}}{\partial q_k}\right\}(q), \tag{4.118}$$

we can further rewrite the Euler–Lagrange equations as

$$\sum_j m_{kj}(q)\ddot{q}_j + \sum_{i,j} c_{ijk}(q)\dot{q}_i\dot{q}_j + \frac{\partial P}{\partial q_k}(q) = \tau_k , \quad k = 1, \ldots, n,$$

or, more compactly,

$$M(q)\ddot{q} + C(q,\dot{q})\dot{q} + \frac{\partial P}{\partial q}(q) = \tau , \tag{4.119}$$

where the (k, j)-th element of the matrix $C(q, \dot{q})$ is defined as

$$c_{kj}(q) = \sum_{i=1}^n c_{ijk}(q)\dot{q}_i . \tag{4.120}$$

In a mechanical system context the forces $C(q, \dot{q})\dot{q}$ in (4.119) correspond to the *centrifugal and Coriolis forces*.

The definition of the Christoffel symbols leads to the following important observation. Adopt the notation $\dot{M}(q)$ for the $n \times n$ matrix with (i, j)-th element given by $\dot{m}_{ij}(q) = \frac{d}{dt}m_{ij}(q) = \sum_k \frac{\partial m_{ij}}{\partial q_k}(q)\dot{q}_k$.

Lemma 4.6.1 *The matrix*

$$\dot{M}(q) - 2C(q,\dot{q}) \tag{4.121}$$

is skew-symmetric for every $q, \dot{q}$.

Proof Leaving out the argument q, the (k, j)-th element of (4.121) is given as

$$\dot{m}_{kj} - 2c_{kj} = \sum_{i=1}^n \left[\frac{\partial m_{kj}}{\partial q_i} - \left\{ \frac{\partial m_{kj}}{\partial q_i} + \frac{\partial m_{ki}}{\partial q_j} - \frac{\partial m_{ij}}{\partial q_k} \right\} \right] \dot{q}_i$$

$$= \sum_{i=1}^n \left[\frac{\partial m_{ij}}{\partial q_k} - \frac{\partial m_{ki}}{\partial q_j} \right] \dot{q}_i$$

which changes sign if we interchange k and j.   □

The skew-symmetry of $\dot{M}(q) - 2C(q,\dot{q})$ is another manifestation of the fact that the forces $C(q, \dot{q})\dot{q}$ in (4.119) are *workless*. Indeed by direct differentiation of the total energy $E(q, \dot{q}) := \frac{1}{2}\dot{q}^T M(q)\dot{q} + P(q)$ along (4.119) one obtains

$$\begin{aligned} \frac{d}{dt}H &= \dot{q}^T M(q)\ddot{q} + \frac{1}{2}\dot{q}^T \dot{M}(q)\dot{q} + \dot{q}^T \frac{\partial P}{\partial q}(q) \\ &= \dot{q}^T \tau + \frac{1}{2}\dot{q}^T \left(\dot{M}(q) - 2C(q,\dot{q}) \right)\dot{q} = \dot{q}^T \tau, \end{aligned} \tag{4.122}$$

in accordance with (4.110).

However, skew-symmetry of $\dot{M}(q) - 2C(q,\dot{q})$ is actually a *stronger* property than energy conservation. In fact, if we choose the matrix $C(q,\dot{q})$ different from the matrix of Christoffel symbols (4.116), i.e., as some other matrix $\tilde{C}(q,\dot{q})$ such that

$$\tilde{C}(q,\dot{q})\dot{q} = C(q,\dot{q})\dot{q}, \qquad \text{for all } q,\dot{q}, \tag{4.123}$$

then still $\dot{q}^T(\dot{M}(q) - 2\tilde{C}(q,\dot{q}))\dot{q} = 0$ (conservation of energy), but in general $\dot{M}(q) - 2\tilde{C}(q,\dot{q})$ will *not* be skew-symmetric anymore.

This observation is underlying the following developments. Start out from Eq. (4.119) for zero potential energy P and the vector of external forces τ denoted by u, that is

$$M(q)\ddot{q} + C(q,\dot{q})\dot{q} = u \tag{4.124}$$

Definition 4.6.2 The *virtual system* associated to (4.124) is defined as the *first-order* system in the state vector $s \in \mathbb{R}^n$

$$\begin{aligned} M(q)\dot{s} + C(q,\dot{q})s &= u \\ y &= s \end{aligned} \tag{4.125}$$

with inputs $u \in \mathbb{R}^n$ and outputs $y \in \mathbb{R}^n$, *parametrized* by the vector $q \in \mathbb{R}^n$ and its time-derivative $\dot{q} \in \mathbb{R}^n$.

Thus for *any* curve $q(\cdot)$ and corresponding values $q(t), \dot{q}(t)$ for all t, we may consider the time-varying system (4.125) with state vector s. Clearly, any solution $q(\cdot)$ of the Euler–Lagrange equations (4.124) for a certain input function $\tau(\cdot)$ generates the solution $s(t) := \dot{q}(t)$ to the virtual system (4.125) for $u = \tau$, but on the other hand *not* every pair $q(t), s(t)$, with $s(t)$ a solution of (4.125) parametrized by $q(t)$, corresponds to a solution of (4.124). In fact, this is only the case if additionally $s(t) = \dot{q}(t)$. This explains the name *virtual system*.

Remarkably, not only the Euler–Lagrange equations (4.124) are lossless with respect to the output $y = \dot{q}$, but also the virtual system (4.125) turns out to be lossless with respect to the output $y = s$, *for every time-function* $q(\cdot)$. This follows from the following computation, crucially relying on the skew-symmetry of $\dot{M}(q) - 2C(q,\dot{q})$. Define the storage function of the virtual system (4.125) as the following function of s, parametrized by q

$$S(s,q) := \frac{1}{2}s^T M(q)s \tag{4.126}$$

Then, by skew-symmetry of $\dot{M} - 2C$, along (4.125)

$$\begin{aligned} \tfrac{d}{dt}S(s,q) &= s^T M(q)\dot{s} + \tfrac{1}{2}s^T \dot{M}(q)s \\ &= -s^T C(q,\dot{q})s + \tfrac{1}{2}s^T \dot{M}(q)s + s^T u = s^T u \end{aligned} \tag{4.127}$$

This is summarized in the following proposition.

Proposition 4.6.3 *For any curve $q(\cdot)$ the virtual system (4.125) with input u and output y is lossless, with parametrized storage function $S(s,q) = \frac{1}{2}s^T M(q)s$.*

This can be directly extended to

$$M(q)\ddot{q} + C(q,\dot{q})\dot{q} + \frac{\partial R}{\partial \dot{q}}(\dot{q}) = \tau, \tag{4.128}$$

with Rayleigh dissipation function $R(\dot{q})$ satisfying $\dot{q}^T \frac{\partial R}{\partial \dot{q}}(\dot{q}) \geq 0$, leading to the associated virtual system

$$\begin{aligned} \dot{s} &= -M^{-1}(q)C(q,\dot{q})s - M^{-1}(q)\frac{\partial R}{\partial s}(s) + M^{-1}(q)u \\ y &= s. \end{aligned} \tag{4.129}$$

Corollary 4.6.4 *For any curve $q(\cdot)$ the virtual system (4.129) is passive with parametrized storage function $S(s,q) := \frac{1}{2}s^T M(q)s$, satisfying $\frac{d}{dt}S(s,q) = -s^T \frac{\partial R}{\partial s}(s) + s^T u \leq s^T u$.*

Example 4.6.5 As an application of Proposition 4.6.3 suppose one wants to asymptotically track a given reference trajectory $q_d(\cdot)$ for a mechanical system (e.g., robot manipulator) with dynamics (4.119). Consider first the preliminary feedback

$$\tau = M(q)\dot{\xi} + C(q,\dot{q})\xi + \frac{\partial P}{\partial q}(q) + \nu \tag{4.130}$$

where

$$\xi := \dot{q}_d - \Lambda(q - q_d) \tag{4.131}$$

for some matrix $\Lambda = \Lambda^T > 0$. Substitution of (4.130) into (4.119) yields the virtual dynamics

$$M(q)\dot{s} + C(q,\dot{q})s = \nu \tag{4.132}$$

with $s := \dot{q} - \xi$. Define the additional feedback

$$\nu = -\hat{\nu} + \tau_e := -Ks + \tau_e, \quad K = K^T > 0, \tag{4.133}$$

corresponding to an input strictly passive map $s \mapsto \hat{\nu}$.

Then by Theorem 2.2.15, part (b), for every $\tau_e \in L_2(\mathbb{R}^n)$ such that s (and thus ν) are in L_{2e}^n (see Fig. 4.3), actually the signal s will be in $L_2(\mathbb{R}^n)$. This fact has an important consequence, since by (4.131) and $s = \dot{q} - \xi$ the error $e = q - q_d$ satisfies

$$\dot{e} = -\Lambda e + s. \tag{4.134}$$

Fig. 4.3 Feedback configuration for tracking

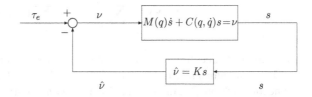

Because we took $\Lambda = \Lambda^T > 0$ it follows from linear systems theory that also $e \in L_2(\mathbb{R}^n)$, and therefore by (4.134) that $\dot{e} \in L_2(\mathbb{R}^n)$. It is well known (see e.g., [83], pp. 186, 237) that this implies[8] $e(t) \to 0$ for $t \to \infty$.

An intrinsic *geometric* interpretation of the skew-symmetry of $\dot{M} - 2C$ and the virtual system (4.125) can be given as follows, within the framework of *Riemannian geometry*. The configuration space $\mathcal{Q}$ of the mechanical system is assumed to be a *manifold* with local coordinates $(q_1, \ldots, q_n)$. Then the generalized mass matrix $M(q) > 0$ defines a *Riemannian metric* $<, >$ on $\mathcal{Q}$ by setting

$$< v, w > := v^T M(q) w \qquad (4.135)$$

for v, w tangent vectors to $\mathcal{Q}$ at the point q. The manifold $\mathcal{Q}$ endowed with the Riemannian metric is called a *Riemannian manifold*.

Furthermore, an affine *connection* ∇ on an arbitrary manifold $\mathcal{Q}$ is a map that assigns to each pair of vector fields X and Y on $\mathcal{Q}$ another vector field $\nabla_X Y$ on $\mathcal{Q}$ such that

(a) $\nabla_X Y$ is bilinear in X and Y
(b) $\nabla_{fX} Y = f \nabla_X Y$
(c) $\nabla_X fY = f \nabla_X Y + (L_X f) Y$

for every smooth function f, where $L_X f$ denotes the directional derivative of f along $\dot{q} = X(q)$, that is, in local coordinates $q = (q_1, \ldots, q_n)$ for $\mathcal{Q}$, $L_X f(q) = \sum_k \frac{\partial f}{\partial q_k}(q) X_k(q)$, where X_k is the k-th component of the vector field X. In particular, as will turn out to be important later on, Property (b) implies that $\nabla_X Y$ at $q \in \mathcal{Q}$ depends on the vector field X only through its value $X(q)$ at q.

In local coordinates q for $\mathcal{Q}$ an affine connection on $\mathcal{Q}$ is determined by n^3 smooth functions

$$\Gamma_{ij}^\ell(q), \quad i, j, \ell = 1, \ldots, n, \qquad (4.136)$$

such that the ℓ-th component of $\nabla_X Y$, $\ell = 1, \ldots, n$, is given as

---

[8]A simple proof runs as follows (with thanks to J.W. Polderman and I.M.Y. Mareels). Take for simplicity $n = 1$. Then, since $\frac{d}{dt} e^2(t) = 2e(t)\dot{e}(t)$, $e^2(t_2) - e^2(t_1) = 2 \int_{t_1}^{t_2} e(t)\dot{e}(t)dt \leq \int_{t_1}^{t_2} [e^2(t) + \dot{e}^2(t)]dt \to 0$ for $t_1, t_2 \to \infty$. Thus for any sequence of time instants $t_1, t_2, \ldots, t_k, \ldots$ with $t_k \to \infty$ for $k \to \infty$ the sequence $e^2(t_i)$ is a Cauchy sequence, implying that $e^2(t_i)$ and thus $e^2(t)$ converges to some finite value for $t_i, t \to \infty$, which has to be zero since $e \in L_2(\mathbb{R})$.

$$(\nabla_X Y)_\ell = \sum_j \frac{\partial Y_\ell}{\partial q_j} X_j + \sum_{i,j} \Gamma_{ij}^\ell X_i Y_j, \tag{4.137}$$

with subscripts denoting the components of the vector fields involved.

The Riemannian metric $<,>$ on $\mathcal{Q}$ obtained from $M(q)$ defines a unique affine connection ∇^M on $\mathcal{Q}$ (called the *Levi-Civita connection*), which in local coordinates is determined by the n^3 *Christoffel symbols* (of the second kind)

$$\Gamma_{ij}^\ell(q) := \sum_{k=1}^n m^{\ell k}(q) c_{ijk}(q), \tag{4.138}$$

with $m^{\ell k}(q)$ the (ℓ, k)-th element of the inverse matrix $M^{-1}(q)$, and $c_{ijk}(q)$ the Christoffel symbols of the first kind as defined in (4.118). Thus in vector notation the affine connection ∇^M is given as

$$\nabla_X^M Y(q) = DY(q)(q)X(q) + M^{-1}(q)C(q, X)Y(q) \tag{4.139}$$

with $DY(q)$ the $n \times n$ Jacobian matrix of Y.

Identifying $s \in \mathbb{R}^n$ with a tangent vector at $q \in \mathcal{Q}$, we conclude that the coordinate-free description of the virtual system (4.125) is given by

$$\begin{aligned} \nabla_{\dot{q}(t)}^M s(t) &= M^{-1}(q(t))u(t) \\ y(t) &= s(t) \end{aligned} \tag{4.140}$$

Thus the state s of the virtual system at any moment t is an element of $T_{q(t)}\mathcal{Q}$. (Recall that $\nabla_X^M s(q)$ depends on the vector field X only through its value $X(q)$. Hence at every time t the expression in the left-hand side of (4.140) depends on the curve $q(\cdot)$ only through the value $\dot{q}(t) \in T_{q(t)}\mathcal{Q}$.)

With regard to the last term $M^{-1}(q)u$ we note that from a geometric point of view, the force u is an element of the *cotangent* space of $\mathcal{Q}$ at q. Since $M^{-1}(q)$ defines a map from the cotangent space to the tangent space, this yields $M^{-1}(q)u \in T_q\mathcal{Q}$. In terms of the Riemannian metric $<,>$ the tangent vector $Z = M^{-1}(q)u \in T_q\mathcal{Q}$ is determined by the requirement that the cotangent vector $< Z, \cdot >$ equals u. This is summarized in the following.

Proposition 4.6.6 *Consider a configuration manifold $\mathcal{Q}$ with Riemannian metric determined by the generalized mass matrix $M(q)$. Let ∇^M be the Levi-Civita connection on $\mathcal{Q}$. Then the virtual system is given by (4.140), where $q(\cdot)$ is any curve on $\mathcal{Q}$ and $s(t) \in T_{q(t)}\mathcal{Q}$ for all t. The virtual system is lossless with parametrized storage function $S(s, q) = \frac{1}{2} < s, s > (q)$.*

Remark 4.6.7 The expression $\nabla_{\dot{q}(t)}^M s(t)$ on the left-hand side of (4.140) is also called the *covariant derivative* of $s(t)$ (with respect to the affine connection ∇^M); sometimes denoted as $\frac{Ds}{dt}(t)$.

We emphasize that one can take *any* curve $q(t)$ in Q with corresponding velocity vector field $\dot{q}(t) = X(q(t))$, and consider the dynamics (4.140) of *any* vector field s *along* this curve $q(t)$ (that is, $s(t)$ being a tangent vector to Q at $q(t)$). If we take s to be *equal to* $\dot{q}$, then (4.140) reduces to

$$\nabla_{\dot{q}}^{M} \dot{q} = M^{-1}(q)\nu \tag{4.141}$$

which is nothing else than the second-order equations (4.124).

Finally, let us come back to the crucial property of skew-symmetry of $\dot{M} - 2C$. This property has the following geometric interpretation. First we note the following obvious lemma.

Lemma 4.6.8 $\dot{M} - 2C$ *is skew-symmetric if and only if* $\dot{M} = C + C^{T}$

Proof $(\dot{M} - 2C) = -(\dot{M} - 2C)^{T}$ iff $2\dot{M} = 2C + 2C^{T}$. □

Given an arbitrary Riemannian metric $<, >$ on Q, an affine connection ∇ on Q is said to be *compatible* with $<, >$ if the following property holds:

$$L_X < Y, Z > = < \nabla_X Y, Z > + < Y, \nabla_X Z > \tag{4.142}$$

for all vector fields X, Y, Z on Q.

Consider now the Riemannian metric $<, >$ determined by the mass matrix M as in (4.135). Furthermore, consider local coordinates $q = (q_1, \ldots, q_n)$ for Q, and let $Y = \frac{\partial}{\partial q_i}, Z = \frac{\partial}{\partial q_j}$. Then (4.142) reduces to (see (4.137))

$$L_X m_{ij} = < \nabla_X \frac{\partial}{\partial q_i}, \frac{\partial}{\partial q_j} > + < \frac{\partial}{\partial q_i}, \nabla_X \frac{\partial}{\partial q_j} > \tag{4.143}$$

with m_{ij} the (i, j)-th element of the mass matrix M. Furthermore, by (4.139) we have

$$\nabla_X \frac{\partial}{\partial q_i} = M^{-1}(q)C(q, X)e_i$$
$$\nabla_X \frac{\partial}{\partial q_j} = M^{-1}(q)C(q, X)e_j \tag{4.144}$$

with e_i, e_j denoting the i-th, respectively j-th, basis vector. Therefore, taking into account the definition of $<, >$ in (4.135), we obtain from (4.143)

$$L_X m_{ij} = (C^{T}(q, X))_{ij} + (C(q, X))_{ij}, \tag{4.145}$$

which we write (replacing L_X by the ˙ operator) as

$$\dot{M}(q) = C^{T}(q, \dot{q}) + C(q, \dot{q}). \tag{4.146}$$

Thus, in view of Lemma 4.6.8, the property of skew-symmetry of the matrix $\dot{M} - 2C$ is nothing else than the *compatibility* of the Levi-Civita connection ∇^M defined by the Christoffel symbols (4.138) with the Riemannian metric $<, >$ defined by $M(q)$.

This observation also implies that one may take *any* other affine connection ∇ (different from the Levi-Civita connection ∇^M), which is compatible with $<, >$ defined by M in order to obtain a lossless virtual system (4.140) (with ∇^M replaced by ∇).

Finally, we note that the Levi-Civita connection ∇^M defined by the Christoffel symbols (4.138) is the *unique* affine connection that is *compatible* with $<, >$ defined by M, as well as is *torsion-free* in the sense that

$$\nabla_X Y - \nabla_X Y = [X, Y] \tag{4.147}$$

for any two vector fields X, Y on Q, where $[X, Y]$ denotes the Lie bracket of X and Y. In terms of the Christoffel symbols (4.138) the condition (4.147) amounts to the *symmetry* condition $\Gamma_{ij}^{\ell} = \Gamma_{ji}^{\ell}$ for all i, j, ℓ, or equivalently, with C_{kj} related to Γ_{ij}^{ℓ} by (4.138) and (4.120), that

$$C(q, X)Y = C(q, Y)X \tag{4.148}$$

for every pair of tangent vectors X, Y.

4.7   Incremental and Shifted Passivity

Recall the definition of *incremental passivity* as given in Definition 2.2.20. A state space version can be given as follows.

Definition 4.7.1 Consider a system as given in (4.1), with input and output spaces $U = Y = \mathbb{R}^m$ and state space $\mathcal{X}$. The system Σ is called *incrementally passive* if there exists a function, called the *incremental storage function*,

$$S : \mathcal{X} \times \mathcal{X} \to \mathbb{R}^+ \tag{4.149}$$

such that

$$
\begin{aligned}
S(x_1(T), x_2(T)) &\leq S(x_1(0), x_2(0)) \\
&+ \int_0^T (u_1(t) - u_2(t))^T (y_1(t) - y_2(t))dt
\end{aligned}
\tag{4.150}
$$

for all $T \geq 0$, and for all pairs of input functions $u_1, u_2 : [0, T] \to \mathbb{R}^m$ and all pairs of initial conditions $x_1(0)$, $x_2(0)$, with resulting pairs of state and output trajectories $x_1, x_2 : [0, T] \to \mathcal{X}$, $y_1, y_2 : [0, T] \to \mathbb{R}^m$.

Remark 4.7.2 Note that if $S(x_1, x_2)$ satisfies (4.150) then so does the function $\frac{1}{2}(S(x_1, x_2) + S(x_2, x_1))$. Hence, without loss of generality, we may assume that the storage function $S(x_1, x_2)$ satisfies $S(x_1, x_2) = S(x_2, x_1)$. Extensions of Definition 4.7.1 to incremental output strict or incremental input strict passivity are immediate.

Definition 4.7.1 directly implies incremental passivity of the input–output map $G_{\bar{x}}$ defined by Σ, for every initial state $\bar{x} \in \mathcal{X}$. This follows from (4.150) by taking *identical* initial conditions $x_1(0) = x_2(0) = \bar{x}$. Hence, the property of incremental passivity defined in Definition 4.7.1 for state space systems is in principle *stronger* than the property defined in Definition 2.2.20 for input–output maps.

As a direct corollary of Theorem 3.1.11 we obtain the following.

Corollary 4.7.3 *The system (4.1) is incrementally passive if and only if*

$$\sup_{u_1(\cdot), u_2(\cdot), T \geq 0} - \int_0^T (u_1(t) - u_2(t))^T (y_1(t) - y_2(t)) dt < \infty \qquad (4.151)$$

for all initial conditions $(x_1(0), x_2(0)) \in \mathcal{X} \times \mathcal{X}$.

The *differential* version of the incremental dissipation inequality (4.149) takes the form

$$S_{x_1}(x_1, x_2) f(x_1, u_1) + S_{x_2}(x_1, x_2) f(x_2, u_2) \leq (u_1 - u_2)^T (y_1 - y_2) \qquad (4.152)$$

for all $x_1, x_2, u_1, u_2, y_1 = h(x_1, u_1), y_2 = h(x_2, u_2)$, where $S_{x_1}(x_1, x_2)$ and $S_{x_2}(x_1, x_2)$ denote row vectors of partial derivatives with respect to x_1, respectively x_2.

An obvious example of an incrementally passive system is a *linear* passive system with quadratic storage function $\frac{1}{2}x^T Q x$. In this case, $S(x_1, x_2) := \frac{1}{2}(x_1 - x_2)^T Q(x_1 - x_2)$ define an incremental storage function, satisfying (4.149). Another example of an incrementally passive system is the *virtual* system defined in (4.125), with incremental storage function given by the parametrized expression (compare with (4.126)) $S(s_1, s_2, q) = \frac{1}{2}(s_1 - s_2)^T M(q)(s_1 - s_2)$. Furthermore, in both cases the system remains incrementally passive in the presence of an extra external (disturbance) input. For example, passivity of $\dot{x} = Ax + Bu$, $y = Cx$ implies incremental passivity of the disturbed system

$$\dot{x} = Ax + Bu + Gd, \quad \dot{d} = Fd, \quad y = Cx \qquad (4.153)$$

for any F, G.

A different type of example of incremental passivity, relying on *convexity*, is given next.

Example 4.7.4 (Primal–dual gradient algorithm) Consider the constrained optimization problem

$$\min_{q; \, Aq = b} C(q), \qquad (4.154)$$

where $C : \mathbb{R}^n \to \mathbb{R}$ is a convex function, and $Aq = b$ are affine constraints, for some $k \times n$ matrix A and vector $b \in \mathbb{R}^k$. The corresponding Lagrangian function is defined as

$$L(q, \lambda) := C(q) + \lambda^T (Aq - b), \quad \lambda \in \mathbb{R}^k, \tag{4.155}$$

which is convex in q and concave in λ. The primal–dual gradient algorithm for solving the optimization problem in continuous time is given as

$$\begin{aligned} \tau_q \dot{q} &= -\frac{\partial L}{\partial q}(q, \lambda) = -\frac{\partial C}{\partial q}(q) - A^T \lambda + u \\ \tau_\lambda \dot{\lambda} &= \frac{\partial L}{\partial \lambda}(q, \lambda) \quad = Aq - b \\ y &= q \,, \end{aligned} \tag{4.156}$$

where τ_q, τ_λ are diagonal positive matrices (determining the time-scales of the algorithm). Furthermore, we have added an input vector $u \in \mathbb{R}^n$ representing possible interaction with other algorithms or dynamics (e.g., if the primal–dual gradient algorithm is carried out in a distributed fashion). The output vector is defined as $y = q \in \mathbb{R}^n$. This defines an incrementally passive system with incremental storage function

$$S(q_1, \lambda_1, q_2, \lambda_2) := \frac{1}{2}(q_1 - q_2)^T \tau_q (q_1 - q_2) + \frac{1}{2}(\lambda_1 - \lambda_2)^T \tau_\lambda (\lambda_1 - \lambda_2) \tag{4.157}$$

Indeed

$$\begin{aligned} \frac{d}{dt} S &= (q_1 - q_2)^T \tau_q (\dot{q}_1 - \dot{q}_2) + (\lambda_1 - \lambda_2)^T \tau_\lambda (\dot{\lambda}_1 - \dot{\lambda}_2) \\ &= -(q_1 - q_2)^T \left(\frac{\partial C}{\partial q}(q_1) - \frac{\partial C}{\partial q}(q_2) \right) + (u_1 - u_2)^T (y_1 - y_2) \\ &\le (u_1 - u_2)^T (y_1 - y_2) \end{aligned} \tag{4.158}$$

since $(q_1 - q_2)^T \left(\frac{\partial C}{\partial q}(q_1) - \frac{\partial C}{\partial q}(q_2) \right) \ge 0$ for all q_1, q_2, by convexity of C.

Finally, a special case of incremental passivity is obtained by letting u_2 to be a *constant* input $\bar{u}$, and x_2 a corresponding *steady-state* $\bar{x}$ satisfying $f(\bar{x}, \bar{u}) = 0$. Defining the corresponding constant output $\bar{y} = h(\bar{x}, \bar{u})$ and denoting u_1, x_1, y_1 simply by u, x, y, this leads to requiring the existence of a storage function $S_{\bar{x}}(x)$ (parametrized[9] by $\bar{x}$) satisfying

$$S_{\bar{x}}(x(T)) \le S_{\bar{x}}(x(0)) + \int_0^T (u(t) - \bar{u})^T (y(t) - \bar{y}) dt \tag{4.159}$$

This existence of a function $S_{\bar{x}}(x) \ge 0$ satisfying (4.159) is called *shifted passivity* (with respect to the steady-state values $\bar{u}, \bar{x}, \bar{y}$). We shall return to the notion of shifted passivity more closely in the treatment of port-Hamiltonian systems in Chap. 6, see especially Sect. 6.5.

---

[9]Note that in this case the subscript $\bar{x}$ does *not* refer to differentiation.

4.8 Notes for Chapter 4

1. The Kalman–Yakubovich–Popov Lemma is concerned with the equivalence between the frequency-domain condition of *positive realness* of the transfer matrix of a linear system and the existence of a solution to the LMI (4.18) or (4.19), and thus to the passivity of a (minimal) input-state-output realization. It was derived by Kalman [154], also bringing together results of Yakubovich and Popov. See Willems [351], Rantzer [257], Brogliato, Lozano, Maschke & Egeland [52]. For the uncontrollable case, see especially Rantzer [257], Camlibel, Belur & Willems [58].

2. Example 4.1.7 is taken from van der Schaft [283].

3. The factorization approach mentioned in Sect. 4.1 is due to Hill & Moylan [123, 126, 225]; see these papers for further developments along these lines.

4. Example 4.2.5 is taken from Dalsmo & Egeland [75, 76].

5. Corollary 4.3.5 is based on Vidyasagar [343], Sastry [267] (in the input–output map setting; see Chap. 2). See also Hill & Moylan [124, 125], Moylan [225] for further developments and generalizations.

6. The treatment of Example 4.3.6 is from Willems [352].

7. Example 4.3.7 is based on van der Schaft & Schumacher [302], where also applications are discussed. For further developments on passive *complementarity* systems see Camlibel, Iannelli & Vasca [59] and the references quoted therein.

8. Proposition 4.3.9 is taken from Kerber & van der Schaft [158].

9. Another interesting extension to the converse passivity theorems discussed in Sect. 4.3 concerns the following scenario. Suppose Σ_1 is such that $\Sigma_1 \|_f \Sigma_2$ is *stable* (in some sense) *for every* passive system Σ_2. Then under appropriate conditions this implies that also Σ_1 is necessarily passive. This is proved, using the Nyquist criterion, for single-input single-output linear systems in Colgate & Hogan [69], and for general nonlinear input–output maps, using the S-procedure lossless theorem, in Khong & van der Schaft [163]. Within a general state space setting the result is formulated and derived in Stramigioli [329], where also other important extensions are discussed. The result is of particular interest for robotic applications, where the "environment" Σ_2 of a controlled robot Σ_1 is usually unknown, but can be assumed to be passive. Hence, overall stability is only guaranteed if Σ_1 is passive; see e.g., Colgate & Hogan [69], Stramigioli [328, 329].

10. The first scenario of network interconnection of passive systems discussed in Sect. 4.4 is emphasized and discussed much more extensively in the textbook Bai, Arcak & Wen [18]. Here also a broad range of applications can be found, continuing on the seminal paper Arcak [10]. See also Arcak, Meissen & Packard

[11] for further developments, as well as Bürger, Zelazo & Allgöwer [55] for a
network flow optimization perspective.

11. Example 4.4.4 can be found in Arcak [10]. See also van der Schaft & Stegink
    [303] for a generalization to "structure-preserving" networks of generators and
    loads.

12. Kirchhoff's matrix tree theorem goes back to the classical work of Kirchhoff on
    resistive electrical circuits [164]; see Bollobas [48] for a succinct treatment (see
    especially Theorem 14 on p. 58), and Mirzaev & Gunawardena [220] and van
    der Schaft, Rao & Jayawardhana [301] for an account in the context of chemical
    reaction networks.
    The *existence* (not the explicit *construction*) of $\gamma \in \mathbb{R}_+^N$ satisfying $L\gamma = 0$
    already follows from the Perron–Frobenius theorem, exploiting the fact that the
    off-diagonal elements of $-L := DK$ are all nonnegative; see Sontag [320]
    (Lemma V.2).

13. The idea to assemble Lyapunov functions from a weighted sum of Lyapunov
    functions of component systems is well known in the literature on large-scale
    systems, see e.g., Michel & Miller [219], Siljak [315], and is sometimes referred
    to as the use of vector Lyapunov functions. Closely related developments to
    the second scenario discussed in Sect. 4.4 can be found in Zhang, Lewis &
    Qu [364]. The exposition here, distinguishing between flow and communication
    Laplacian matrices, is largely based on van der Schaft [287]. The interconnection
    of passive systems through a symmetric Laplacian matrix can be already found in
    Chopra & Spong [66].

14. Remark 4.4.13 generalizes the definition of *effective resistance* for symmetric
    Laplacians, which is well known; see e.g., Bollobas [48]. Note that in case of a
    symmetric Laplacian $R_{ij} = R_{ji}$.

15. The third scenario of network interconnection of passive systems as discussed
    in Sect. 4.4 is based on Arcak & Sontag [12], to which we refer for additional
    references and developments on the secant condition.

16. Section 4.5, as well as the first part of Sect. 4.5 is mainly based on the survey
    paper Ortega & Spong [243], for which we refer to additional references. See also
    the book Ortega, Loria, Nicklasson & Sira-Ramirez [239], as well as Arimoto
    [13]. Example 4.6.5 is due to Slotine & Li [316].

17. (Cf. Remark 4.5.2). If the map from $\dot{q}$ to p is not invertible one is led to con-
    strained Hamiltonian dynamics as considered by Dirac [81, 82]. Under regularity
    conditions the constrained Hamiltonian dynamics is Hamiltonian with respect
    to the Poisson structure defined as the *Dirac bracket*. See van der Schaft [271]
    for an input–output decoupling perspective.

18. Background on the Riemannian geometry in Sect. 4.6 can be found, e.g., in
    Boothby [49], Abraham & Marsden [1]. For related work, see Li & Horowitz
    [180].

19. The concept of the *virtual system* defined in Definition 4.6.2 and the proof of its passivity (in fact, losslessness) is due to Slotine and coworkers, see e.g., Wang & Slotine [344], Jouffroy & Slotine [153], Manchester & Slotine [193].

20. Incremental passivity is also closely related to *differential passivity*, as explored in Forni & Sepulchre [100], Forni, Sepulchre & van der Schaft [101], van der Schaft [285]. Following the last reference, the notion of differential passivity involves the notion of the *variational systems* of Σ, defined as follows (cf. Crouch & van der Schaft [73]). Consider a one-parameter family of input-state-output trajectories $(x(t, \epsilon), u(t, \epsilon), y(t, \epsilon))$, $t \in [0, T]$, of Σ parametrized by $\epsilon \in (-c, c)$, for some constant $c > 0$. Denote the nominal trajectory by $x(t, 0) = x(t), u(t, 0) = u(t)$ and $y(t, 0) = y(t), t \in [0, T]$. Then the infinitesimal *variations*

$$\delta x(t) = \frac{\partial x}{\partial \epsilon}(t, 0), \quad \delta u(t) = \frac{\partial u}{\partial \epsilon}(t, 0), \quad \delta y(t) = \frac{\partial y}{\partial \epsilon}(t, 0)$$

satisfy

$$\begin{aligned}\dot{\delta x}(t) &= \frac{\partial f}{\partial x}(x(t), u(t))\delta x(t) + \frac{\partial f}{\partial x}(x(t), u(t))\delta u(t) \\ \delta y(t) &= \frac{\partial h}{\partial x}(x(t), u(t))\delta x(t) + \frac{\partial f}{\partial x}(x(t), u(t))\delta u(t)\end{aligned} \tag{4.160}$$

The system (4.160) (parametrized by $u(\cdot), x(\cdot), y(\cdot)$) is called the *variational system*, with variational state $\delta x(t) \in T_{x(t)}\mathcal{X}$, variational inputs $\delta u \in \mathbb{R}^m$, and variational outputs $\delta y \in \mathbb{R}^m$.

Suppose now that the original system Σ is *incrementally passive*. Identify $u(\cdot), x(\cdot), y(\cdot)$ with $u_2(\cdot), x_2(\cdot), y_2(\cdot)$ in (4.150), and $(x(t, \epsilon), u(t, \epsilon), y(t, \epsilon))$ for $\epsilon \neq 0$ with $u_1(\cdot), x_1(\cdot), y_1(\cdot)$. Dividing both sides of (4.150) by ϵ^2, and taking the limit for $\epsilon \to 0$, yields under appropriate assumptions

$$\bar{S}(x(T), \delta x(T)) \leq \bar{S}(x(0), \delta x(0)) + \int_0^T (\delta u(t))^T \delta y(t)dt \tag{4.161}$$

where

$$\bar{S}(x(t), \delta x(t)) := \lim_{\epsilon \to 0} \frac{S(x(t, \epsilon), x(t))}{\epsilon^2} \tag{4.162}$$

The thus obtained Eq. (4.161) amounts to the *definition* of differential passivity adopted in Forni & Sepulchre [100], van der Schaft [285].

21. For the numerous applications of the theory of passive systems to *adaptive control* we refer, e.g., to Brogliato, Lozano, Maschke & Egeland [52], and Astolfi, Karagiannis & Ortega [16], and the references quoted therein.

Chapter 5
Passivity by Feedback

In the previous Chaps. 2 and 4, we have seen the importance of the notion of passivity, both for analysis and for control. This motivates to consider the problem of *transforming* a non-passive system into a passive system by the application of state feedback.

We will give necessary as well as sufficient conditions for the solvability of this problem. The main idea in the proof of the sufficiency part is to transform the system into the feedback interconnection of two passive systems. This idea is further explored in Sect. 5.2 for the stabilization of cascaded systems, and leading in Sect. 5.3 to the technique of stabilization of systems with triangular structure known as backstepping.

5.1 Feedback Equivalence to a Passive System

Consider throughout this chapter affine input-state-output systems without feedthrough terms, denoted as

$$\Sigma : \begin{aligned} \dot{x} &= f(x) + g(x)u, \quad x \in \mathcal{X}, \ u \in \mathbb{R}^m \\ y &= h(x), \qquad\qquad y \in \mathbb{R}^m \end{aligned} \tag{5.1}$$

with an *equal* number of inputs and outputs. Furthermore, let us consider the set of *regular state feedback* laws

$$(\alpha, \beta) : u = \alpha(x) + \beta(x)v,$$
$$\alpha(x) \in \mathbb{R}^m, \quad \beta(x) \in \mathbb{R}^{m \times m}, \ \det \beta(x) \neq 0, \tag{5.2}$$

leading to the closed-loop systems

© Springer International Publishing AG 2017
A. van der Schaft, *L_2-Gain and Passivity Techniques in Nonlinear Control*,
Communications and Control Engineering, DOI 10.1007/978-3-319-49992-5_5

$$\Sigma_{\alpha,\beta}: \begin{array}{l} \dot{x} = \big[f(x) + g(x)\alpha(x)\big] + g(x)\beta(x)v \\ y = h(x) \end{array} \qquad (5.3)$$

with new inputs $v \in \mathbb{R}^m$. The system Σ is said to be *feedback equivalent* to $\Sigma_{\alpha,\beta}$.

The problem we want to address is the following:

Problem

Under what conditions on Σ is it possible to find a feedback law (α, β) as in (5.2) such that $\Sigma_{\alpha,\beta}$ is a passive system, i.e., *when is Σ feedback equivalent to a passive system?*

We will first derive, modulo some technicalities, two *necessary* conditions for feedback equivalence to a passive system. Subsequently, we will show that these two conditions are sufficient as well; at least for a *locally* defined feedback transformation.

Suppose Σ is feedback equivalent to a passive system, i.e., there exists a regular feedback (α, β) such that $\Sigma_{\alpha,\beta}$ defined in (5.3) is passive with storage function $S \geq 0$. Then by definition

$$S(x(t_1)) - S(x(t_0)) \leq \int_{t_0}^{t_1} v^T(t)y(t)dt \qquad (5.4)$$

along all solutions $(x(t), v(t), y(t))$, $t \in [t_0, t_1]$, of (5.3), and all $t_1 \geq t_0$. Therefore, if we consider the *constrained* system, defined by setting $y = h(x)$ to zero, that is,

$$\Sigma_{\alpha,\beta}^c: \begin{array}{l} \dot{x} = \big[f(x) + g(x)\alpha(x)\big] + g(x)\beta(x)v \\ 0 = h(x), \end{array} \qquad (5.5)$$

then $S(x(t_1)) - S(x(t_0)) \leq 0$, for all solutions $(x(t), v(t))$, $t \in [t_0, t_1]$, of this constrained system. On the other hand, as follows from the definition of regular state feedback in (5.2), in particular the requirement $\det \beta(x) \neq 0$, the state space part of the solutions $(x(t), v(t))$ of $\Sigma_{\alpha,\beta}$ equals the state space part of solutions $(x(t), u(t))$ of the original system Σ. Indeed, if $(x(t), u(t))$ is solution of Σ, then $(x(t), v(t)) = \beta^{-1}(x(t))[u(t) - \alpha(x(t))])$ is a solution of $\Sigma_{\alpha,\beta}^c$, and conversely if $(x(t), v(t))$ is a solution of $\Sigma_{\alpha,\beta}$ then $(x(t), u(t) = \alpha(x(t)) + \beta(x(t))v(t))$ is a solution of Σ. Hence, also for the *original* system Σ with constraints $y = 0$

$$\Sigma^c: \begin{array}{l} \dot{x} = f(x) + g(x)u \\ 0 = h(x) \end{array} \qquad (5.6)$$

it follows that necessarily

$$S(x(t_1)) - S(x(t_0)) \leq 0 \qquad (5.7)$$

along all solutions $(x(t), u(t))$ of Σ^c. The constrained system Σ^c in (5.6) is called the *zero-output constrained dynamics* of Σ.

Summarizing, we have obtained the following necessary condition for feedback equivalence to a passive system.

Proposition 5.1.1 *Suppose Σ is feedback equivalent to a passive system $\Sigma_{(\alpha,\beta)}$ with storage function S. Then (5.7) holds for all solutions $(x(t), u(t))$, $t \in [t_0, t_1]$, of the zero-output constrained dynamics Σ^c.*

A second necessary condition can be derived, under some extra technical conditions, as follows. We give two versions, requiring slightly different technical assumptions.

Denote by $h_x(x)$ the $m \times n$-Jacobian matrix of h, that is, the i-th row of $h_x(x)$ is the gradient vector $h_{ix}(x)$ of the i-th component function h_i.

Lemma 5.1.2 *Suppose Σ is passive with a C^2 storage function S that is positive definite at x^* and has nondegenerate Hessian matrix $S_{xx}(x^*) := \frac{\partial^2 S}{\partial x^2}(x^*)$. Furthermore, assume that rank $g(x^*) = m$. Then $h_x(x)g(x)$ has rank m in a neighborhood of x^*.*

Proof By positive-definiteness of S at x^* it follows that $\frac{\partial S}{\partial x}(x^*) = 0$. Hence, since $h(x) = g^T(x)\frac{\partial S}{\partial x}(x)$,

$$h_x(x^*) = g^T(x^*)S_{xx}(x^*)$$

and thus

$$h_x(x^*)g(x^*) = g^T(x^*)S_{xx}(x^*)g(x^*) \tag{5.8}$$

By positive-definiteness of S the non-degenerate Hessian matrix $S_{xx}(x^*)$ is positive definite, and thus, since rank $g(x^*) = m$, also $h_x(x^*)g(x^*)$ has maximal rank m, implying that it has rank m in a neighborhood of x^*. $\qquad\square$

Lemma 5.1.3 *Suppose Σ is passive with a C^1 storage function S which is positive definite at x^*. Assume the $m \times m$-matrix $h_x(x)g(x)$ has constant rank in a neighborhood of x^*. Then $h_x(x)g(x)$ has rank m in a neighborhood of x^*.*

Proof Suppose rank $h_x(x)g(x) < m$ in a neighborhood of x^*. Then there exists a smooth function $u(x) \in \mathbb{R}^m$, defined on a neighborhood of x^*, such that

$$h_x(x)g(x)u(x) = 0 \tag{5.9}$$

while $\gamma(x) := g(x)u(x) \neq 0$. As before, positive-definiteness of S at x^* implies $\frac{\partial S}{\partial x}(x^*) = 0$. Hence, $h^T(x) = S_x(x)g(x)$ satisfies $h(x^*) = 0$. Denote the solutions of $\dot{x} = \gamma(x)$, $x(0) = x_0$, by $x_\gamma^t(x_0)$, and consider the function $f(t) = S(x_\gamma^t(x^*))$. By the Mean Value Theorem there exists for any $t > 0$ some s with $0 \leq s \leq t$ such that

$$S(x_\gamma^t(x^*)) = S(x^*) + u^T(x_\gamma^s(x^*))h(x_\gamma^s(x^*))t \tag{5.10}$$

since $f'(t) = S_x(x_\gamma^t(x^*))\gamma(x_\gamma^t(x^*)) = h^T(x_\gamma^t(x^*))u(x_\gamma^t(x^*))$. From (5.9) it follows that $h(x_\gamma^s(x^*)) = h(x^*) = 0$, and thus by (5.10) $S(x_\gamma^t(x^*)) = 0$. However, by positive-definiteness of S this implies $x_\gamma^t(x^*) = 0$, which is in contradiction with $\gamma(x^*) \neq 0$. $\qquad\square$

Finally, we note that the full-rank property of $h_x(x)g(x)$ is invariant under feedback (α, β), since by such a feedback $h_x(x)g(x)$ is transformed into $h_x(x)g(x)\beta(x)$ with $\det \beta(x) \neq 0$. Hence, modulo technical conditions spelled out in the previous two Lemma's, we conclude that a second necessary condition for feedback equivalence to a passive system is that rank $h_x(x)g(x) = m$.

The property of $h_x(x)g(x)$ being full rank in a neighborhood of x^* has several consequences. First, it implies that rank $h_x(x) = m$ around x^*, implying that the constrained set

$$\mathcal{X}_c := \{x \in \mathcal{X} \mid h(x) = 0\} \tag{5.11}$$

is actually a *submanifold* of $\mathcal{X}$ (of co-dimension m) in a neighborhood of x^*. Furthermore, it implies that the zero-output constrained dynamics Σ^c has a *unique* solution $(x(t), u(t))$, $t \geq 0$, for every initial condition $x(0) = x_c \in \mathcal{X}_c$. Indeed, computing the time-derivative $\dot{y}$ of $y(t) = h(x(t))$ for Σ yields

$$\dot{y} = h_x(x)f(x) + h_x(x)g(x)u. \tag{5.12}$$

Hence, by defining

$$u_c(x) := -[h_x(x)g(x)]^{-1} h_x(x)f(x) \tag{5.13}$$

the closed-loop dynamics $\dot{x} = f(x) + g(x)u^*(x)$ satisfies $\dot{y} = 0$, while conversely $\dot{y} = 0$ *implies* (5.13). Thus every solution $x(t)$ of the closed-loop dynamics starting in $\mathcal{X}_c$ ($h(x(0)) = 0$) satisfies $h(x(t)) = 0$, $t \geq 0$, and thus *remains* in $\mathcal{X}_c$. Hence $\mathcal{X}_c$ is an *invariant submanifold* for the closed-loop dynamics, and the closed-loop dynamics can be *restricted* to $\mathcal{X}_c$. Thus, under the assumption that $h_x(x)g(x)$ has rank m, the zero-output constrained dynamics Σ^c is equivalently given by the *explicit* system on $\mathcal{X}_c$

$$\dot{x} = f(x) + g(x)u_c(x), \qquad x \in \mathcal{X}_c \tag{5.14}$$

Summarizing, we have arrived at the following necessary conditions for local feedback equivalence to a passive system.

Proposition 5.1.4 *Suppose Σ is locally about x^* feedback equivalent to a passive system with C^2 storage function S, which is positive definite at x^*. Assume that either $h_x(x)g(x)$ has constant rank on a neighborhood of x^*, or that $S_{xx}(x^*)$ is positive definite and rank $g(x^*) = m$. Then rank $h_x(x)g(x) = m$ in a neighborhood of x^*, and*

$$S_x(x)\left[f(x) + g(x)u^*(x)\right] \leq 0, \qquad x \in \mathcal{X}_c, \tag{5.15}$$

with u_c defined by (5.13).

Proof Suppose there exists, locally about x^*, a feedback (α, β) such that $\Sigma_{\alpha,\beta}$ is passive. Since rank $g(x^*) = \text{rank } h_x(x)g(x)\beta(x)$ it follows from Lemma's 5.1.3, respectively, Lemma 5.1.2, that $h_x(x)g(x)$ has rank m on a neighborhood of x^*. Since $y = 0$ on $\mathcal{X}_c$, the inequality (5.15) directly follows from the dissipation inequality (5.4). $\qquad \qquad \square$

Loosely speaking (disregarding technical assumptions), Proposition 5.1.4 expresses that necessary conditions for (local) feedback equivalence of Σ to a passive system are that (locally) $h_x(x)g(x)$ has rank m, and that the resulting zero-output constrained dynamics Σ^c of Σ on $\mathcal{X}_c$ given by (5.14), with $u_c(x)$ defined by (5.13), satisfies (5.15) for a certain function $S_c : \mathcal{X}_c \to \mathbb{R}^+$.

Next step is to show that these conditions are actually *sufficient* as well.

Theorem 5.1.5 *Consider the system Σ. Suppose* rank $h_x(x^*)g(x^*) = m$ *at a certain point x^* with $h(x^*) = 0$, and let $S_c : \mathcal{X}_c \to \mathbb{R}^+$ satisfy (5.15) locally around x^*, where $\mathcal{X}_c$ and u_c are defined as in (5.11), respectively (5.13). Then Σ is locally around x^* feedback equivalent to a passive system with storage function $S : \mathcal{X} \to \mathbb{R}^+$ (locally defined around x^*). Furthermore, if S_c is positive definite at x^*, then so is S.*

Proof We will give the proof under the additional assumption that the Lie brackets[1] $[g_i, g_j]$ of the vector fields defined by the columns $g_1, \ldots, g_m$ of $g(x)$ are contained in span $\{g_1(x), \ldots, g_m(x)\}$; see [56] for the general case. Since rank $h_x(x) = m$ for all x in a neighborhood of x^* the functions $y_1 = h_1(x), \ldots, y_m = h_m(x)$, can be taken as partial local coordinate functions for $\mathcal{X}$ around x^*. Furthermore, since rank $h_x(x)g(x) = m$ and by the additional assumption Lie bracket condition, we can find $n - m$ complementary local coordinates z around x^* such that in the new coordinates (z, y) for $\mathcal{X}$ the first $n - m$ rows of the matrix g are zero.[2] Hence, after suitable feedback the dynamics of (5.7) takes, locally around x^*, the following form

$$\begin{aligned} \dot{z} &= f_1(z, y) \\ \dot{y} &= v \end{aligned} \qquad (5.16)$$

Furthermore, z define local coordinates for $\mathcal{X}_c$, and the dynamics $\dot{z} = f_1(z, 0) =: f_c(z)$ is a coordinate expression of (5.14). Expressing S_c in these coordinates we have by (5.15)

$$S_{cz}(z)f_c(z) \leq 0 \qquad (5.17)$$

Since $f_1(z, y) - f_c(z)$ is zero for $y = 0$ we can write (see e.g., the Notes at the end of this chapter)

$$f_1(z, y) = f_c(z) + p(z, y)y \qquad (5.18)$$

for some smooth matrix $p(z, y)$. Then the following system, defined locally around x^* on $\mathcal{X}_c$,

$$\Sigma_z : \quad \begin{aligned} \dot{z} &= f_c(z) + p(z, y)y \\ w &= p^T(z, y)\frac{\partial S_c}{\partial z}(z) \end{aligned} \qquad (5.19)$$

---

[1] In local coordinates x the Lie bracket vector field $[g_i, g_j]$ is given by the expression $[g_i, g_j](x) = \frac{\partial g_j}{\partial x}(x)g_i(x) - \frac{\partial g_i}{\partial x}(x)g_j(x)$.
[2] Since dim span $\{g_1(x), \ldots, g_m(x)\} = m$ near x^* and $[g_i, g_i](x) \in$ span $\{g_1(x), \ldots, g_m(x)\}$ it follows from Frobenius' theorem that there exist $n - m$ partial local coordinates $z = (z_1, \ldots, z_{n-m})$ such that $\frac{\partial z_i}{\partial x} g_j = 0$ [135, 233].

is a passive system with storage function S_c, with respect to the inputs y and outputs w, since by (5.17) $\frac{dS_c}{dt} \leq w^T y$. Furthermore, the integrator system $\dot{y} = v$ is trivially a passive system with respect to the inputs v and outputs y, and with storage function $\frac{1}{2}\|y\|^2$. Hence the standard feedback interconnection

$$v = -w + e = -p^T(z, y)\frac{\partial S_c}{\partial z}(z) + e, \qquad e \in \mathbb{R}^m \qquad (5.20)$$

will result by Proposition 4.3.1 into a passive system, with respect to the inputs e and outputs y, and with storage function

$$S(z, y) := S_c(z) + \frac{1}{2}\|y\|^2 \qquad (5.21)$$

on $\mathcal{X}$. If additionally S_c is assumed to be positive definite at x^*, then also $S(z, y)$ is positive definite at x^*. □

5.2   Stabilization of Cascaded Systems

The main idea in proving Theorem 5.1.5 was to transform Σ into a standard feedback interconnection of two passive systems, one being the system (5.19), and the other being the integrator system $\dot{y} = v$. This idea can be generalized in a number of directions. Consider the system

$$\begin{aligned} \dot{z} &= f(z, \xi) & , \quad f(0, 0) = 0 \\ \dot{\xi} &= \Phi(z, \xi, u), \ u \in \mathbb{R}^m, \quad \Phi(0, 0, 0) = 0 \end{aligned} \qquad (5.22)$$

The feedback stabilization problem is to find a feedback $u = \alpha(z, \xi)$ such that $(z, \xi) = (0, 0)$ is an (asymptotically) stable equilibrium of the closed-loop system.

Theorem 5.2.1 *Consider the system (5.22). Suppose there exists an output map $y = h(z, \xi) \in \mathbb{R}^m$ such that*

(i) the system (5.22) is passive with respect to this output map, with storage function $S(\xi)$,
(ii)

$$f(z, \xi) = f_0(z, \xi) + \sum_{j=1}^m y_j f_j(z, \xi) \qquad (5.23)$$

for some functions $f_0(z, \xi)$ and $f_j(z, \xi), j = 1, \ldots, m$, with $f_0(z, \xi)$ satisfying

$$V_z(z)f_0(z, \xi) \leq 0, \quad \text{for all } z, \xi, \qquad (5.24)$$

for some function $V(z) \geq 0$.

Then the feedback

$$u_j = -V_z(z)f_j(z, \xi) + v_j, \qquad j = 1, \ldots, m, \tag{5.25}$$

transforms (5.22) into a passive system with respect to the inputs $v \in \mathbb{R}^m$ and outputs $y = h(z, \xi)$, having storage function $V(z) + S(\xi)$. Furthermore, the feedback

$$u_j = -V_z(z)f_j(z, \xi) - y_j + v_j, \ j = 1, \ldots, m, \tag{5.26}$$

renders the system (5.22) output strictly passive.

Proof Define the system

$$\Sigma_z : \begin{aligned} \dot{z} &= f_0(z, \xi) + \sum_{j=1}^{m} p_j f_j(z, \xi) \\ w_j &= V_z(z)f_j(z, \xi) \qquad j = 1, \ldots, m \end{aligned} \tag{5.27}$$

This system, with inputs p_j and outputs $w_j, j = 1, \ldots, m$, and *modulated* by ξ, is passive with storage function V, since by (5.24)

$$\frac{dV}{dt} \leq \sum_{j=1}^{m} p_j w_j \tag{5.28}$$

Hence, the standard feedback interconnection of Σ_z with Σ_ξ defined by $u = -w + v, p = y$ results in a passive system with storage function $V(z) + S(\xi)$, while $u = -w - y + v, p = y$ results in output strict passivity. □

Corollary 5.2.2 *If $V(z) + S(\xi)$ is positive definite at $(z, \xi) = (0, 0)$, then the feedback (5.25) renders the equilibrium $(0, 0)$ stable, while the feedback (5.26) renders the equilibrium asymptotically stable assuming a zero-state detectability property. See Chap. 3 for further generalizations.*

Remark 5.2.3 Let the output functions $y = h(z, \xi)$ be chosen, as well as the drift vector field $f_0(z, \xi)$. Then there exist functions $f_j(z, \xi), j = 1, \ldots, m$, satisfying (5.23) if and only if the function $f(z, \xi) - f_0(z, \xi)$ is zero whenever $h(z, \xi)$ is zero; see the Notes at the end of this chapter.

The range of applicability of Theorem 5.2.1 is substantially enlarged by relaxing the condition (i) to the condition that the system (5.22) can be *rendered* passive by a preliminary feedback $u = \Psi(z, \xi)$, as formulated in the following corollary.

Corollary 5.2.4 *Replace condition (i) in Theorem 5.2.1 by*

(i) ′ *There exists a preliminary feedback $u = \Psi(z, \xi) + u'$ such that the system (5.22) is passive with respect to the output map $y = h(z, \xi)$ and storage function $S(\xi)$.*

Then after this preliminary feedback the results of Theorem 5.2.1 continue to hold with u replaced by u'.

Example 5.2.5 (*Examples 4.2.4 and 4.2.5 continued*) Consider the combined dynamics of the orientation and the angular velocities of a rigid body spinning around its center of mass as discussed before, cf. (4.43), (4.41),

$$\dot{R} = RS(\omega) \tag{5.29}$$

$$I\dot{\omega} = -S(\omega)I\omega + u \tag{5.30}$$

with I the diagonal matrix containing the moments of inertia along the principal axes. Recall from Example 4.2.5 that by representing the rotation matrix $R \in SO(3)$ in its Euler parameters (ε, η), the dynamics (5.29) takes the form, cf. (4.48),

$$\begin{bmatrix} \dot{\varepsilon} \\ \dot{\eta} \end{bmatrix} = \frac{1}{2} \begin{bmatrix} \eta I_3 + S(\varepsilon) \\ -\varepsilon^T \end{bmatrix} \omega, \tag{5.31}$$

evolving on the three-dimensional unit sphere S^3 in $\mathbb{R}^4$. In Example 4.2.5 it was noted that the dynamics (5.31), with inputs ω and outputs ε, is lossless with respect to the storage function $V(\varepsilon, \eta) = \varepsilon^T \varepsilon + (1 - \eta)^2 = 2(1 - \eta)$. Moreover, in Example 4.2.4 it was observed that the dynamics (5.30) with inputs u and outputs ω is lossless with respect to the storage function $T(\omega) = \frac{1}{2}\omega^T M \omega$ (kinetic energy). Hence, the total set of Eqs. (5.29), (5.30) can be regarded as the cascade (series) interconnection of two lossless systems. Therefore, closing the loop by setting

$$u = -\varepsilon + e \tag{5.32}$$

(with e an external input), one obtains a lossless system with inputs e and outputs ω, with total storage function $S(\eta, \omega) := T(\omega) + V(\eta)$.

5.3 Stabilization by Backstepping

An important case where condition (i)' of Corollary 5.2.4 is always satisfied occurs if the ξ-dynamics in (5.22) is of the form

$$\dot{\xi} = a(z, \xi) + b(z, \xi)u$$

with rank $b(z, \xi) = m$ everywhere. In fact, if rank $b(z, \xi) = m$ then the ξ-dynamics can be feedback transformed into any dynamics.

   This motivates the consideration of systems with the following *triangular* structure

$$\dot{z} = f(z, \xi_1)$$

$$\dot{\xi}_1 = a_1(z, \xi_1) + b_1(z, \xi_1)\xi_2$$

$$\dot{\xi}_2 = a_2(z, \xi_1, \xi_2) + b_2(z, \xi_1, \xi_2)\xi_3 \tag{5.33}$$

$$\vdots$$

$$\dot{\xi}_k = a_k(z, \xi_1, \ldots, \xi_k) + b_k(z, \xi_1, \ldots, \xi_k)u$$

with $\xi_i \in \mathbb{R}^m$, where it is assumed that rank $b_i = m$, $i = 1, \ldots, k$, everywhere.

Let $z = 0, \xi_1 = \cdots = \xi_k = 0$ be an equilibrium of (5.33), that is, $f(0, 0) = 0$, $a_1(0, 0) = \cdots = a_k(0, 0) = 0$. In order to asymptotically stabilize this equilibrium, we start by assuming that there exists a *virtual* feedback

$$\xi_1 = \alpha_0(z) \tag{5.34}$$

such that $z = 0$ is an asymptotically stable equilibrium of

$$\dot{z} = f(z, \alpha_0(z)), \tag{5.35}$$

having a Lyapunov function V that is positive definite at $z = 0$. Defining the corresponding virtual output

$$y_1 := \xi_1 - \alpha_0(z) \tag{5.36}$$

consideration of Theorem 5.1.5 then leads to rewriting the z-dynamics as

$$\dot{z} = f(z, \alpha_0(z) + y_1) = f(z, \alpha_0(z)) + g(z, y_1)y_1 \tag{5.37}$$

Furthermore, the ξ_1-dynamics can be replaced by the dynamics for $y_1 = \xi - \alpha_0(z)$, given as

$$\dot{y}_1 = \dot{\alpha}_0(z) + a_1(z, \xi_1) + b_1(z, \xi_1)\xi_2 , \tag{5.38}$$

which is regarded as a system with virtual input ξ_2. This last system can be rendered output strictly passive with respect to the virtual output y_1 and the storage function $\frac{1}{2}\|y_1\|^2$ by setting

$$\xi_2 = b_1^{-1}(z, \xi_1)[\dot{\alpha}_0(z) - a_1(z, \xi_1) - y_1 + u], \tag{5.39}$$

with u a new input. Define

$$\alpha_1(z, \xi_1) := b_1^{-1}(z, \xi_1)[\dot{\alpha}_0(z) - a_1(z, \xi_1) - y_1 - (V_z(z)g(z, y))^T] \tag{5.40}$$

Hence by application of Theorem 5.1.5, the virtual feedback

$$\xi_2 = \alpha_1(z, \xi_1) + b_1^{-1}(z, \xi_1)v, \tag{5.41}$$

with v another new input, will result in the system

$$\dot{z} = f(z, \alpha_0(z)) + g(z, y_1)y_1$$
$$\dot{y}_1 = -y_1 - (V_z(z)g(z, y_1))^T + v, \tag{5.42}$$

which is output strictly passive with storage function

$$S_1(z, \xi) := V(z) + \frac{1}{2}\|y_1\|^2 \tag{5.43}$$

Alternatively, returning to the original form

$$\Sigma_2 : \begin{cases} \dot{z} = f(z, \xi_1) \\ \dot{\xi}_1 = a_1(z, \xi_1) + b_1(z, \xi_1)\xi_2 \end{cases} \tag{5.44}$$

we conclude that by defining

$$y_2 := \xi_2 - \alpha_1(z), \tag{5.45}$$

and substituting $\xi_2 = \alpha_1(z, \xi_1) + y_2$ in (5.47) we obtain a system

$$\dot{z} = f(z, \xi_1)$$
$$\dot{\xi}_1 = a_1(z, \xi_1) + b_1(z, \xi_1)\alpha_1(z, \xi_1) + b_1(z, \xi_1)y_2 \tag{5.46}$$

where $z = 0, \xi_1 = 0$ is an asymptotically stable equilibrium for $y_2 = 0$, with Lyapunov function $S_1(z, \xi_1)$. This means that once more we can apply Theorem 5.1.5 to the extended dynamics

$$\Sigma_2 : \begin{cases} \dot{z} = f(z, \xi_1) \\ \dot{\xi}_1 = a_1(z, \xi_1) + b_1(z, \xi_1)\xi_2 \\ \dot{\xi}_2 = a_2(z, \xi_1, \xi_2) + b_2(z, \xi_1, \xi_2)\xi_3 \end{cases} \tag{5.47}$$

with virtual input ξ_3 and virtual output $y_2 = \xi_2 - \alpha_1(z, \xi_1)$. Since its zero-output constrained dynamics is by construction asymptotically stable with Lyapunov function S_1, it again follows that (5.47) can be transformed into a passive system by the virtual feedback

$$\xi_3 = \alpha_2(z, \xi_1, \xi_2) + b_2^{-1}(z, \xi_1, \xi_2)e_3$$
$$\alpha_2(z, \xi_1, \xi_2) = b_2^{-1}(z, \xi_1, \xi_2)[-a_2(z, \xi_1, \xi_2) + \dot{\alpha}_1(z, \xi_1) - y_2 - b_1(z, \xi_1)y_1] \tag{5.48}$$

This leads to the *recursion*

$$y_i = \xi_i - \alpha_{i-1}(z, \xi_1, \ldots, \xi_{i-1})$$
$$\alpha_i(z, \xi_1, \ldots, \xi_i) = \qquad\qquad i = 1, 2, \ldots, k \qquad (5.49)$$
$$b_i^{-1}(-a_i + \dot\alpha_{i-1} - y_i - b_{i-1}(z, \xi_1, \ldots, \xi_{i-1})y_{i-1})$$

resulting in a feedback transformed system with Lyapunov function

$$S := S_k = V(z) + \frac{1}{2}\|y_1\|^2 + \cdots + \frac{1}{2}\|y_k\|^2 \qquad (5.50)$$

satisfying

$$\frac{dS}{dt} \le -\|y_1\|^2 - \|y_2\|^2 - \cdots - \|y_k\|^2 \qquad (5.51)$$

This procedure is commonly called (exact) *backstepping*.

The class of system (5.33) to which the backstepping procedure applies can be described as follows. The ξ-dynamics of (5.33) (with $\xi = (\xi_1, \ldots, \xi_k)$) is a *feedback linearizable*, that is, there exists a feedback $u = \alpha(\xi) + \beta(\xi)v$, $\det\beta(\xi) \ne 0$, such that in suitable new coordinates $\tilde\xi$ the system is a *linear* system (with input v). Hence, the backstepping procedure applies to the cascade of a system (5.33) which is asymptotically stabilizable (by virtual feedback $\xi_1 = \alpha_0(x)$), and a feedback linearizable system.

Necessary and sufficient geometric conditions for feedback linearizability have been obtained in geometric control theory, see e.g., the textbooks [135, 233] for a coverage. These conditions imply that the class of feedback linearizable systems is an important but, mathematically speaking, a *thin* subset of the set of all systems (see [335]).

In practice, the recursively defined feedbacks α_i, $i = 0, 1, \ldots, k$, in (5.49) tend to become rather complex, primarily due to the appearance of $\dot\alpha_i$ in the definition of α_{i+1}. On the other hand, the procedure can be made more flexible by not insisting on Lyapunov functions of the precise form $S_i = V + \frac{1}{2}\|y_1\|^2 + \cdots + \frac{1}{2}\|y_i\|^2$, and by generalizing (5.51). This flexibility can be also exploited for avoiding exact cancelation of terms involving unknown parameters and, in general, for improving the characteristics of the resulting feedback $\alpha_k(z_1, \ldots, z_k)$. We refer again to [312] and the references quoted in there.

5.4 Notes for Chapter 5

1. Sect. 5.1 is based on Byrnes, Isidori & Willems [56].

2. The factorizations in (5.18) and in Remark 5.2.3 are based on the following fact (see, e.g., Nijmeijer & van der Schaft [233], Lemma 2.23, for a proof): Let

$f : \mathbb{R}^n \to \mathbb{R}$ be a C^∞ function with $f(0) = 0$. Then $f(x_1, \ldots, x_n) = \sum_{i=1}^{n} x_i g_i(x)$, for certain C^∞ functions g_i satisfying $g_i(0) = \frac{\partial f}{\partial x_i}(0)$.

3. The first part of Sect. 5.2 is largely based on Ortega [237], as well as on Sussmann & Kokotovic [332]. The backstepping procedure is detailed in Sepulchre, Jankovic and Kokotović [312], and Krstić, Kanellakopoulos and Kokotović [173]; where many extensions and refinements can be found. Related work is presented in Marino and Tomei [195].

4. The example in Sect. 5.2 is taken from Dalsmo and Egeland [75], Dalsmo [74].

Chapter 6
Port-Hamiltonian Systems

As described in the previous Chaps. 3 and 4, (cyclo-)passive systems are defined by the existence of a storage function (nonnegative in case of passivity) satisfying the dissipation inequality with respect to the supply rate $s(u, y) = u^T y$. In contrast, port-Hamiltonian systems, the topic of the current chapter, are endowed with the property of (cyclo-)passivity as a *consequence* of their system formulation. In fact, port-Hamiltonian systems arise from first principles physical modeling. They are defined in terms of a *Hamiltonian* function together with two *geometric structures* (corresponding, respectively, to power-conserving interconnection and energy dissipation), which are such that the Hamiltonian function automatically satisfies the dissipation inequality.

6.1 Input-State-Output Port-Hamiltonian Systems

An important subclass of port-Hamiltonian systems, especially for control purposes, is defined as follows.

Definition 6.1.1 An *input-state-output port-Hamiltonian system* with n-dimensional state space manifold $\mathcal{X}$, input and output spaces $U = Y = \mathbb{R}^m$, and Hamiltonian $H : \mathcal{X} \to \mathbb{R}$, is given as[1]

$$
\begin{aligned}
\dot{x} &= [J(x) - R(x)] \frac{\partial H}{\partial x}(x) + g(x)u \\
y &= g^T(x) \frac{\partial H}{\partial x}(x)
\end{aligned}
\tag{6.1}
$$

where the $n \times n$ matrices $J(x), R(x)$ satisfy $J(x) = -J^T(x)$ and $R(x) = R^T(x) \geq 0$. By the properties of $J(x), R(x)$, it immediately follows that

---

[1] As before, $\frac{\partial H}{\partial x}(x)$ denotes the column vector of partial derivatives of H.

© Springer International Publishing AG 2017
A. van der Schaft, *L2-Gain and Passivity Techniques in Nonlinear Control*,
Communications and Control Engineering, DOI 10.1007/978-3-319-49992-5_6

$$\frac{dH}{dt}(x(t)) = \frac{\partial^T H}{\partial x}(x(t))\dot{x}(t) =$$
$$-\frac{\partial^T H}{\partial x}(x(t))R(x(t))\frac{\partial H}{\partial x}(x(t)) + y^T(t)u(t) \le u^T(t)y(t), \tag{6.2}$$

implying, cf. Definition 4.1.1, *cyclo-passivity* and *passivity* if $H \ge 0$.

The Hamiltonian H is is equal to the *total stored* energy of the system, while $u^T y$ is the externally supplied power. In the definition of a port-Hamiltonian system, two geometric structures on the state space $\mathcal{X}$ play a role: the internal *interconnection* structure given by $J(x)$, which by skew-symmetry is *power-conserving*, and a *resistive* structure given by $R(x)$, which by nonnegativity is responsible for internal *dissipation* of energy. For a further discussion on the mathematical theory underlying these geometric structures, as well as the port-based modeling origins of port-Hamiltonian systems, we refer to the Notes at the end of this chapter.

A useful extension of Definition 6.1.1 to systems with *feedthrough terms* is given as follows.

Definition 6.1.2 An *input-state-output port-Hamiltonian system with feedthrough terms* is specified by an n-dimensional state space manifold $\mathcal{X}$, input and output spaces $U = Y = \mathbb{R}^m$, Hamiltonian $H : \mathcal{X} \to \mathbb{R}$, and dynamics

$$\dot{x} = [J(x) - R(x)]\frac{\partial H}{\partial x}(x) + [G(x) - P(x)]u$$
$$y = [G(x) + P(x)]^T \frac{\partial H}{\partial x}(x) + [M(x) + S(x)]u, \tag{6.3}$$

where the matrices $J(x), M(x), R(x), P(x), S(x)$ satisfy the skew-symmetry conditions $J(x) = -J^T(x), M(x) = -M^T(x)$, and the nonnegativity condition

$$\begin{bmatrix} R(x) & P(x) \\ P^T(x) & S(x) \end{bmatrix} \ge 0, \quad x \in \mathcal{X} \tag{6.4}$$

In this case, the power balance (6.2) takes the following form (using skew-symmetry of $J(x), M(x)$, and exploiting the nonnegativity condition (6.4))

$$\frac{d}{dt}H(x) = \frac{\partial^T H}{\partial x}(x)\left([J(x) - R(x)]\frac{\partial H}{\partial x}(x) + [G(x) - P(x)]u\right) =$$
$$-\left[\frac{\partial^T H}{\partial x}(x)\ u^T\right]\begin{bmatrix} R(x) & P(x) \\ P^T(x) & S(x) \end{bmatrix}\begin{bmatrix} \frac{\partial H}{\partial x}(x) \\ u \end{bmatrix} + y^T u \le u^T y \tag{6.5}$$

leading to the same conclusion regarding (cyclo-)passivity as above.

Remark 6.1.3 Note that by (6.4) $P = 0$ whenever $S = 0$ (no feedthrough).

Both (6.3) and (6.5) correspond to a *linear* resistive structure. The extension to *nonlinear* energy dissipation is given next.

Definition 6.1.4 An *input-state-output port-Hamiltonian system with nonlinear resistive structure* is given as

$$\dot{x} = J(x)z - \mathcal{R}(x, z) + g(x)u, \quad z = \frac{\partial H}{\partial x}(x)$$
$$y = g^T(x)z \tag{6.6}$$

where $J(x) = -J^T(x)$, and the *resistive mapping* $\mathcal{R}(x, \cdot) : \mathbb{R}^n \to \mathbb{R}^n$ satisfies

$$z^T \mathcal{R}(x, z) \geq 0, \quad \text{for all } z \in \mathbb{R}^n, \; x \in \mathcal{X} \tag{6.7}$$

Remark 6.1.5 Geometrically $z = \frac{\partial H}{\partial x}(x) \in T_x^* \mathcal{X}$, with $T_x^* \mathcal{X}$, the co-tangent space of $\mathcal{X}$ at $x \in \mathcal{X}$, while $\dot{x} \in T_x \mathcal{X}$, the tangent space at $x \in \mathcal{X}$. Hence, the resistive mapping $\mathcal{R}$ is defined geometrically as a *vector bundle map* $\mathcal{R} : T^* \mathcal{X} \to T \mathcal{X}$.

Similarly to (6.2) we obtain

$$\frac{d}{dt} H = \frac{\partial^T H}{\partial x}(x) \dot{x} = -\frac{\partial^T H}{\partial x}(x) \mathcal{R} \left(x, \frac{\partial H}{\partial x}(x) \right) + y^T u \leq u^T y, \tag{6.8}$$

showing again (cyclo-)passivity. We leave the extension to systems with feedthrough terms to the reader.

Example 6.1.6 Consider a mass–spring–damper system (mass m, spring constant k, momentum p, spring extension q, external force F) subject to ideal Coulomb friction

$$\begin{bmatrix} \dot{q} \\ \dot{p} \end{bmatrix} = \begin{bmatrix} 0 & 1 \\ -1 & 0 \end{bmatrix} \begin{bmatrix} kq \\ \frac{p}{m} \end{bmatrix} - \begin{bmatrix} 0 \\ c \operatorname{sign} \frac{p}{m} \end{bmatrix} + \begin{bmatrix} 0 \\ F \end{bmatrix}, \tag{6.9}$$

where sign is the multivalued function defined by

$$\operatorname{sign} v = \begin{cases} 1 & , \quad v > 0 \\ [-1, 1] & , \quad v = 0 \\ -1 & , \quad v < 0 \end{cases} \tag{6.10}$$

and $c > 0$ is a constant. This defines an input-state-output port-Hamiltonian system with nonlinear resistive structure defined by the multivalued function $c \operatorname{sign}$. Note that strictly speaking, this entails a further generalization of Definition 6.1.4 since the Coulomb friction mapping (6.10) is multivalued. The Hamiltonian $H(q, p) = \frac{1}{2m} p^2 + \frac{1}{2} k q^2$ satisfies

$$\frac{d}{dt} H = -\frac{p}{m} \operatorname{sign} \frac{p}{m} + F \frac{p}{m} \leq F \frac{p}{m} \tag{6.11}$$

Example 6.1.7 The dynamics of a detailed-balanced mass action kinetics chemical reaction network can be written as, see the Notes at the end of this chapter for further information,

$$\begin{aligned} \dot{x} &= -Z\mathcal{L} \operatorname{Exp} \left(Z^T \operatorname{Ln} \frac{x}{x^*} \right) + S_b u \\ y &= S_b^T \operatorname{Ln} \frac{x}{x^*} \end{aligned} \tag{6.12}$$

where $x \in \mathbb{R}^n$ is the vector of chemical species concentrations, u is the vector of boundary fluxes, and y is the vector of boundary chemical potentials. Furthermore, x^* is a thermodynamic equilibrium, Z is the complex composition matrix, S_b specifies which are the boundary chemical species, and $\mathcal{L}$ is a symmetric Laplacian matrix (see Definition 4.4.6) on the graph of chemical complexes, with weights determined by the kinetic reaction constants. Exp and Ln denote the component-wise exponential and logarithm mappings, i.e., $(\text{Exp}\,(x))_i = \exp x_i$, $(\text{Ln}\,(x))_i = \ln x_i$, $i = 1, \ldots, n$. Similarly, $\frac{x}{x^*}$ denotes component-wise division of the vector x by the vector x^*. The Hamiltonian is given by the Gibbs' free energy, which (up to constants) is equal to

$$H(x) = \sum_{i=1}^{n} x_i \ln \frac{x_i}{x_i^*} + \sum_{i=1}^{n} (x_i^* - x_i), \tag{6.13}$$

corresponding to the chemical potentials $z_i = \frac{\partial H}{\partial x_i}(x) = \ln \frac{x_i}{x_i^*}$. Since [299]

$$\gamma^T \mathcal{L} \, \text{Exp} \, \gamma \geq 0 \tag{6.14}$$

for all vectors γ, this defines an input-state-output port-Hamiltonian system, with $J = 0$ and nonlinear resistive structure given by the mapping $z \mapsto Z^T \mathcal{L} \, \text{Exp}\, Z^T z$.

Finally, a *linear* input-state-output port-Hamiltonian system with feedthrough terms is given by the following specialization of Definition 6.1.2

$$\begin{aligned} \dot{x} &= [J - R]\,Qx + [G - P]\,u \\ y &= [G + P]^T \, Qx + [M + S]\,u \end{aligned} \tag{6.15}$$

with quadratic Hamiltonian $H(x) = \frac{1}{2}x^T Qx$, $Q = Q^T$, and constant matrices J, M, R, P, S satisfying $J = -J^T$, $M = -M^T$ and

$$\begin{bmatrix} R & P \\ P^T & S \end{bmatrix} \geq 0 \tag{6.16}$$

Since subtracting a constant from the Hamiltonian function H does not change the system, the condition $H \geq 0$ can be replaced by H being bounded from below. Hence based on (6.2), (6.5), (6.8), we can summarize the characterization of (cyclo-)passivity of input-state-output port-Hamiltonian systems as follows.

Proposition 6.1.8 *Any input-state-output port-Hamiltonian system given by one of the expressions (6.1), (6.3), (6.6), (6.15) is cyclo-passive, and passive if H is bounded from below, respectively, $Q \geq 0$. Furthermore, if the (nonlinear) resistive structure is absent, then the system is lossless in case H is bounded from below.*

In the modeling of physical systems, the port-Hamiltonian formulation directly follows from the physical structure of the system; see Sects. 6.2 and 6.3 and the Notes at the end of this chapter for further information. On the other hand, one may still wonder when the converse of Proposition 6.1.8 holds, i.e., when and how a passive system can be written as a port-Hamiltonian system. In the linear case, this question can be answered as follows. Consider the passive linear system (for simplicity without feedthrough terms)

$$\dot{x} = Ax + Bu$$
$$y = Cx \tag{6.17}$$

with *positive-definite* storage function $\frac{1}{2}x^T Q x$, i.e.,

$$A^T Q + QA \leq 0, \quad B^T Q = C, \quad Q > 0 \tag{6.18}$$

Now decompose AQ^{-1} into its skew-symmetric and symmetric part as

$$AQ^{-1} = J - R, \quad J = -J^T, R = R^T \tag{6.19}$$

Then $A^T Q + QA \leq 0$ implies $R \geq 0$, and $\dot{x} = Ax + Bu$, $y = Cx$, can be rewritten into port-Hamiltonian form $\dot{x} = (J - R)Qx + Bu$, $y = B^T Qx$. The same result can be shown to hold for linear passive systems with $Q \geq 0$ under the additional assumption $\ker Q \subset \ker A$. In this case, one defines F such that $A = FQ$, and factorizes F into its skew-symmetric and symmetric part.

On the other hand, since in general the storage matrix Q of a passive system is not unique, also the interconnection and resistive structure matrices J and R as obtained in the above port-Hamiltonian formulation are not unique. Hence, if Q is not unique then there exist essentially *different* port-Hamiltonian formulations of the same linear passive system $\dot{x} = Ax + Bu$, $y = Cx$.

For *nonlinear* systems, the conversion from passive to port-Hamiltonian systems is more subtle. For example,

$$\dot{x} = f(x) + g(x)u$$
$$y = h(x) \tag{6.20}$$

is lossless with storage function $H \geq 0$ iff

$$\frac{\partial^T H}{\partial x}(x)f(x) = 0$$
$$g^T(x)\frac{\partial H}{\partial x}(x) = h(x) \tag{6.21}$$

Nevertheless, the first equality in (6.21) does not imply that there exists a skew-symmetric matrix $J(x)$ such that $f(x) = J(x)\frac{\partial H}{\partial x}(x)$, as illustrated by the next example.

Example 6.1.9 Consider the system

$$\begin{bmatrix} \dot{x}_1 \\ \dot{x}_2 \end{bmatrix} = \begin{bmatrix} x_1 \\ -x_2 \end{bmatrix} + \begin{bmatrix} 0 \\ 1 \end{bmatrix} u \tag{6.22}$$
$$y = x_1^2 x_2$$

which is lossless with respect to the storage function $H(x_1, x_2) = \frac{1}{2}x_1^2 x_2^2$. However, it is easy to see that there does not exist a 2×2 matrix $J(x) = -J^T(x)$, depending smoothly on $x = (x_1, x_2)$, such that

$$\begin{bmatrix} x_1 \\ -x_2 \end{bmatrix} = J(x) \begin{bmatrix} x_1 x_2^2 \\ x_1^2 x_2 \end{bmatrix}$$

Hence, the system is *not* a port-Hamiltonian system with respect to $H(x_1, x_2)$.

6.2 Mechanical Systems

The port-Hamiltonian formulation of standard mechanical systems directly follows from classical mechanics. Consider as in Proposition 4.5.1, the Hamiltonian representation of fully actuated Euler–Lagrange equations in n configuration coordinates $q = (q_1, \ldots, q_n)$ given by the $2n$-dimensional system

$$\begin{aligned}
\dot{q} &= \tfrac{\partial H}{\partial p}(q, p), & p &= (p_1, \ldots, p_n) \\
\dot{p} &= -\tfrac{\partial H}{\partial q}(q, p) + u, & u &= (u_1, \ldots, u_n) \\
y &= \tfrac{\partial H}{\partial p}(q, p) \ (= \dot{q}), & y &= (y_1, \ldots, y_n)
\end{aligned} \tag{6.23}$$

with u the vector of (generalized) external forces and y the vector of (generalized) velocities. The state space of (6.23) with local coordinates (q, p) is called the *phase space*. In most mechanical systems, the Hamiltonian $H(q, p)$ is the sum of a positive kinetic energy and a potential energy

$$H(q, p) = \frac{1}{2} p^T M^{-1}(q) p + P(q) \tag{6.24}$$

It was shown in Proposition 4.5.1 that along every trajectory of (6.23)

$$H(q(t_1), p(t_1)) = H(q(t_0), p(t_0)) + \int_{t_0}^{t_1} u^T(t) y(t) dt, \tag{6.25}$$

expressing that the increase in internal energy H equals the *work* supplied to the system ($u^T y$ is generalized force times generalized velocity, i.e., power). Hence, the system (6.23) is an input-state-output port-Hamiltonian system, which is lossless if H is bounded from below. The system description (6.23) can be further generalized to

$$\begin{aligned}
\dot{q} &= \tfrac{\partial H}{\partial p}(q, p), & (q, p) &= (q_1, \ldots, q_n, p_1, \ldots, p_n) \\
\dot{p} &= -\tfrac{\partial H}{\partial q}(q, p) + B(q)u, & u &\in \mathbb{R}^m \\
y &= B^T(q) \tfrac{\partial H}{\partial p}(q, p) \ (= B^T(q)\dot{q}), & y &\in \mathbb{R}^m,
\end{aligned} \tag{6.26}$$

where $B(q)$ is an input force matrix, with $B(q)u$ denoting the generalized forces resulting from the control inputs $u \in \mathbb{R}^m$. If $m < n$, we speak of an *underactuated* mechanical system. Also for (6.26) we obtain the power balance

$$\frac{dH}{dt}(q(t), p(t)) = u^T(t)y(t) \tag{6.27}$$

A further generalization is obtained by extending (6.26) to input-state-output port-Hamiltonian systems

$$\dot{x} = J(x)\frac{\partial H}{\partial x}(x) + g(x)u, \quad J(x) = -J^T(x), \ x \in \mathcal{X}$$
$$y = g^T(x)\frac{\partial H}{\partial x}(x), \tag{6.28}$$

where $\mathcal{X}$ is an $\bar{n}$-dimensional state space manifold, and $J(x)$ is state-dependent skew-symmetric matrix. Note that (6.23) and (6.26) correspond to the full rank and constant skew-symmetric matrix J given by

$$J = \begin{bmatrix} 0 & I_n \\ -I_n & 0 \end{bmatrix} \tag{6.29}$$

Models (6.28) arise, for example, by *symmetry reduction* of (6.23) or (6.26). A classical example is Euler's equations for the dynamics of the angular velocities of a rigid body.

Example 6.2.1 (*Euler's equations*; *Example* 4.2.4 *continued*) Consider a rigid body spinning around its center of mass in the absence of gravity. In Example 4.2.4, we already encountered Euler's equations for the dynamics of the angular velocities. The Hamiltonian formulation is obtained by considering the body angular momenta $p = (p_x, p_y, p_z)$ along the three principal axes, and the Hamiltonian given by the kinetic energy

$$H(p) = \frac{1}{2}\left(\frac{p_x^2}{I_x} + \frac{p_y^2}{I_y} + \frac{p_z^2}{I_z}\right), \tag{6.30}$$

where I_x, I_y, I_z are the principal moments of inertia. The vector p of angular momenta is related to the vector ω of angular velocities as $p = I\omega$, where I is the diagonal matrix with positive diagonal elements I_x, I_y, I_z. Euler's equations are now given as

$$\begin{bmatrix} \dot{p}_x \\ \dot{p}_y \\ \dot{p}_z \end{bmatrix} = \underbrace{\begin{bmatrix} 0 & -p_z & p_y \\ p_z & 0 & -p_x \\ -p_y & p_x & 0 \end{bmatrix}}_{J(p)} \begin{bmatrix} \frac{\partial H}{\partial p_x} \\ \frac{\partial H}{\partial p_y} \\ \frac{\partial H}{\partial p_z} \end{bmatrix} + \begin{bmatrix} b_x \\ b_y \\ b_z \end{bmatrix} u \tag{6.31}$$

In the scalar input case, the last term bu denotes the torque around an axis with coordinates $b = (b_x\ b_y\ b_z)^T$, with corresponding collocated output given as

$$y = b_x \frac{p_x}{I_x} + b_y \frac{p_y}{I_y} + b_z \frac{p_z}{I_z}, \tag{6.32}$$

which is the velocity around the same axis $(b_x\ b_y\ b_z)^T$.

In many cases (including the one obtained by symmetry reduction from the canonical J given in (6.29)), the dependence of the matrix J on the state x will satisfy the *integrability* conditions

$$\sum_{l=1}^{n} \left[J_{lj}(x) \frac{\partial J_{ik}}{\partial x_l}(x) + J_{li}(x) \frac{\partial J_{kj}}{\partial x_l}(x) + J_{lk}(x) \frac{\partial J_{ji}}{\partial x_l}(x) \right] = 0, \tag{6.33}$$

for $i, j, k = 1, \ldots, n$. These integrability conditions are also referred to as the *Jacobi-identity*. If these integrability conditions are met, we can construct by Darboux's theorem (see e.g., [347]), around any point x_0 where the rank of the matrix $J(x)$ is constant, local coordinates

$$\tilde{x} = (q, p, s) = (q_1, \ldots, q_l, p_1, \ldots, p_k, s_1, \ldots s_l), \tag{6.34}$$

with $2k$ the rank of J and $n = 2k + l$, such that J in these coordinates takes the form

$$J = \begin{bmatrix} 0 & I_k & 0 \\ -I_k & 0 & 0 \\ 0 & 0 & 0 \end{bmatrix} \tag{6.35}$$

The coordinates (q, p, s) are also called *canonical* coordinates, and J satisfying (6.33) is called a *Poisson structure matrix*. Otherwise, it is called an *almost-Poisson structure*.

Example 6.2.2 (*Example* 6.2.1 *continued*) It can be directly checked that the skew-symmetric matrix $J(p)$ defined in (6.31) satisfies the Jacobi-identity (6.33). This also follows from the fact that $J(p)$ is the canonical *Lie–Poisson structure* matrix on the dual of the Lie algebra $so(3)$ corresponding to the configuration space $SO(3)$ of the rigid body; see the Notes at the end of this chapter for further information.

The rest of this section will be devoted to mechanical systems with *kinematic constraints*, which is an important class of systems in applications (for example in robotics). Consider a mechanical system with n degrees of freedom, locally described by n configuration variables

$$q = (q_1, \ldots, q_n) \tag{6.36}$$

Expressing the kinetic energy as $\frac{1}{2} \dot{q}^T M(q) \dot{q}$, with $M(q) > 0$ being the generalized mass matrix, we define in the usual way the Lagrangian function

$L(q, \dot{q}) = \frac{1}{2}\dot{q}^T M(q)\dot{q} - P(q)$, where P is the potential energy. Suppose now that there are constraints on the generalized velocities $\dot{q}$, described as

$$A^T(q)\dot{q} = 0, \tag{6.37}$$

with $A(q)$ an $n \times k$ matrix of rank k everywhere. This means that there are k independent kinematic constraints. Classically, the constraints (6.37) are called *holonomic* if it is possible to find new configuration coordinates $\bar{q} = (\bar{q}_1, \ldots, \bar{q}_n)$ such that the constraints are equivalently expressed as

$$\dot{\bar{q}}_{n-k+1} = \dot{\bar{q}}_{n-k+2} = \cdots = \dot{\bar{q}}_n = 0 \,, \tag{6.38}$$

in which case it is possible to eliminate the configuration variables $\bar{q}_{n-k+1}$, $\ldots, \bar{q}_n$, since the kinematic constraints (6.38) are equivalent to the *geometric constraints* constraints

$$\bar{q}_{n-k+1} = c_{n-k+1}, \ldots, \bar{q}_n = c_n \,, \tag{6.39}$$

for constants $c_{n-k+1}, \ldots, c_n$ determined by the initial conditions. Then the system reduces to an *unconstrained* system in the remaining configuration coordinates $(\bar{q}_1, \ldots, \bar{q}_{n-k})$. If it is *not* possible to find coordinates $\bar{q}$ such that (6.38) holds (that is, if we are not able to *integrate* the kinematic constraints as above), then the kinematic constraints are called *nonholonomic*.

The equations of motion for the mechanical system with Lagrangian $L(q, \dot{q})$ and kinematic constraints (6.37) are given by the *constrained Euler–Lagrange equations*

$$\frac{d}{dt}\left(\frac{\partial L}{\partial \dot{q}}(q, \dot{q})\right) - \frac{\partial L}{\partial q}(q, \dot{q}) = A(q)\lambda + B(q)u, \quad \lambda \in \mathbb{R}^k, u \in \mathbb{R}^m$$
$$A^T(q)\dot{q} = 0, \tag{6.40}$$

where $B(q)u$ are the external forces applied to the system, for some $n \times m$ matrix $B(q)$, while $A(q)\lambda$ are the *constraint forces*. The Lagrange multipliers $\lambda(t)$ are uniquely determined by the requirement that the constraints $A^T(q(t))\dot{q}(t) = 0$ are satisfied for all t.

Defining as before (cf. (4.107)) the generalized momenta

$$p = \frac{\partial L}{\partial \dot{q}}(q, \dot{q}) = M(q)\dot{q}, \tag{6.41}$$

the constrained Euler–Lagrange equations (6.40) transform into *constrained Hamiltonian equations*

$$\dot{q} = \frac{\partial H}{\partial p}(q, p)$$

$$\dot{p} = -\frac{\partial H}{\partial q}(q, p) + A(q)\lambda + B(q)u$$

$$y = B^T(q)\frac{\partial H}{\partial p}(q, p) \qquad (6.42)$$

$$0 = A^T(q)\frac{\partial H}{\partial p}(q, p)$$

with $H(q, p) = \frac{1}{2}p^T M^{-1}(q)p + P(q)$ the total energy. Thus, the kinematic constraints appear as *algebraic constraints* on the phase space, and the *constrained* state space is given as the following subset of the phase space

$$\mathcal{X}_c = \left\{ (q, p) \mid A^T(q)\frac{\partial H}{\partial p}(q, p) = 0 \right\} \qquad (6.43)$$

The algebraic constraints $A^T(q)\frac{\partial H}{\partial p}(q, p) = 0$ and constraint forces $A(q)\lambda$ can be *eliminated* in the following way. Since rank $A(q) = k$, there exists locally an $n \times (n - k)$ matrix $S(q)$ of rank $n - k$ such that

$$A^T(q)S(q) = 0 \qquad (6.44)$$

Now define $\tilde{p} = (\tilde{p}^1, \tilde{p}^2) = (\tilde{p}_1, \dots, \tilde{p}_{n-k}, \tilde{p}_{n-k+1}, \dots, \tilde{p}_n)$ as

$$\begin{aligned} \tilde{p}^1 &:= S^T(q)p, \ \tilde{p}^1 \in \mathbb{R}^{n-k} \\ \tilde{p}^2 &:= A^T(q)p, \ \tilde{p}^2 \in \mathbb{R}^k \end{aligned} \qquad (6.45)$$

It is readily checked that $(q, p) \mapsto (q, \tilde{p}^1, \tilde{p}^2)$ is a coordinate transformation. Indeed, by (6.44) the rows of $S^T(q)$ are orthogonal to the rows of $A^T(q)$. In the new coordinates the constrained system (6.42) takes the form [293], $*$ denoting unspecified elements,

$$\begin{bmatrix} \dot{q} \\ \dot{\tilde{p}}^1 \\ \dot{\tilde{p}}^2 \end{bmatrix} = \begin{bmatrix} 0_n & S(q) & * \\ -S^T(q) & (-p^T[S_i, S_j](q))_{i,j} & * \\ * & * & * \end{bmatrix} \begin{bmatrix} \frac{\partial \tilde{H}}{\partial q} \\ \frac{\partial \tilde{H}}{\partial \tilde{p}^1} \\ \frac{\partial \tilde{H}}{\partial \tilde{p}^2} \end{bmatrix} +$$

$$\begin{bmatrix} 0 \\ 0 \\ A^T(q)A(q) \end{bmatrix} \lambda + \begin{bmatrix} 0 \\ B_c(q) \\ \overline{B}(q) \end{bmatrix} u \qquad (6.46)$$

$$A^T(q)\frac{\partial H}{\partial p} = A^T(q)A(q)\frac{\partial \tilde{H}}{\partial \tilde{p}^2} = 0$$

with $\tilde{H}(q, \tilde{p})$ the Hamiltonian H expressed in the new coordinates $q, \tilde{p}$. Here S_i denotes the i-th column of $S(q)$, $i = 1, \dots, n - k$, and $[S_i, S_j]$ is the *Lie bracket* of S_i and S_j, in local coordinates q given as (see e.g., [1, 233])

$$[S_i, S_j](q) = \frac{\partial S_j}{\partial q}(q)S_i(q) - \frac{\partial S_i}{\partial q}S_j(q) \tag{6.47}$$

with $\frac{\partial S_j}{\partial q}, \frac{\partial S_i}{\partial q}$ denoting the $n \times n$ Jacobian matrices.

The constraints $A^T(q)\frac{\partial H}{\partial p}(q, p) = 0$ are equivalently given as $\frac{\partial \tilde{H}}{\partial \tilde{p}^2}(q, \tilde{p}) = 0$, and by non-degeneracy of the kinetic energy $\frac{1}{2}p^T M^{-1}(q)p$ these equations can be solved for $\tilde{p}^2$. Since λ only influences the $\tilde{p}^2$-dynamics, the constrained dynamics is thus determined by the dynamics of q and $\tilde{p}^1$ alone (which together serve as coordinates for the constrained state space $\mathcal{X}_c$), given as

$$\begin{bmatrix} \dot{q} \\ \dot{\tilde{p}}^1 \end{bmatrix} = J_c(q, \tilde{p}^1) \begin{bmatrix} \frac{\partial H_c}{\partial q}(q, \tilde{p}^1) \\ \frac{\partial H_c}{\partial \tilde{p}^1}(q, \tilde{p}^1) \end{bmatrix} + \begin{bmatrix} 0 \\ B_c(q) \end{bmatrix} u \tag{6.48}$$

Here $H_c(q, \tilde{p}^1)$ equals $\tilde{H}(q, \tilde{p})$ with $\tilde{p}^2$ satisfying $\frac{\partial \tilde{H}}{\partial \tilde{p}^2}(q, \tilde{p}^1, \tilde{p}^2) = 0$, and where the skew-symmetric matrix $J_c(q, \tilde{p}^1)$ is given as the left-upper part of the structure matrix in (6.46), that is

$$J_c(q, \tilde{p}^1) = \begin{bmatrix} O_n & S(q) \\ -S^T(q) & \left(-p^T[S_i, S_j](q)\right)_{i,j} \end{bmatrix}, \tag{6.49}$$

where p is expressed as function of $q, \tilde{p}$, with $\tilde{p}^2$ eliminated from $\frac{\partial \tilde{H}}{\partial \tilde{p}^2} = 0$. Finally, in the coordinates $q, \tilde{p}$, the output map is given as

$$y = \begin{bmatrix} B_c^T(q) & \bar{B}^T(q) \end{bmatrix} \begin{bmatrix} \frac{\partial \tilde{H}}{\partial \tilde{p}^1}(q, \tilde{p}^1) \\ \frac{\partial \tilde{H}}{\partial \tilde{p}^2}(q, \tilde{p}^1) \end{bmatrix} \tag{6.50}$$

which reduces on the constrained state space $\mathcal{X}_c$ to

$$y = B_c^T(q)\frac{\partial \tilde{H}}{\partial \tilde{p}^1}(q, \tilde{p}^1) \tag{6.51}$$

Summarizing, (6.48) and (6.51) define an *input-state-output port-Hamiltonian system* on $\mathcal{X}_c$, with Hamiltonian H_c given by the constrained total energy, and with structure matrix J_c given by (6.49).

The skew-symmetric matrix J_c defined on $\mathcal{X}_c$ is an *almost-Poisson structure* since it does not necessarily the integrability conditions (6.33). In fact, J_c satisfies the integrability conditions (6.33), and thus defines a Poisson structure on $\mathcal{X}_c$, if and only if the kinematic constraints (6.37) are *holonomic*. In fact, if the constraints are holonomic then the coordinates s as in (6.34) can be taken equal to the "integrated constraint functions" $\bar{q}_{n-k+1}, \ldots, \bar{q}_n$ of (6.39).

Example 6.2.3 (Rolling coin) Let x, y be the Cartesian coordinates of the point of contact of a vertical coin with the plane. Furthermore, φ denotes the heading angle

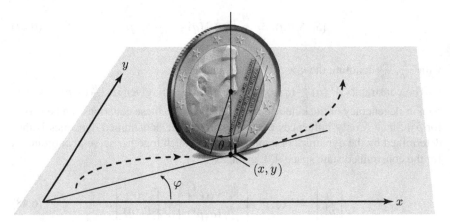

Fig. 6.1 The geometry of the rolling coin

of the coin on the plane, and θ the angle of Willem Alexander's head; cf. Fig. 6.1. With all constants set to unity, the constrained Lagrangian equations of motion are

$$
\begin{aligned}
\ddot{x} &= \lambda_1 \\
\ddot{y} &= \lambda_2 \\
\ddot{\theta} &= -\lambda_1 \cos\varphi - \lambda_2 \sin\varphi + u_1 \\
\ddot{\varphi} &= u_2
\end{aligned}
\tag{6.52}
$$

where u_1 is the control torque about the rolling axis, and u_2 the control torque about the vertical axis. The rolling constraints are

$$
\dot{x} = \dot{\theta}\cos\varphi, \quad \dot{y} = \dot{\theta}\sin\varphi
\tag{6.53}
$$

(rolling without slipping). The energy is $H = \frac{1}{2}p_x^2 + \frac{1}{2}p_y^2 + \frac{1}{2}p_\theta^2 + \frac{1}{2}p_\varphi^2$, and the kinematic constraints can be rewritten as $p_x = p_\theta \cos\varphi, p_y = p_\theta \sin\varphi$. Define according to (6.45) new p-coordinates

$$
\begin{aligned}
p_1 &= p_\varphi \\
p_2 &= p_\theta + p_x \cos\varphi + p_y \sin\varphi \\
p_3 &= p_x - p_\theta \cos\varphi \\
p_4 &= p_y - p_\theta \sin\varphi
\end{aligned}
\tag{6.54}
$$

The constrained state space $\mathcal{X}_c$ is given by $p_3 = p_4 = 0$, and the dynamics on $\mathcal{X}_c$ is computed as

$$
\begin{bmatrix} \dot{x} \\ \dot{y} \\ \dot{\theta} \\ \dot{\varphi} \\ \dot{p}_1 \\ \dot{p}_2 \end{bmatrix} = \begin{bmatrix} & & & & 0 & \cos\varphi \\ & & O_4 & & 0 & \sin\varphi \\ & & & & 0 & 1 \\ & & & & 1 & 0 \\ 0 & 0 & 0 & -1 & 0 & 0 \\ -\cos\varphi & -\sin\varphi & -1 & 0 & 0 & 0 \end{bmatrix} \begin{bmatrix} \frac{\partial H_c}{\partial x} \\ \frac{\partial H_c}{\partial y} \\ \frac{\partial H_c}{\partial \theta} \\ \frac{\partial H_c}{\partial \varphi} \\ \frac{\partial H_c}{\partial p_1} \\ \frac{\partial H_c}{\partial p_2} \end{bmatrix} \tag{6.55}
$$

$$
+ \begin{bmatrix} 0 & 0 \\ 0 & 0 \\ 0 & 0 \\ 0 & 0 \\ 0 & 1 \\ 1 & 0 \end{bmatrix} \begin{bmatrix} u_1 \\ u_2 \end{bmatrix}
$$

$$
\begin{bmatrix} y_1 \\ y_2 \end{bmatrix} = \begin{bmatrix} 0 & 1 \\ 1 & 0 \end{bmatrix} \begin{bmatrix} \frac{\partial H_c}{\partial p_1} \\ \frac{\partial H_c}{\partial p_2} \end{bmatrix} = \begin{bmatrix} \frac{1}{2}p_2 \\ p_1 \end{bmatrix}
$$

where $H_c(x, y, \theta, \varphi, p_1, p_2) = \frac{1}{2}p_1^2 + \frac{1}{4}p_2^2$. (Note that $\frac{\partial H_c}{\partial p_2} = \frac{1}{2}p_2 = p_\theta$.) It can be verified that the structure matrix J_c in (6.55) does *not* satisfy the integrability conditions, in accordance with the fact that the rolling constraints (rolling without slipping) are *nonholonomic*.

6.3 Port-Hamiltonian Models of Electromechanical Systems

This section will contain a collection of characteristic examples of port-Hamiltonian systems arising in electromechanical systems, illustrating the use of port-Hamiltonian models for *multi-physics* systems. In most of the examples the interaction between the *mechanical* and the *electrical* part of the system will take place through the Hamiltonian function, which will depend in a non-separable way on state variables belonging to the mechanical and variables belonging to the electrical domain.

Example 6.3.1 (Capacitor microphone [230]) Consider the capacitor microphone depicted in Fig. 6.2.

The capacitance $C(q)$ of the capacitor is varying as a function of the displacement q of the right plate (with mass m), which is attached to a spring (with spring constant $k > 0$) and a damper (with constant $d > 0$), and affected by a mechanical force F (air pressure arising from sound). Furthermore, E is a voltage source. The equations of motion can be written as the port-Hamiltonian system

Fig. 6.2 Capacitor
microphone

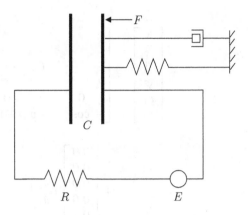

$$
\begin{bmatrix} \dot{q} \\ \dot{p} \\ \dot{Q} \end{bmatrix} = \left(\begin{bmatrix} 0 & 1 & 0 \\ -1 & 0 & 0 \\ 0 & 0 & 0 \end{bmatrix} - \begin{bmatrix} 0 & 0 & 0 \\ 0 & d & 0 \\ 0 & 0 & 1/R \end{bmatrix} \right) \begin{bmatrix} \frac{\partial H}{\partial q} \\ \frac{\partial H}{\partial p} \\ \frac{\partial H}{\partial Q} \end{bmatrix}
$$
$$
+ \begin{bmatrix} 0 \\ 1 \\ 0 \end{bmatrix} F + \begin{bmatrix} 0 \\ 0 \\ 1/R \end{bmatrix} E \qquad (6.56)
$$
$$
y_1 = \frac{\partial H}{\partial p} = \dot{q}
$$
$$
y_2 = \frac{1}{R} \frac{\partial H}{\partial Q} = I
$$

where p is the momentum, R the resistance of the resistor, I the current through the
voltage source, and the Hamiltonian H is the total energy

$$
H(q, p, Q) = \frac{1}{2m} p^2 + \frac{1}{2} k (q - \bar{q})^2 + \frac{1}{2C(q)} Q^2, \qquad (6.57)
$$

with $\bar{q}$ denoting the rest length of the spring. Note that the electric energy $\frac{1}{2C(q)} Q^2$
not only depends on the electric charge Q, but also on the q-variable belonging to
the mechanical part of the system. Furthermore

$$
\frac{d}{dt} H = -c \dot{q}^2 - R I^2 + F \dot{q} + E I \leq F \dot{q} + E I, \qquad (6.58)
$$

with $F \dot{q}$ the mechanical power and EI the electrical power supplied to the system.
In the application as a microphone the voltage over the resistor will be used (after
amplification) as a measure for the mechanical force F. Finally, we note that the
same model can be used for an electrical *micro-actuator*. In this case, the system
is controlled at its electrical side in order to produce a certain desired force at its
mechanical side. This physical phenomenon of *bilateral operation* will be also evi-
dent in the following examples.

Fig. 6.3 Magnetically
levitated ball

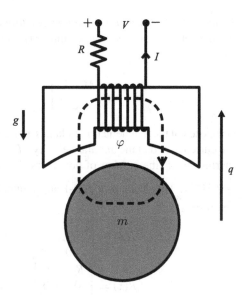

Example 6.3.2 (Magnetically levitated ball) Consider the dynamics of an iron ball
that is levitated by the magnetic field of a controlled inductor as schematically
depicted in Fig. 6.3. The port-Hamiltonian description of this system (with q the
height of the ball, p the vertical momentum, and φ the magnetic flux linkage of the
inductor) is given as

$$
\begin{bmatrix} \dot{q} \\ \dot{p} \\ \dot{\varphi} \end{bmatrix} = \begin{bmatrix} 0 & 1 & 0 \\ -1 & 0 & 0 \\ 0 & 0 & -R \end{bmatrix} \begin{bmatrix} \dfrac{\partial H}{\partial q} \\ \dfrac{\partial H}{\partial p} \\ \dfrac{\partial H}{\partial \varphi} \end{bmatrix} + \begin{bmatrix} 0 \\ 0 \\ 1 \end{bmatrix} V
$$

(6.59)

$$
I = \frac{\partial H}{\partial \varphi}
$$

Although at first instance the mechanical and the magnetic part of the system look
decoupled, they are actually coupled via the Hamiltonian

$$
H(q, p, \varphi) = mgq + \frac{p^2}{2m} + \frac{\varphi^2}{2L(q)},
$$

(6.60)

where the inductance $L(q)$ depends on the height q. In fact, the magnetic energy $\frac{\varphi^2}{2L(q)}$
depends both on the flux φ and the mechanical variable q. As a result, the right-hand
side of the second equation (describing the evolution of the mechanical momentum
variable p) depends on the magnetic variable φ, and conversely the right-hand side
of the third equation (describing the evolution of the magnetic variable φ) depends
on the mechanical variable q.

Example 6.3.3 (*Permanent magnet synchronous motor* [242]) A state vector for a permanent magnet synchronous motor (in rotating reference (dq) frame) is defined as

$$x = M \begin{bmatrix} i_d \\ i_q \\ \omega \end{bmatrix}, \qquad M = \begin{bmatrix} L_d & 0 & 0 \\ 0 & L_q & 0 \\ 0 & 0 & \frac{j}{n_p} \end{bmatrix} \tag{6.61}$$

composed of the magnetic flux linkages and mechanical momentum (with i_d, i_q being the currents, and ω the angular velocity), L_d, L_q stator inductances, j the moment of inertia, and n_p the number of pole pairs. The Hamiltonian $H(x)$ is given as $H(x) = \frac{1}{2}x^T M^{-1} x$. This leads to a port-Hamiltonian formulation with $J(x)$, $R(x)$ and $g(x)$ determined as

$$J(x) = \begin{bmatrix} 0 & L_0 x_3 & 0 \\ -L_0 x_3 & 0 & -\Phi_{q0} \\ 0 & \Phi_{q0} & 0 \end{bmatrix},$$

$$R(x) = \begin{bmatrix} R_S & 0 & 0 \\ 0 & R_S & 0 \\ 0 & 0 & 0 \end{bmatrix}, \quad g(x) = \begin{bmatrix} 1 & 0 & 0 \\ 0 & 1 & 0 \\ 0 & 0 & -\frac{1}{n_p} \end{bmatrix}, \tag{6.62}$$

with R_S the stator winding resistance, Φ_{q0} a constant term due to interaction of the permanent magnet and the magnetic material in the stator, and $L_0 := L_d n_p / j$. The three inputs are the stator voltages $\left(v_d, v_q\right)^T$ and the (constant) load torque. Outputs are i_d, i_q, and ω. The system can also operate as a *dynamo*, converting mechanical power into electrical power.

Example 6.3.4 (*Synchronous machine*) The standard eight-dimensional model for the synchronous machine, as described, e.g., in [177], can be written in port-Hamiltonian form as (see [98] for details)

$$\begin{bmatrix} \dot{\psi}_s \\ \dot{\psi}_r \\ \dot{p} \\ \dot{\theta} \end{bmatrix} = \begin{bmatrix} -R_s & 0_{33} & 0_{31} & 0_{31} \\ 0_{33} & -R_r & 0_{31} & 0_{31} \\ 0_{13} & 0_{13} & -d & -1 \\ 0_{13} & 0_{13} & 1 & 0 \end{bmatrix} \begin{bmatrix} \frac{\partial H}{\partial \psi_s} \\ \frac{\partial H}{\partial \psi_r} \\ \frac{\partial H}{\partial p} \\ \frac{\partial H}{\partial \theta} \end{bmatrix} +$$

$$\begin{bmatrix} I_3 & 0_{31} & 0_{31} \\ 0_{33} & e_1 & 0_{31} \\ 0_{13} & 0 & 1 \\ 0_{13} & 0 & 0 \end{bmatrix} \begin{bmatrix} V_s \\ V_f \\ \tau \end{bmatrix} \tag{6.63}$$

$$\begin{bmatrix} I_s \\ I_f \\ \omega \end{bmatrix} = \begin{bmatrix} I_3 & 0_{33} & 0_{31} & 0_{31} \\ 0_{13} & e_1^T & 0 & 0 \\ 0_{13} & 0_{13} & 1 & 0 \end{bmatrix} \begin{bmatrix} \frac{\partial H}{\partial \psi_s} \\ \frac{\partial H}{\partial \psi_r} \\ \frac{\partial H}{\partial p} \\ \frac{\partial H}{\partial \theta} \end{bmatrix},$$

where 0_{lk} denotes the $l \times k$ zero matrix, I_3 denotes the 3×3 identity matrix, and e_1 is the first basis vector of $\mathbb{R}^3$. This defines a port-Hamiltonian input-state-output system with Poisson structure matrix $J(x)$ given by the constant matrix

$$J = \begin{bmatrix} 0_{66} & 0_{62} \\ 0_{26} & \begin{matrix} 0 & -1 \\ 1 & 0 \end{matrix} \end{bmatrix}, \tag{6.64}$$

and resistive structure matrix $R(x)$, which is also constant, having diagonal blocks

$$R_s = \begin{bmatrix} r_s & 0 & 0 \\ 0 & r_s & 0 \\ 0 & 0 & r_s \end{bmatrix}, \quad R_r = \begin{bmatrix} r_f & 0 & 0 \\ 0 & r_{kd} & 0 \\ 0 & 0 & r_{kq} \end{bmatrix}, \quad d, \quad 0, \tag{6.65}$$

denoting, respectively, the *stator resistances, rotor resistances*, and *mechanical friction*. The state variables x of the synchronous machine comprise of

- $\psi_s \in \mathbb{R}^3$, the *stator fluxes*,
- $\psi_r \in \mathbb{R}^3$, the *rotor fluxes*: the first one corresponding to the *field winding* and the remaining two to the *damper windings*,
- p, the *angular momentum* of the rotor,
- θ, the *angle* of the rotor.

Moreover, $V_s \in \mathbb{R}^3$, $I_s \in \mathbb{R}^3$ are the three-phase *stator terminal voltages and currents*, V_f, I_f are the rotor *field winding voltage and current*, and τ, ω are the mechanical *torque* and *angular velocity*.

The synchronous machine is designed depending on two possible modes of operation: synchronous generator or synchronous motor. In the first case, mechanical power is converted to electrical power (supplied to an electrical transmission network); see Fig. 6.4 for a schematic view. Conversely, in the synchronous motor case electrical power is drawn from the power grid in order to deliver mechanical power.

The *Hamiltonian H* (total stored energy of the synchronous machine) is the sum of the magnetic energy of the machine and the kinetic energy of the rotating rotor, given as the sum of the two nonnegative terms

Fig. 6.4 The state and port variables of the synchronous generator

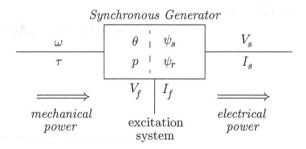

$$H(\psi_s, \psi_r, p, \theta) = \tfrac{1}{2} \begin{bmatrix} \psi_s^T & \psi_r^T \end{bmatrix} L^{-1}(\theta) \begin{bmatrix} \psi_s \\ \psi_r \end{bmatrix} + \tfrac{1}{2J_r} p^2 \tag{6.66}$$
$$= \text{ magnetic energy } H_m + \text{ kinetic energy } H_k,$$

where J_r is the rotational inertia of the rotor, and $L(\theta)$ is an 6×6 *inductance* matrix. In the *round rotor* case (no saliency; cf. [177, 192])

$$L(\theta) = \begin{bmatrix} L_{ss} & L_{sr}(\theta) \\ L_{sr}^T(\theta) & L_{rr} \end{bmatrix} \tag{6.67}$$

where

$$L_{ss} = \begin{bmatrix} L_{aa} & -L_{ab} & -L_{ab} \\ -L_{ab} & L_{aa} & -L_{ab} \\ -L_{ab} & -L_{ab} & L_{aa} \end{bmatrix}, \quad L_{rr} = \begin{bmatrix} L_{ffd} & L_{akd} & 0 \\ L_{akd} & L_{kkd} & 0 \\ 0 & 0 & L_{kkq} \end{bmatrix} \tag{6.68}$$

while

$$L_{sr}(\theta) = \begin{bmatrix} \cos\theta & \cos\theta & -\sin\theta \\ \cos(\theta - \tfrac{2\pi}{3}) & \cos(\theta - \tfrac{2\pi}{3}) & -\sin(\theta - \tfrac{2\pi}{3}) \\ \cos(\theta + \tfrac{2\pi}{3}) & \cos(\theta + \tfrac{2\pi}{3}) & -\sin(\theta + \tfrac{2\pi}{3}) \end{bmatrix} \times$$
$$\begin{bmatrix} L_{afd} & 0 & 0 \\ 0 & L_{akd} & 0 \\ 0 & 0 & L_{akq} \end{bmatrix} \tag{6.69}$$

A crucial feature of the magnetic energy term H_m in the Hamiltonian H is its dependency on the mechanical rotor angle θ; see the formula (6.69) for $L_{sr}(\theta)$. This dependence is responsible for the interaction between the mechanical domain of the generator (the mechanical motion of the rotor) and the electromagnetic domain (the dynamics of the magnetic fields in the rotor and stator), and thus for the functioning of the synchronous machine as an energy-conversion device, transforming mechanical power into electrical power, or conversely (Fig. 6.4).

The synchronous machine is connected to its environment by three types of ports; see Fig. 6.4. In the case of operation as a synchronous *generator*, the scalar *mechanical port* with power variables τ, ω is to be interconnected to a *prime mover*, such as a

Fig. 6.5 DC motor

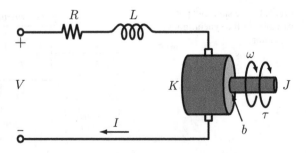

turbine. This port is also used for control purposes, e.g., via so-called *droop control*. Second, there are three *stator terminal ports*, with vectors of power variables V_s, I_s. Third, there is the port with scalar power variables V_f, I_f, which is responsible for the magnetization of the rotor, and which is controlled by an *excitation system*.

Example 6.3.5 (*DC motor*) The system depicted in Fig. 6.5 consists of five ideal modeling subsystems: an inductor L with state φ (flux), a rotational inertia J with state p (angular momentum), a resistor R and friction b, and a gyrator K. The Hamiltonian (corresponding to the linear inductor and inertia) reads as $H(p, \varphi) = \frac{1}{2L}\varphi^2 + \frac{1}{2J}p^2$. The linear resistive relations are $V_R = -RI$, $\tau_d = -b\omega$, with $R, b > 0$ and τ_d a damping torque. The equations of the gyrator (converting magnetic power into mechanical, and conversely) are

$$V_K = -K\omega, \quad \tau = KI \tag{6.70}$$

with K the gyrator constant. The subsystems are interconnected by the equations $V_L + V_R + V_K + V = 0$ (and equal currents), as well as $\tau_J + \tau_d + \tau = 0$ (with common angular velocity), leading to the port-Hamiltonian input-state-output system

$$\begin{bmatrix} \dot{\varphi} \\ \dot{p} \end{bmatrix} = \begin{bmatrix} -R & -K \\ K & -b \end{bmatrix} \begin{bmatrix} \dfrac{\varphi}{L} \\ \dfrac{p}{J} \end{bmatrix} + \begin{bmatrix} 1 \\ 0 \end{bmatrix} V$$

$$I = \begin{bmatrix} 1 & 0 \end{bmatrix} \begin{bmatrix} \dfrac{\varphi}{L} \\ \dfrac{p}{J} \end{bmatrix}. \tag{6.71}$$

Note that, as in the case of the synchronous machine, the system can operate in two modes: either as a motor (converting electrical power into mechanical power) or as a dynamo (converting rotational motion and mechanical power into electrical current and power).

6.4   Properties of Port-Hamiltonian Systems

A crucial property of a port-Hamiltonian system is *cyclo-passivity*, and *passivity* if the Hamiltonian satisfies $H \geq 0$. Apart from this, the port-Hamiltonian formulation also reveals other structural properties. The first one is the existence of *conserved quantities*, which are determined by the structure matrices $J(x)$, $R(x)$.

Definition 6.4.1 A *Casimir* function for an input-state-output port-Hamiltonian system (6.1) or (6.3) is any function $C : \mathcal{X} \to \mathbb{R}$ satisfying

$$\frac{\partial^T C}{\partial x}(x) [J(x) - R(x)] = 0, \quad x \in \mathcal{X} \tag{6.72}$$

It follows that for $u = 0$

$$\frac{d}{dt}C = \frac{\partial^T C}{\partial x}(x)\,[J(x) - R(x)]\,\frac{\partial H}{\partial x}(x) = 0, \tag{6.73}$$

and thus a Casimir function is a *conserved quantity* of the system for $u = 0$, *independently* of the Hamiltonian H. Note furthermore that if $C_1, \ldots, C_r$ are Casimirs, then also the composed function $\Phi(C_1, \ldots, C_r)$ is a Casimir for any $\Phi : \mathbb{R}^r \to \mathbb{R}$. Finally, the existence of Casimirs $C_1, \ldots, C_r$ entails the following invariance property of the dynamics: any subset

$$\{x \mid C_1(x) = c_1, \ldots, C_r(x) = c_r\} \tag{6.74}$$

for arbitrary constants $c_1, \ldots, c_r$ is an invariant subset of the dynamics.

Proposition 6.4.2 $C : \mathcal{X} \to \mathbb{R}$ *is a Casimir function for* (6.1) *or* (6.3) *if and only if*

$$\frac{\partial^T C}{\partial x}(x)J(x) = 0 \text{ and } \frac{\partial^T C}{\partial x}(x)R(x) = 0, \quad x \in \mathcal{X} \tag{6.75}$$

Proof The "if" implication is obvious. For the converse we note that (6.72) implies by skew-symmetry of $J(x)$

$$0 = \frac{\partial^T C}{\partial x}(x)\,[J(x) - R(x)]\,\frac{\partial C}{\partial x}(x) = -\frac{\partial^T C}{\partial x}(x)R(x)\frac{\partial C}{\partial x}(x), \tag{6.76}$$

and therefore, in view of $R(x) \geq 0$, $\frac{\partial^T C}{\partial x}(x)R(x) = 0$, and thus also $\frac{\partial^T C}{\partial x}(x)$ $J(x) = 0$. $\qquad\square$

Hence, the vectors $\frac{\partial C}{\partial x}(x)$ of partial derivatives of the Casimirs C are contained in the intersection of the kernels of the matrices $J(x)$ and $R(x)$ for any $x \in \mathcal{X}$, implying that the maximal number of independent number of Casimirs is always bounded from above by $\dim (\ker J(x) \cap \ker R(x))$. Equality, however, need not be true because of *lack of integrability* of $J(x)$ and/or $R(x)$; see Example 6.4.4 below and the Notes at the end of this chapter.

Example 6.4.3 (*Example* 6.2.1 *continued*) Consider Euler's equations for the angular momenta of a rigid body, with J being given by (6.31) and $R = 0$. It follows that $C(p_1, p_2, p_3) = p_x^2 + p_y^2 + p_z^2$ (the squared total angular momentum) is a Casimir function.

Example 6.4.4 (*Example* 6.2.3 *continued*) The pde's (6.72) for the existence of a Casimir function take the form

$$\frac{\partial C}{\partial p_1} = \frac{\partial C}{\partial p_2} = \frac{\partial C}{\partial \phi} = 0$$

$$\frac{\partial C}{\partial x}\cos\phi + \frac{\partial C}{\partial y}\sin\phi + \frac{\partial C}{\partial \theta} = 0 \tag{6.77}$$

This can be seen *not* to possess a non-trivial solution C, due to the non-holonomicity of the kinematic constraints.

The definition of Casimir C for (6.1) can be further strengthened by requiring that $\frac{d}{dt}C = 0$ for all input values u. This leads to the *stronger* condition

$$\frac{\partial^T C}{\partial x}(x)J(x) = 0, \quad \frac{\partial^T C}{\partial x}(x)R(x) = 0, \quad \frac{\partial^T C}{\partial x}(x)g(x) = 0, \quad x \in \mathcal{X} \qquad (6.78)$$

A second property of the dynamics of port-Hamiltonian systems, which is closely connected to the structure matrix $J(x)$ and its integrability conditions (6.33) is *volume-preservation*. Indeed, consider the case $R(x) = 0$, and let us assume that (6.33) is satisfied with rank $J(x) = \dim \mathcal{X} = n$, implying the existence of local coordinates (q, p) such that (see (6.35))

$$J = \begin{bmatrix} 0 & I_k \\ -I_k & 0 \end{bmatrix} \qquad (6.79)$$

with $n = 2k$. Define the *divergence* of any set of differential equations

$$\dot{x}_i = X_i(x_1, \ldots, x_n), \quad i = 1, \ldots, n, \qquad (6.80)$$

in a set of local coordinates $x_1, \ldots, x_n$ as

$$\operatorname{div}(X)(x) = \sum_{i=1}^{n} \frac{\partial X_i}{\partial x_i}(x) \qquad (6.81)$$

Denote the solution trajectories of (6.80) from $x(0) = x_0$ by $x(t; x_0) = X^t(x_0)$, $t \geq 0$. Then it is a standard fact that the maps $X^t : \mathbb{R}^n \to \mathbb{R}^n$ are *volume-preserving*, that is,

$$\det\left[\frac{\partial X^t}{\partial x}(x)\right] = 1, \quad \text{for all } x, t \geq 0, \qquad (6.82)$$

if and only if $\operatorname{div}(X)(x) = 0$ for all x. Returning to the Hamiltonian dynamics

$$\dot{x} = J(x)\frac{\partial H}{\partial x}(x), \qquad (6.83)$$

with J given by (6.79) it is easily verified that the divergence in the (q, p)-coordinates is everywhere zero, and hence the solutions of (6.83) preserve the standard volume in (q, p)-space. In case rank $J(x) < \dim \mathcal{X}$ and there exist local coordinates (q, p, s) as in (6.35), then the divergence is still zero, and it follows that the Hamiltonian dynamics (6.83) preserves the standard volume in (q, p, s)-space, with the additional property that on any (invariant) level set

Fig. 6.6 LC-circuit

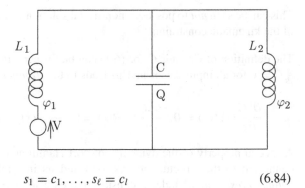

$$s_1 = c_1, \ldots, s_\ell = c_l \tag{6.84}$$

the volume in (q, p)-coordinates is preserved.

Example 6.4.5 (LC-circuit) Consider the LC-circuit (see Fig. 6.6) consisting of two inductors with magnetic energies $H_1(\varphi_1), H_2(\varphi_2)$ (φ_1 and φ_2 being the magnetic flux linkages), and a capacitor with electric energy $H_3(Q)$ (Q being the charge). If the elements are linear then $H_1(\varphi_1) = \frac{1}{2L_1}\varphi_1^2$, $H_2(\varphi_2) = \frac{1}{2L_2}\varphi_2^2$ and $H_3(Q) = \frac{1}{2C}Q^2$. Furthermore, $V = u$ denotes a voltage source. Using Kirchhoff's current and voltage laws one immediately arrives at the port-Hamiltonian system formulation

$$\begin{bmatrix} \dot{Q} \\ \dot{\varphi}_1 \\ \dot{\varphi}_2 \end{bmatrix} = \underbrace{\begin{bmatrix} 0 & 1 & -1 \\ -1 & 0 & 0 \\ 1 & 0 & 0 \end{bmatrix}}_{J} \begin{bmatrix} \frac{\partial H}{\partial Q} \\ \frac{\partial H}{\partial \varphi_1} \\ \frac{\partial H}{\partial \varphi_2} \end{bmatrix} + \begin{bmatrix} 0 \\ 1 \\ 0 \end{bmatrix} u \tag{6.85}$$

$$y = \frac{\partial H}{\partial \varphi_1} \quad (= \text{current through first inductor})$$

with $H(Q, \varphi_1, \varphi_2) := H_1(\varphi_1) + H_2(\varphi_2) + H_3(Q)$ the total energy. Clearly, the matrix J is skew-symmetric, and since J is constant it trivially satisfies (6.33). The quantity $\varphi_1 + \varphi_2$ (total flux linkage) can be seen to be a Casimir function. The volume in $(Q, \varphi_1, \varphi_2)$-space is preserved.

Finally, let us comment on the implications of the port-Hamiltonian structure for the use of *Brockett's necessary condition* for asymptotic stabilizability. Loosely speaking, Brockett's necessary condition [51] tells us that a necessary condition for asymptotic stabilizability of a nonlinear system $\dot{x} = f(x, u), f(0, 0) = 0$, using *continuous* state feedback is that the image of the map $(x, u) \mapsto f(x, u)$, for x and u arbitrarily close to zero, should contain a neighborhood of the origin. Application to input-state-output port-Hamiltonian systems leads to the following necessary condition for asymptotic stabilizability.

Proposition 6.4.6 *Consider the input-state-output port-Hamiltonian system (6.3) with equilibrium x_0. A necessary condition for asymptotic stabilizability around x_0 is that for every $\varepsilon > 0$*

$$\cup_{\{x; \|x-x_0\|<\varepsilon\}} (\operatorname{im}[J(x) - R(x)] + \operatorname{im}[G(x) - P(x)]) = \mathbb{R}^n \tag{6.86}$$

As an application of this result, we note that the port-Hamiltonian system (6.48) arising from a mechanical system with kinematic constraints does never satisfy the necessary condition (6.86). Indeed, by the specific forms of J, $R = 0$, and g, see (6.48) and (6.49),

$$\operatorname{im}[J(x) - R(x)] + \operatorname{im}[G(x) - P(x)] \subset \operatorname{im} \begin{bmatrix} S(q) \\ 0 \end{bmatrix} + \operatorname{im} \begin{bmatrix} 0 \\ I_{n-k} \end{bmatrix} \tag{6.87}$$

where rank $S(q) = n - k$, $q \in \mathcal{Q}$. After possibly reordering the rows of $S(q)$ we may without loss of generality assume that

$$S(q) = \begin{bmatrix} S_1(q) \\ S_2(q) \end{bmatrix} \tag{6.88}$$

with the $(n - k) \times (n - k)$ matrix $S_2(q)$ of full rank $n - k$ in a neighborhood of the equilibrium position vector of interest, and therefore the rows of S_1 depending on the rows of S_2. It follows that vectors of the form $\begin{bmatrix} * \\ 0 \end{bmatrix}$, with 0 the $(n - k)$-dimensional zero-vector, can not be in the image of $S(q)$, and hence not in $\operatorname{im}[J(x) - R(x)] + \operatorname{im}[G(x) - P(x)]$. Hence,

Corollary 6.4.7 *Mechanical systems with kinematic constraints (6.48) are not asymptotically stabilizable using continuous feedback.*

For *holonomic* kinematic constraints this is not surprising, since in this case we should first eliminate the conserved quantities $\bar{q}_{n-k+1} \ldots, \bar{q}_n$ as in (6.39) from the system (6.48). However, since for *nonholonomic* kinematic constraints such an elimination is *not* possible, the above observation indeed entails an important obstruction for asymptotic stabilization[2] of mechanical systems with nonholonomic constraints.

For a further discussion of the dynamical properties of port-Hamiltonian systems, we refer to the extensive literature on this topic; see the references quoted in the Notes at the end of this chapter. Still another use of the port-Hamiltonian structure will be provided separately in the next section.

6.5 Shifted Passivity of Port-Hamiltonian Systems

In many cases of interest, the desired set-point of a port-Hamiltonian system is *not* equal to the minimum of the Hamiltonian function H (an equilibrium of the system for zero-input), but instead is a *steady-state* value corresponding to a nonzero constant

---

[2]However, asymptotic feedback stabilization using *discontinuous* or *time-varying* feedback may still be possible.

input. (We already encountered the same scenario in Chap. 4 in the context of passive systems.) This motivates the following developments.

Proposition 6.5.1 *Consider an input-state-output port-Hamiltonian system with feedthrough terms (6.3), together with a constant input $\bar{u}$ with corresponding steady-state $\bar{x}$ determined by*

$$0 = [J(\bar{x}) - R(\bar{x})] \frac{\partial H}{\partial x}(\bar{x}) + [G(\bar{x}) - P(\bar{x})]\bar{u} \tag{6.89}$$

Denote

$$\bar{y} = [G(\bar{x}) + P(\bar{x})]^T \frac{\partial H}{\partial x}(\bar{x}) + [M(\bar{x}) + S(\bar{x})]\bar{u} \tag{6.90}$$

Suppose we can find coordinates x in which the system matrices $J(x), M(x), R(x), P(x), S(x), G(x)$ are all constant. Then the system can be rewritten as

$$\begin{aligned}
\dot{x} &= [J - R] \frac{\partial \widehat{H}_{\bar{x}}}{\partial x}(x) + [G - P](u - \bar{u}) \\
y - \bar{y} &= [G + P]^T \frac{\partial \widehat{H}_{\bar{x}}}{\partial x}(x) + [M + S](u - \bar{u})
\end{aligned} \tag{6.91}$$

with respect to the shifted Hamiltonian[3] defined as

$$\widehat{H}_{\bar{x}}(x) := H(x) - \frac{\partial^T H}{\partial x}(\bar{x})(x - \bar{x}) - H(\bar{x}) \tag{6.92}$$

If H is convex in the coordinates x, then $\widehat{H}_{\bar{x}}$ has a minimum at $x = \bar{x}$ (with value 0), and the port-Hamiltonian system is passive with respect to the shifted supply rate $s(u,y) = (u - \bar{u})^T(y - \bar{y})$, with storage function $\widehat{H}_{\bar{x}}$.

Proof Observe that

$$\frac{\partial \widehat{H}_{\bar{x}}}{\partial x}(x) = \frac{\partial H}{\partial x}(x) - \frac{\partial H}{\partial x}(\bar{x}) \tag{6.93}$$

Adopting the shorthand notation $z = \frac{\partial H}{\partial x}(x)$ and $\bar{z} = \frac{\partial H}{\partial x}(\bar{x})$, we obtain

$$\begin{aligned}
\tfrac{d}{dt}\widehat{H}_{\bar{x}} &= (z - \bar{z})^T [(J - R)z + (G - P)u] \\
&= (z - \bar{z})^T [(J - R)(z - \bar{z}) + (G - P)(u - \bar{u})] \\
&= -\left[(z - \bar{z})^T \ (u - \bar{u})^T\right] \begin{bmatrix} R & P \\ P^T & S \end{bmatrix} \begin{bmatrix} (z - \bar{z}) \\ (u - \bar{u}) \end{bmatrix} + (u - \bar{u})^T(y - \bar{y}) \\
&\le (u - \bar{u})^T(y - \bar{y}),
\end{aligned} \tag{6.94}$$

showing passivity with respect to the shifted supply rate $(u - \bar{u})^T(y - \bar{y})$. Finally, $\widehat{H}_{\bar{x}}(\bar{x}) = 0$ and convexity of H is equivalent to

---

[3]Note that the function $\widehat{H}_{\bar{x}}$ admits the following geometric interpretation. Consider the surface in $\mathbb{R}^{n+1}$ defined by H, and the tangent plane at the point $(\bar{x}, H(\bar{x})) \in \mathbb{R}^n$ to this surface. Then $\widehat{H}_{\bar{x}}(x)$ is the vertical distance above the point $x \in \mathbb{R}^n$ from this tangent plane to the surface.

$$H(x) \geq \frac{\partial^T H}{\partial x}(\bar{x})(x - \bar{x}) + H(\bar{x}), \text{ for all } x, \bar{x}, \tag{6.95}$$

implying that $\widehat{H}_{\bar{x}}(x) \geq 0, \ x \in \mathcal{X}.$ □

Recall from Chap. 4, cf. (4.159), that the property of being passive with respect to the shifted supply rate $(u - \bar{u})^T (y - \bar{y})$ is referred to as *shifted passivity*. It follows from Proposition 6.5.1 that with *constant* system matrices J, M, R, P, S, G and a *convex* Hamiltonian H the input-state-output port-Hamiltonian system with feedthrough term is shifted passive with respect to *any* constant $\bar{u}$ for which there exists a steady-state $\bar{x}$ (and corresponding $\bar{y}$).

Remark 6.5.2 The function $\widehat{H}_{\bar{x}}(x)$, regarded as a function of x *and* $\bar{x}$, is known in convex analysis as the *Bregman divergence* or *Bregman distance*. It also appears as the *availability function* in thermodynamics (dating back to the classical work of Gibbs [113]), and was introduced in the present context in [146]. Note that the definition of $\widehat{H}_{\bar{x}}$ (as well as the notion of a convex function) depends on the choice of coordinates x for the state space $\mathcal{X}$.

Example 6.5.3 Consider the model of a power network formulated in Example 4.4.4; see (4.87). Identifying the Hamiltonian with the storage function already defined[4] in (4.88)

$$H(q, p) = \frac{1}{2} p^T J^{-1} p - \sum_{j=1}^{M} \gamma_j \cos q_j, \tag{6.96}$$

the system takes the port-Hamiltonian form

$$\begin{bmatrix} \dot{q} \\ \dot{p} \end{bmatrix} = \begin{bmatrix} 0 & D^T \\ -D & -A \end{bmatrix} \begin{bmatrix} \frac{\partial H}{\partial q}(q, p) \\ \frac{\partial H}{\partial p}(q, p) \end{bmatrix} + \begin{bmatrix} 0 \\ u \end{bmatrix} \tag{6.97}$$

$$y = \frac{\partial H}{\partial p}(q, p)$$

The steady-state $(\bar{q}, \bar{p})$ corresponding to constant input $\bar{u}$ is determined by

$$0 = D^T \frac{\partial H}{\partial p}(\bar{q}, \bar{p})$$
$$0 = D \frac{\partial H}{\partial q}(\bar{q}, \bar{p}) + A \frac{\partial H}{\partial p}(\bar{q}, \bar{p}) + \bar{u} \tag{6.98}$$

Assuming the graph to be *connected* the first equation leads to $\frac{\partial H}{\partial p}(\bar{q}, \bar{p}) = \mathbb{1} \omega_*$, with $\omega_* \in \mathbb{R}$ a common frequency deviation. Furthermore, by premultiplying the second equation by the row-vector $\mathbb{1}^T$ of all ones,

$$0 = \omega_* \sum_{i=1}^{N} A_i + \sum_{i=1}^{N} \bar{u}_i, \quad \bar{p} = J \mathbb{1} \omega_*, \tag{6.99}$$

---

[4]The matrix J in the Hamiltonian refers to the inertia of the generators; not to be confused with the Poisson structure.

determining ω_*, and thus $\bar{p}$, as a function of the total generated and consumed power $\sum \bar{u}_i$. Finally, the steady-state vector $\bar{q}$ of phase angle differences is determined by

$$0 = D\Gamma \text{Sin} \, \bar{q} - A\mathbb{1} \frac{\sum_{i=1}^{N} \bar{u}_i}{\sum_{i=1}^{N} A_i} + \bar{u} \qquad (6.100)$$

(Note that by boundedness of the mapping Sin this does not have a solution for large $\bar{u}$.) Defining the shifted Hamiltonian as in (6.92) yields

$$\begin{aligned} \widehat{H}_{(\bar{q},\bar{p})}(q, p) = {} & \tfrac{1}{2}(p - \bar{p})^T J^{-1}(p - \bar{p}) - \sum_{j=1}^{M} \gamma_j \cos q_j \\ & - \sum_{j=1}^{M} \gamma_j \sin \bar{q}_j (q_j - \bar{q}_j) + \sum_{j=1}^{M} \gamma_j \cos \bar{q}_j \end{aligned} \qquad (6.101)$$

It follows that the system is shifted passive with respect to the shifted supply rate $(u - \bar{u})^T (y - \bar{y})$ and storage function $\widehat{H}_{(\bar{q},\bar{p})}$, with $\bar{y} = \frac{\partial H}{\partial p}(\bar{q}, \bar{p}) = J^{-1}\bar{p} = \mathbb{1}\omega_*$.

If no coordinates exist in which the matrices $J(x), M(x), R(x), P(x), S(x), G(x)$ are all constant, the analysis for nonzero $\bar{u}$ becomes much harder. Define the combined interconnection and resistive structure matrix

$$K(x) := \begin{bmatrix} -J(x) + R(x) & -G(x) + P(x) \\ G^T(x) + P^T(x) & M(x) + S(x) \end{bmatrix} \qquad (6.102)$$

Proposition 6.5.4 *Consider an input-state-output port-Hamiltonian system with feedthrough terms (6.3), and a steady-state triple $\bar{u}, \bar{x}, \bar{y}$. Then the system is shifted passive with storage function $\widehat{H}_{\bar{x}}$ if*[5]

$$\left[\frac{\partial^T H}{\partial x}(x) - \frac{\partial^T H}{\partial x}(\bar{x}) \; u^T - \bar{u}^T \right] \left(K(x) \begin{bmatrix} \frac{\partial H}{\partial x}(x) \\ u \end{bmatrix} - K(\bar{x}) \begin{bmatrix} \frac{\partial H}{\partial x}(\bar{x}) \\ \bar{u} \end{bmatrix} \right) \geq 0 \qquad (6.103)$$

for all x, u.

Proof By direct computation of $\frac{d}{dt}\widehat{H}_{\bar{x}}$; see [97]. $\square$

Furthermore, shifting with respect to constant inputs $\bar{u}$ can still be done if the input matrix $g(x)$ of the port-Hamiltonian system (6.1) satisfies the following *integrability condition* with respect to the combined geometric structure $J(x) - R(x)$. Assume that for each j-th column $g_j(x)$ of the input matrix $g(x)$ there exists a function $C_j : \mathcal{X} \to \mathbb{R}$ such that

$$g_j(x) = -[J(x) - R(x)] \frac{\partial F_j}{\partial x}(x), \quad j = 1, \dots, m \qquad (6.104)$$

Then for any constant $\bar{u}$, the dynamics of the port-Hamiltonian system (6.1) can be rewritten as

---

[5]Since $K(x) + K^T(x) \geq 0$ the condition (6.103) is automatically satisfied in case K does not depend on x.

$$\dot{x} = [J(x) - R(x)] \frac{\partial \widetilde{H}}{\partial x}(x) + g(x)(u - \bar{u}), \qquad (6.105)$$

with

$$\widetilde{H}(x) := H(x) - \sum_{j=1}^{m} F_j(x)\bar{u}_j \qquad (6.106)$$

(In case of constant matrices J, R, g one verifies that $\widetilde{H}(x) = \widehat{H}_{\bar{x}}$, where $\bar{x}$ is the steady-state corresponding to $\bar{u}$.) However, in general this does *not* imply passivity with respect to the shifted supply rate $(u - \bar{u})^T (y - \bar{y})$, with $\bar{y}$ the steady-state output value. Nevertheless, the condition *is* useful especially in case part of the inputs can be considered as constant "disturbances" $\bar{u}$, with remaining other inputs v corresponding to the dynamics

$$\dot{x} = [J(x) - R(x)] \frac{\partial H}{\partial x}(x) + b(x)v + g(x)\bar{u} \qquad (6.107)$$

for some input matrix $b(x)$. In this case, satisfaction of (6.104) allows one to rewrite the system as

$$\dot{x} = [J(x) - R(x)] \frac{\partial \widetilde{H}}{\partial x}(x) + b(x)v \qquad (6.108)$$

which is port-Hamiltonian with respect to the inputs v and corresponding outputs $z = b^T(x)\frac{\partial \widetilde{H}}{\partial x}(x)$.

Finally, let us come back to the notion of the *steady-state input–output relation* as defined in Chap. 4, cf. (4.31). In the port-Hamiltonian case, one obtains the following result

Proposition 6.5.5 *Consider an input-state-output port-Hamiltonian system with feedthrough terms (6.3). Its steady-state input–output relation is given as*

$$\{(\bar{u}, \bar{y}) \mid \exists \bar{x} \text{ s.t. } 0 = [J(\bar{x}) - R(\bar{x})]\frac{\partial H}{\partial x}(\bar{x}) + [G(\bar{x}) - P(\bar{x})]\bar{u}, \\ \bar{y} = [G(\bar{x}) + P(\bar{x})]^T \frac{\partial H}{\partial x}(\bar{x}) + [M(\bar{x}) + S(\bar{x})]\bar{u}\} \qquad (6.109)$$

In particular, if $[J(\bar{x}) - R(\bar{x})]$ is invertible, the steady-state input–output relation is given as the graph of the mapping (from $\bar{u}$ to $\bar{y}$)

$$\bar{y} = -[G(\bar{x}) + P(\bar{x})]^T (J(\bar{x}) - R(\bar{x}))^{-1} [G(\bar{x}) - P(\bar{x})]\bar{u} + [M(\bar{x}) + S(\bar{x})]\bar{u} \qquad (6.110)$$

which is linear in case the matrices J, R, G, P, M, S are all constant.

Note that the matrix in (6.110) is equal to the Schur complement of the matrix $K(\bar{x})$ defined in (6.102) with respect to its left-upper block. Since the symmetric part of $K(\bar{x})$ is ≥ 0, this Schur complement inherits the same positivity property.

Proposition 6.5.5 can be extended to input-state-output port-Hamiltonian systems with *nonlinear resistive structure* as in Definition 6.1.4. For example, the steady-state input–output relation corresponding to the port-Hamiltonian system

$$\dot{x} = -\mathcal{R}(z) + u, \quad y = z, \quad z = \frac{\partial H}{\partial x}(x) \tag{6.111}$$

with the nonlinear resistive mapping $\mathcal{R}$ satisfying (6.7), is given by

$$\{(\bar{u}, \bar{y}) \mid \bar{u} = \mathcal{R}(\bar{y})\} \tag{6.112}$$

provided for each $\bar{u}$ there exists a steady-state $\bar{x}$ such that $\bar{u} = \mathcal{R}(\frac{\partial H}{\partial x}(\bar{x}))$.

6.6 Dirac Structures

In Chap. 3, the definition of *dissipativity* was extended to differential-algebraic equation (DAE) systems $F(\dot{x}, x, w) = 0$, with w denoting the vector of external variables (inputs *and* outputs).

Similarly, in this and the next section we show how the definition of input-state-output port-Hamiltonian systems can be extended to the DAE case. This extension is crucial from a modeling point of view, since first principles modeling of physical systems often leads to DAE systems. This stems from the fact that in many modeling approaches the system under consideration is naturally regarded as obtained from interconnecting simpler subsystems. These interconnections often give rise to algebraic constraints between the state space variables of the subsystems; thus leading to DAE systems.

The key to define port-Hamiltonian DAE systems is the geometric notion of a *Dirac structure*, formalizing the concept of a power-conserving interconnection, and generalizing the notion of an (almost-)Poisson structure matrix $J(x)$ as encountered before.

Let us return to the basic setting of passivity (see Chap. 2), starting with a finite-dimensional linear space and its dual, with the duality product defining power. Thus, let $\mathcal{F}$ be an ℓ-dimensional linear space, and denote its dual (the space of linear functions on $\mathcal{F}$) by $\mathcal{E} := \mathcal{F}^*$. We call $\mathcal{F}$ the space of *flows* f, and $\mathcal{E}$ the space of *efforts* e. On the product space $\mathcal{F} \times \mathcal{E}$, *power* is defined by

$$< e \mid f >, \quad (f, e) \in \mathcal{F} \times \mathcal{E}, \tag{6.113}$$

where $< e \mid f >$ denotes the *duality product*, that is, the linear function $e \in \mathcal{E} = \mathcal{F}^*$ acting on $f \in \mathcal{F}$.

Remark 6.6.1 Recall from Chap. 2 that if $\mathcal{F}$ is endowed with an *inner-product* structure $<, >$, then $\mathcal{E} = \mathcal{F}^*$ can be *identified* with $\mathcal{F}$ in such a way that $< e \mid f > = < e, f >$, $f \in \mathcal{F}$, $e \in \mathcal{E} \simeq \mathcal{F}$.

Example 6.6.2 Let $\mathcal{F}$ be the space of generalized *velocities*, and $\mathcal{E} = \mathcal{F}^*$ the space of generalized *forces*, then $< e \mid f >$ is mechanical power. Similarly, let $\mathcal{F}$ be the space of *currents*, and $\mathcal{E} = \mathcal{F}^*$ be the space of *voltages*, then $< e \mid f >$ is electrical power.

In multi-body systems one considers the space of *twists* $\mathcal{F} = se(3)$ (the Lie algebra of the matrix special Euclidian group $SE(3)$), with $\mathcal{E} = \mathcal{F}^* = se^*(3)$ the space of *wrenches*.

As already introduced in Sect. 2.4, there exists on $\mathcal{F} \times \mathcal{E}$ a canonically defined symmetric bilinear form

$$\ll (f_1, e_1), (f_2, e_2) \gg := < e_1 \mid f_2 > + < e_2 \mid f_1 > \qquad (6.114)$$

for $f_i \in \mathcal{F}$, $e_i \in \mathcal{E}, i = 1, 2$. Now consider a subspace

$$\mathcal{D} \subset \mathcal{F} \times \mathcal{E} \qquad (6.115)$$

and its orthogonal companion $\mathcal{D}^{\perp\!\!\!\perp}$ with respect to the bilinear form $\ll , \gg$ on $\mathcal{F} \times \mathcal{E}$, defined as

$$\mathcal{D}^{\perp\!\!\!\perp} = \{(f, e) \in \mathcal{F} \times \mathcal{E} \mid \ll (f, e), (\tilde{f}, \tilde{e}) \gg = 0 \text{ for all } (\tilde{f}, \tilde{e}) \in \mathcal{D}\} \qquad (6.116)$$

Clearly, if $\mathcal{D}$ has dimension d, then the subspace $\mathcal{D}^{\perp\!\!\!\perp}$ has dimension $2 \dim \mathcal{F} - d$ (since $\ll , \gg$ is a non-degenerate form on $\mathcal{F} \times \mathcal{E}$, and furthermore $\dim \mathcal{F} \times \mathcal{E} = 2 \dim \mathcal{F}$).

Definition 6.6.3 A subspace $\mathcal{D} \subset \mathcal{F} \times \mathcal{E}$ is a (constant) Dirac structure if

$$\mathcal{D} = \mathcal{D}^{\perp\!\!\!\perp} \qquad (6.117)$$

It immediately follows that the dimension of any Dirac structure $\mathcal{D}$ is equal to $\dim \mathcal{F}$. Furthermore, let $(f, e) \in \mathcal{D} = \mathcal{D}^{\perp\!\!\!\perp}$. Then by (6.114)

$$0 = \ll (f, e), (f, e) \gg = 2 < e \mid f > = 0 \qquad (6.118)$$

Hence, a Dirac structure $\mathcal{D}$ defines a *power-conserving* relation between the variables $(f, e) \in \mathcal{F} \times \mathcal{E}$. Conversely, we obtain

Proposition 6.6.4 *Let $\mathcal{F}$ be a finite-dimensional linear space. Then $\mathcal{D} \subset \mathcal{F} \times \mathcal{E}$ is a Dirac structure if and only if $< e \mid f >= 0$ for all $(f, e) \in \mathcal{D}$, and $\mathcal{D}$ is a maximal subspace with this property. In particular, for any subspace $\mathcal{D} \subset \mathcal{F} \times \mathcal{E}$ satisfying $< e \mid f >= 0$ for all $(f, e) \in \mathcal{D}$ we have $\dim \mathcal{D} \leq \dim \mathcal{F}$, while $\mathcal{D}$ satisfying $< e \mid f >= 0$ for all $(f, e) \in \mathcal{D}$ is a Dirac structure if and only if $\dim \mathcal{D} = \dim \mathcal{F}$.*

Proof First, consider any subspace $\mathcal{D} \subset \mathcal{F} \times \mathcal{E}$ satisfying $< e \mid f >= 0$ for all $(f, e) \in \mathcal{D}$. Let $(f_1, e_1), (f_2, e_2) \in \mathcal{D}$. Then also $(f_1 + f_2, e_1 + e_2) \in \mathcal{D}$, and thus

$$\begin{aligned} 0 &= < e_1 + e_2 \mid f_1 + f_2 > = \\ &\quad < e_1 \mid f_2 > + < e_2 \mid f_1 > + < e_1 \mid f_1 > + < e_2 \mid f_2 > = \qquad (6.119) \\ &\quad < e_1 \mid f_2 > + < e_2 \mid f_1 > = \ll (f_1, e_1), (f_2, e_2) \gg \end{aligned}$$

Hence, $\mathcal{D} \subset \mathcal{D}^{\perp\perp}$. In view of (6.118), we have thus proved that $< e \mid f >= 0$ for all $(f, e) \in \mathcal{D}$ if and only if $\mathcal{D} \subset \mathcal{D}^{\perp\perp}$. Furthermore, $\mathcal{D} \subset \mathcal{D}^{\perp\perp}$ implies $\dim \mathcal{D} \leq \dim \mathcal{D}^{\perp\perp} = 2 \dim \mathcal{F} - \dim \mathcal{D}$, and hence $\dim \mathcal{D} \leq \dim \mathcal{F}$. Conversely, if $\dim \mathcal{D} = \dim \mathcal{F}$ then $\mathcal{D} = \mathcal{D}^{\perp\perp}$, and $\mathcal{D}$ is a Dirac structure. Hence, we have proved the second claim, and the "only if" direction of the first claim using (6.118). For the "if" direction of the first claim we use again that $< e \mid f >= 0$ for all $(f, e) \in \mathcal{D}$ implies $\mathcal{D} \subset \mathcal{D}^{\perp\perp}$. Now suppose that $\mathcal{D} \subsetneq \mathcal{D}^{\perp\perp}$. Then we can non-trivially *extend* $\mathcal{D}$ to a subspace $\mathcal{D}'$ such that $\mathcal{D}' \subset \mathcal{D}'^{\perp\perp}$, and thus $\mathcal{D}$ is not maximal. $\qquad\square$

Remark 6.6.5 The condition $\dim \mathcal{D} = \dim \mathcal{F}$ is intimately related to the statement that a physical interconnection can *not* determine at the same time both the flow and effort (e.g., current *and* voltage, or velocity *and* force).

Constant Dirac structures admit different *matrix representations*.

Proposition 6.6.6 *Let $\mathcal{D} \subset \mathcal{F} \times \mathcal{E}$, with $\dim \mathcal{F} = \ell$, be a constant Dirac structure. Take linear coordinates for $\mathcal{F}$ and dual coordinates for $\mathcal{E} = \mathcal{F}^*$, resulting in $\mathcal{F} \simeq \mathbb{R}^m \simeq \mathcal{E}$. Then $\mathcal{D}$ can be represented in any of the following ways.*

1. *(Kernel and Image representation)*

$$\mathcal{D} = \{(f, e) \in \mathcal{F} \times \mathcal{E} \mid Ff + Ee = 0\} \tag{6.120}$$

for $\ell \times \ell$ matrices[6] F and E satisfying

$$\begin{aligned} &(i)\ \ EF^T + FE^T = 0 \\ &(ii)\ \text{rank}\,[F \vdots E] = \ell \end{aligned} \tag{6.121}$$

Equivalently in image representation,

$$\mathcal{D} = \{(f, e) \in \mathcal{F} \times \mathcal{E} \mid \exists \lambda \in \mathbb{R}^\ell\ s.t.\ f = E^T \lambda,\ e = F^T \lambda\} \tag{6.122}$$

Conversely, for any $\ell \times \ell$ matrices F and E satisfying (6.121), the subspaces (6.120) and (6.122) are Dirac structures.
2. *(Constrained input–output representation)*

$$\mathcal{D} = \{(f, e) \in \mathcal{F} \times \mathcal{E} \mid \exists \lambda\ s.t.\ f = Je + G\lambda,\ G^T e = 0\} \tag{6.123}$$

for an $\ell \times \ell$ skew-symmetric matrix J, and a matrix G such that $\text{im}\,G = \{f \mid (f, 0) \in \mathcal{D}\}$. Furthermore, $\ker J = \{e \mid (0, e) \in \mathcal{D}\}$. Conversely, for any G and skew-symmetric J the subspace (6.123) is a Dirac structure.
3. *(Hybrid input–output representation).*
   Let $\mathcal{D}$ be given as in (6.120). Suppose $\text{rank}\,F = \ell^1 \leq \ell$. Select ℓ^1 independent

---

[6]We may also allow F and E to be $l' \times l$ matrices with $l' \geq l$, and satisfying (6.121). This is called a *relaxed* kernel representation.

columns of F, and group them into a matrix F^1. Write (possibly after permutations) $F = [F^1 \vdots F^2]$, and correspondingly $E = [E^1 \vdots E^2]$, $f = \begin{bmatrix} f^1 \\ f^2 \end{bmatrix}$, $e = \begin{bmatrix} e^1 \\ e^2 \end{bmatrix}$. Then the matrix $[F^1 \vdots E^2]$ can be shown to be invertible, and

$$\mathcal{D} = \left\{ \begin{bmatrix} f^1 \\ f^2 \end{bmatrix}, \begin{bmatrix} e^1 \\ e^2 \end{bmatrix} \mid \begin{bmatrix} f^1 \\ e^2 \end{bmatrix} = J \begin{bmatrix} e^1 \\ f^2 \end{bmatrix} \right\} \tag{6.124}$$

where $J := -[F^1 \vdots E^2]^{-1}[F^2 \vdots E^1]$ is skew-symmetric. Conversely, for any skew-symmetric J the subspace (6.124) is a Dirac structure.

4. *(Canonical coordinate representation)*
   There exist linear coordinates (q, p, r, s) for $\mathcal{F}$ such that in these coordinates and dual coordinates for $\mathcal{E} = \mathcal{F}^*$, $(f, e) = (f_q, f_p, f_r, f_s, e_q, e_p, e_r, e_s) \in \mathcal{D}$ if and only if

$$\begin{aligned} f_q &= e_p, \quad f_p = -e_q \\ f_r &= 0, \quad e_s = 0 \end{aligned} \tag{6.125}$$

Proof (1) It is directly checked that (6.122) defines a Dirac structure. Since by (6.121) im $\begin{bmatrix} E^T \\ F^T \end{bmatrix}$ = ker $[F \vdots E]$, also (6.120) defines the same Dirac structure. Conversely, any ℓ-dimensional subspace $\mathcal{D}$ can be written as $\mathcal{D} = $ im $\begin{bmatrix} E^T \\ F^T \end{bmatrix}$ for some

$\ell \times \ell$ matrices F, E satisfying rank $[F \vdots E] = \ell$. If $\mathcal{D}$ is a Dirac structure then $0 = e^T f = (F^T \lambda)^T E^T \lambda = \lambda^T FE^T \lambda$ for all $\lambda \in \mathbb{R}^\ell$. This is equivalent to $EF^T + FE^T = 0$.
(2) Consider $\mathcal{D}$ given by (6.123) with $J = -J^T$. Then $e^T f = e^T(Je + G\lambda) = e^T Je + e^T G\lambda = (G^T e)^T \lambda = 0$. Hence, $\mathcal{D} \subset \mathcal{D}^{\perp\!\perp}$. Let now $(\tilde{f}, \tilde{e})$ be such that $0 = \ll (f, e), (\tilde{f}, \tilde{e}) \gg$ for all $(f, e) \in \mathcal{D}$, i.e., $f = Je + G\lambda$, $G^T e = 0$. Then

$$0 = e^T \tilde{f} + \tilde{e}^T f = e^T \tilde{f} + \tilde{e}^T (Je + G\lambda)$$

for all λ and e with $G^T e = 0$. First take $e = 0$. Then $0 = \tilde{e}^T G\lambda$ for all λ, implying that $G^T \tilde{e} = 0$. Hence, $0 = e^T \tilde{f} + \tilde{e}^T Je = e^T(\tilde{f} - J\tilde{e})$ for all e with $G^T e = 0$, implying that $\tilde{f} = J\tilde{e} + G\tilde{\lambda}$, for some $\tilde{\lambda}$. Thus $\mathcal{D}^{\perp\!\perp} \subset \mathcal{D}$, and therefore $\mathcal{D}^{\perp\!\perp} = \mathcal{D}$. On the other hand, take any Dirac structure $\mathcal{D} \subset \mathcal{F} \times \mathcal{E}$. Define the following subspace of $\mathcal{E}$

$$\mathcal{E}_{\mathcal{D}} = \{ e \in \mathcal{E} \mid \exists f \text{ s.t. } (f, e) \in \mathcal{D} \} \tag{6.126}$$

It can be checked that

$$\mathcal{E}_{\mathcal{D}}^\perp = \{ f \in \mathcal{F} \mid (f, 0) \in \mathcal{D} \}, \tag{6.127}$$

where $^\perp$ denotes orthogonality with respect to the duality product $< \mid >$. Furthermore, select any subspace $\bar{\mathcal{E}}_{\mathcal{D}}$ complementary to $\mathcal{E}_{\mathcal{D}} \subset \mathcal{E}$, i.e.,

$$\mathcal{E} = \mathcal{E}_D \oplus \bar{\mathcal{E}}_D$$

Define any matrix G such that $\mathcal{E}_D = \ker G^T$. Define the linear map $J : \mathcal{E} \to \mathcal{F}$ as follows. Define J to be zero on $\bar{\mathcal{E}}_D$. In view of (6.127), there exists for any $e \in \mathcal{E}_D$ a unique $f \in \bar{\mathcal{E}}_D^\perp$ such that $(f, e) \in \mathcal{D}$. Define $Je = f$. Since $(f, e) \in \mathcal{D}$ we have $e^T f = 0$, implying skew-symmetry of J. It is readily checked that $\mathcal{D}$ is given as in (6.123).
(3) By skew-symmetry of J it directly follows that $\mathcal{D}$ defined by (6.124) is a Dirac structure. With regard to all remaining statements, see [47].
(4) See [72].   □

Remark 6.6.7 One may also convert any matrix representation into any other one. For example, start from a Dirac structure $\mathcal{D}$ given in constrained input–output representation (6.123). Define $G^\perp$ as a matrix of maximal rank such that $G^\perp G = 0$ and with independent rows. Then $\mathcal{D}$ is equivalently given in kernel representation as

$$\mathcal{D} = \{(f, e) \mid \begin{bmatrix} -G^\perp \\ 0 \end{bmatrix} f + \begin{bmatrix} G^\perp J \\ G^T \end{bmatrix} e = 0\} \tag{6.128}$$

Example 6.6.8 The combination of Kirchhoff's current and voltage laws for an electrical circuit constitute an example of a constrained input–output representation (6.123) of a Dirac structure. Let $\mathcal{F}$ be the space of currents I through the edges of the circuit graph, and $\mathcal{E} = \mathcal{F}^*$ the space of voltages V across the edges. Let D be the $N \times M$ incidence matrix of the circuit graph (N nodes/vertices, M branches/edges). Then Kirchhoff's current and voltage laws define the Dirac structure

$$\mathcal{D} := \{(I, V) \in \mathbb{R}^M \times \mathbb{R}^M \mid DI = 0, \exists \lambda \in \mathbb{R}^N \text{ s.t. } V = D^T \lambda\}, \tag{6.129}$$

which is in constrained input–output representation (6.123), with $J = 0$ and $G = D^T$. Defining a matrix E such that $\operatorname{im} D^T = \ker E$ one obtains the relaxed kernel representation $DI = 0, EV = 0$.

Given a Dirac structure $\mathcal{D} \subset \mathcal{F} \times \mathcal{E}$, one can define the following subspaces of $\mathcal{F}$, respectively, $\mathcal{E}$,

$$\begin{aligned}
\mathcal{G}_0 &:= \{f \in \mathcal{F} \mid (f, 0) \in \mathcal{D}\} \\
\mathcal{G}_1 &:= \{f \in \mathcal{F} \mid \exists e \in \mathcal{E} \text{ s.t. } (f, e) \in \mathcal{D}\} \\
\mathcal{P}_0 &:= \{e \in \mathcal{E} \mid (0, e) \in \mathcal{D}\} \\
\mathcal{P}_1 &:= \{e \in \mathcal{E} \mid \exists f \in \mathcal{F} \text{ s.t. } (f, e) \in \mathcal{D}\}
\end{aligned} \tag{6.130}$$

It can be readily checked that

$$\begin{aligned}
\mathcal{P}_0 &= \mathcal{G}_1^\perp := \{e \in \mathcal{E} \mid < e \mid f >= 0, \ \forall f \in \mathcal{G}_1\} \\
\mathcal{P}_1 &= \mathcal{G}_0^\perp := \{e \in \mathcal{E} \mid < e \mid f >= 0, \ \forall f \in \mathcal{G}_0\}
\end{aligned} \tag{6.131}$$

With $\mathcal{D}$ expressed in kernel/image representation (6.120), (6.122) one obtains

$$\mathcal{G}_1 = \operatorname{im} E^T, \quad \mathcal{P}_0 = \ker E$$
$$\mathcal{P}_1 = \operatorname{im} F^T, \quad \mathcal{G}_0 = \ker F \tag{6.132}$$

The subspace $\mathcal{G}_1$ expresses the set of *admissible flows* f, and $\mathcal{P}_1$ the set of *admissible efforts* e. The first subspace will turn out to be instrumental in the determination of the *Casimirs* of a port-Hamiltonian DAE system in the next section, and the second subspace in the characterization of its *algebraic constraints*.

Another key property of Dirac structures is the fact that the *composition* of Dirac structures is again a Dirac structure. In the next section, this will lead to the fundamental property that any power-conserving interconnection of port-Hamiltonian DAE systems defines another port-Hamiltonian DAE system. We will start by showing that the composition of two Dirac structures is again a Dirac structure. This readily implies that the power-conserving interconnection of any number of Dirac structures is a Dirac structure.

Thus let us consider a Dirac structure $\mathcal{D}_A \subset \mathcal{F}_1 \times \mathcal{F}_2 \times \mathcal{E}_1 \times \mathcal{E}_2$, and another Dirac structure $\mathcal{D}_B \subset \mathcal{F}_2 \times \mathcal{F}_3 \times \mathcal{E}_2 \times \mathcal{E}_3$. The space $\mathcal{F}_2$ is the space of *shared flow* variables, and $\mathcal{E}_2$ is the space of *shared effort* variables; see Fig. 6.7.

Consider the interconnection equations (the minus sign included for a consistent power flow convention)

$$f_A = -f_B \in \mathcal{F}_2, \quad e_A = e_B \in \mathcal{E}_2 \tag{6.133}$$

Then the *composition* $\mathcal{D}_A \circ \mathcal{D}_B$ of the Dirac structures $\mathcal{D}_A$ and $\mathcal{D}_B$ is defined as

$$\mathcal{D}_A \circ \mathcal{D}_B := \Big\{ (f_1, e_1, f_3, e_3) \in \mathcal{F}_1 \times \mathcal{E}_1 \times \mathcal{F}_3 \times \mathcal{E}_3 \mid \exists (f_2, e_2) \in \mathcal{F}_2 \times \mathcal{E}_2 \\ \text{s.t. } (f_1, e_1, f_2, e_2) \in \mathcal{D}_A \text{ and } (-f_2, e_2, f_3, e_3) \in \mathcal{D}_B \Big\} \tag{6.134}$$

The next theorem is proved in [63].

Theorem 6.6.9 *Let* $\mathcal{D}_A \subset \mathcal{F}_1 \times \mathcal{E}_1 \times \mathcal{F}_2 \times \mathcal{E}_2$ *and* $\mathcal{D}_B \subset \mathcal{F}_2 \times \mathcal{E}_2 \times \mathcal{F}_3 \times \mathcal{E}_3$ *be Dirac structures. Then* $\mathcal{D}_A \circ \mathcal{D}_B \subset \mathcal{F}_1 \times \mathcal{E}_1 \times \mathcal{F}_3 \times \mathcal{E}_3$ *is a Dirac structure.*

(We refer to the next Sect. 6.7, see in particular (6.165), how this extends to the composition of *multiple* Dirac structures.) The following explicit expression can be given for the composition of two Dirac structures in terms of their kernel/image representation.

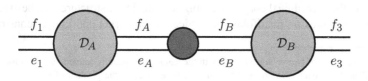

Fig. 6.7 The composition of $\mathcal{D}_A$ and $\mathcal{D}_B$

Proposition 6.6.10 *Consider Dirac structures* $\mathcal{D}_A \subset \mathcal{F}_1 \times \mathcal{E}_1 \times \mathcal{F}_2 \times \mathcal{E}_2, \mathcal{D}_B \subset \mathcal{F}_2 \times \mathcal{E}_2 \times \mathcal{F}_3 \times \mathcal{E}_3$, *given in combined kernel representation*

$$
\begin{bmatrix} F_1 & E_1 & F_{2A} & E_{2A} & 0 & 0 \\ 0 & 0 & -F_{2B} & E_{2B} & F_3 & E_3 \end{bmatrix} \begin{bmatrix} f_1 \\ e_1 \\ f_2 \\ e_2 \\ f_3 \\ e_3 \end{bmatrix} = 0 \tag{6.135}
$$

Then define

$$
M = \begin{bmatrix} F_{2A} & E_{2A} \\ -F_{2B} & E_{2B} \end{bmatrix} \tag{6.136}
$$

and let L_A, L_B *be matrices such*

$$
L = \begin{bmatrix} L_A & L_B \end{bmatrix}, \qquad \ker L = \operatorname{im} M
$$

Then a relaxed kernel representation of $\mathcal{D}_A \circ \mathcal{D}_B$ *is obtained by premultiplying* (6.135) *by the matrix L, resulting in*

$$
L_A F_1 f_1 + L_A E_1 e_1 + L_B F_3 f_3 + L_B E_3 e_3 = 0
$$

In many cases of interest, the notion of a *constant* Dirac structure $\mathcal{D} \subset \mathcal{F} \times \mathcal{E}$, with $\mathcal{F}$ and $\mathcal{E} = \mathcal{F}^*$ linear spaces, is not sufficient for modeling purposes. We already observed this for input-state-output port-Hamiltonian systems, where the matrices J, R, P, S, M in (6.3) were allowed to be *state-dependent*. Furthermore, in many examples the state space $\mathcal{X}$ is not a linear space, but instead a manifold. In particular, this often occurs for 3-D mechanical systems. In such cases, the notion of a constant Dirac structure given in Definition 6.6.3 needs to be extended to the following definition of *Dirac structures on manifolds*.

Definition 6.6.11 Let $\mathcal{X}$ be a manifold. A Dirac structure $\mathcal{D}$ on $\mathcal{X}$ is a vector sub-bundle of the Whitney sum[7] $T\mathcal{X} \oplus T^*\mathcal{X}$ such that

$$
\mathcal{D}(x) \subset T_x \mathcal{X} \times T_x^* \mathcal{X}
$$

is for every $x \in \mathcal{X}$ a constant Dirac structure as before.

Simply put, a Dirac structure on a manifold $\mathcal{X}$ is point-wise (for every $x \in \mathcal{X}$) a constant Dirac structure $\mathcal{D}(x) \subset T_x \mathcal{X} \times T_x^* \mathcal{X}$.

Most of the preceding theory concerning constant Dirac structures can be extended to Dirac structures on manifolds. In particular, the kernel and image, constrained

---

[7]The Whitney sum of two vector bundles with the same base space is defined as the vector bundle whose fiber above each element of this common base space is the product of the fibers of each individual vector bundle.

input–output, and hybrid input–output representation of Proposition 6.6.6 carry over to the case of a Dirac structure on a manifold; the difference being that the matrices involved may be *depending on x*, and that the representations may exist only *locally* on the state space manifold $\mathcal{X}$.

In particular, given a Dirac structure $\mathcal{D}$ on a manifold $\mathcal{X}$ and any point $x_0 \in \mathcal{X}$ there exists a coordinate neighborhood of x_0 such that for x within this coordinate neighborhood

$$\mathcal{D}(x) = \{(f, e) \in T_x\mathcal{X} \times T_x^*\mathcal{X} \mid F(x)f + E(x)e = 0\} \tag{6.137}$$

for $\ell \times \ell$ matrices $F(x)$ and $E(x)$ satisfying

$$E(x)F^T(x) + F(x)E^T(x) = 0, \quad \text{rank}[F(x) \vdots E(x)] = \ell, \tag{6.138}$$

or equivalently,

$$\mathcal{D}(x) = \{(f, e) \in T_x\mathcal{X} \times T_x^*\mathcal{X} \mid f = E^T(x)\lambda, \ e = F^T(x)\lambda, \lambda \in \mathbb{R}^\ell\} \tag{6.139}$$

Conversely, for any $\ell \times \ell$ matrices $F(x)$ and $E(x)$ satisfying (6.138), the subspaces (6.137) and (6.139) define locally a Dirac structure on $\mathcal{X}$.

Furthermore, $\mathcal{D}$ may be locally represented as

$$\mathcal{D}(x) = \{(f, e) \in T_x\mathcal{X} \times T_x^*\mathcal{X} \mid \exists \lambda \text{ s.t. } f = J(x)e + G(x)\lambda, \ G^T(x)e = 0\}, \tag{6.140}$$

for an $\ell \times \ell$ skew-symmetric matrix $J(x)$, and a matrix $G(x)$. Conversely, for any $G(x)$ and skew-symmetric $J(x)$ (6.140) defines locally a Dirac structure.

Finally, starting from (6.137) we may locally split the flows f and e, and correspondingly $F(x), E(x)$, in such a way that

$$\mathcal{D}(x) = \left\{(f, e) \in T_x\mathcal{X} \times T_x^*\mathcal{X} = \left\{\begin{bmatrix} f^1 \\ f^2 \end{bmatrix}, \begin{bmatrix} e^1 \\ e^2 \end{bmatrix} \middle| \begin{bmatrix} f^1 \\ e^2 \end{bmatrix} = J(x) \begin{bmatrix} e^1 \\ f^2 \end{bmatrix}\right\}\right\} \tag{6.141}$$

where $J(x) := -[F(x)^1 \vdots E(x)^2]^{-1}[F(x)^2 \vdots E(x)^1]$ is skew-symmetric. Conversely, for any skew-symmetric $J(x)$ as above (6.141) defines locally a Dirac structure.

On the other hand, the canonical coordinate representation (6.125) is *not* always possible for a Dirac structure $\mathcal{D}$ on a manifold $\mathcal{X}$. In fact, analogously to the *integrability conditions* (6.33) characterizing $J(x)$ to be a *Poisson structure* for which canonical coordinates as in (6.35) can be found, one can formulate integrability conditions on $\mathcal{D}$ which (together with a constant rank assumption) are necessary and sufficient for the local existence of canonical coordinates representing $\mathcal{D}$ as in (6.125). We refer to the Notes at the end of this chapter for further information.

The subspaces G_0, G_1, P_0, P_1 defined for a constant Dirac structure in (6.130) generalize for a Dirac structure $\mathcal{D}$ on $\mathcal{X}$ to the *distributions*, respectively, *co-distributions*, on $\mathcal{X}$

$$
\begin{aligned}
\mathcal{G}_0(x) &:= \{f \in T_x\mathcal{X} \mid (f, 0) \in \mathcal{D}(x)\} \\
\mathcal{G}_1(x) &:= \{f \in T_x\mathcal{X} \mid \exists e \in T_x^*\mathcal{X} \text{ s.t. } (f, e) \in \mathcal{D}(x)\} \\
\mathcal{P}_0(x) &:= \{e \in T_x^*\mathcal{X} \mid (0, e) \in \mathcal{D}(x)\} \\
\mathcal{P}_1(x) &:= \{e \in T_x^*\mathcal{X} \mid \exists f \in T_x\mathcal{X} \text{ s.t. } (f, e) \in \mathcal{D}(x)\}
\end{aligned}
\tag{6.142}
$$

The integrability of the distributions G_0, G_1 and co-distributions P_0, P_1 on $\mathcal{X}$ is *implied* by the integrability of the Dirac structure $\mathcal{D}$; see again the Notes at the end of this chapter.

Also, the theory regarding composition of constant Dirac structures can be extended to Dirac structures on manifolds; we refer to the next section for the appropriate setting.

6.7 Port-Hamiltonian DAE Systems

From a network modeling perspective (see also the Notes at the end of this chapter), lumped-parameter physical systems are naturally described by a set of ideal *energy-storing* elements, a set of *energy-dissipating* or *resistive* elements, and a set of *external ports* by which interaction with the environment can take place. All of them are interconnected to each other by a *power-conserving interconnection*, see Fig. 6.8.

This power-conserving interconnection includes ideal *power-conserving* elements such as (in the electrical domain) transformers, gyrators, or (in the mechanical domain) transformers, kinematic pairs, and kinematic constraints. Power-conserving elements do not store energy, nor dissipate energy, but instead *route* the energy flow.

Associated with the energy-storing elements are state variables $x_1, \ldots, x_n$, being coordinates for some n-dimensional state space manifold $\mathcal{X}$, and a total energy $H : \mathcal{X} \to \mathbb{R}$. The power-conserving interconnection is formalized by a *Dirac structure*

Fig. 6.8 Port-Hamiltonian
DAE system

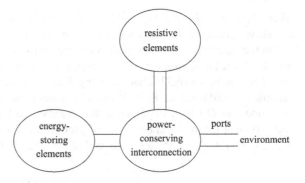

relating the flows and efforts of the energy-storing, energy-dissipating elements and external ports

$$\mathcal{D}(x) \subset T_x \mathcal{X} \times T_x^* \mathcal{X} \times \mathcal{F}_R \times \mathcal{E}_R \times \mathcal{F}_P \times \mathcal{E}_P, \quad x \in \mathcal{X}, \tag{6.143}$$

where $(f_S, e_S) \in T_x \mathcal{X} \times T_x^* \mathcal{X}$ are the flows and efforts of the energy-storing elements, $(f_R, e_R) \in \mathcal{F}_R \times \mathcal{E}_R$ are the flows and efforts of the energy-dissipating elements, and finally $(f_P, e_P) \in \mathcal{F}_P \times \mathcal{E}_P$ are the flows and efforts of the external ports.

Remark 6.7.1 Geometrically, $\mathcal{D}$ is a Dirac structure on the manifold $\mathcal{X} \times \mathcal{F}_R \times \mathcal{F}_P$, which is invariant under translation along F_R, F_P directions, and therefore only depending on $x \in \mathcal{X}$. See also [41, 218].

In the case of a *linear* state space $\mathcal{X}$ and a *constant* Dirac structure $\mathcal{D}$, the expression (6.143) simplifies to

$$\mathcal{D} \subset \mathcal{F}_S \times \mathcal{E}_S \times \mathcal{F}_R \times \mathcal{E}_R \times \mathcal{F}_P \times \mathcal{E}_P \tag{6.144}$$

where $\mathcal{F}_S = \mathcal{X}, \mathcal{E}_S = \mathcal{X}^*$. The equally dimensioned vectors of flow variables and effort variables of the energy-storing elements are given as

$$\dot{x}(t) = \frac{dx}{dt}(t), \quad \frac{\partial H}{\partial x}(x(t)), \quad t \in \mathbb{R}, \tag{6.145}$$

which are equated with f_S, e_S by[8]

$$\begin{aligned} f_S &= -\dot{x} \\ e_S &= \frac{\partial H}{\partial x}(x) \end{aligned} \tag{6.146}$$

Furthermore, f_R, e_R are related by a (static) *energy-dissipating* (resistive) relation, which can be any subset $\mathcal{R} \subset \mathcal{F}_R \times \mathcal{E}_R$, satisfying the property

$$e_R^T f_R \leq 0, \quad \text{for all } (f_R, e_R) \in \mathcal{R} \tag{6.147}$$

This leads to the following.

Definition 6.7.2 A port-Hamiltonian DAE system is defined by a Dirac structure $\mathcal{D}$ as in (6.143), a Hamiltonian $H : \mathcal{X} \to \mathbb{R}$, and an energy-dissipating relation $\mathcal{R} \subset \mathcal{F}_R \times \mathcal{E}_R$ satisfying (6.147). The dynamics is given by the requirement that for all $t \in \mathbb{R}$

$$\begin{aligned} \left(-\frac{dx}{dt}(t), \frac{\partial H}{\partial x}(x(t)), f_R(t), e_R(t), f_P(t), e_P(t)\right) &\in \mathcal{D}(x(t)) \\ (f_R(t), e_R(t)) &\in \mathcal{R} \end{aligned} \tag{6.148}$$

It is directly verified that this definition includes the definitions of input-state-output port-Hamiltonian systems as given before, cf. (6.1), (6.3), (6.6), (6.15), as special

---

[8]The minus sign is inserted in order to have a consistent power flow convention.

cases (with inputs u and outputs y given by (f_P, e_P). However, in general Definition 6.7.2 entails *algebraic constraints* on the state variables x.

By the power-conservation property of a Dirac structure (6.118) and (6.147), any port-Hamiltonian DAE system satisfies the energy-balance

$$
\begin{aligned}
\frac{dH}{dt}(x(t)) &= < \frac{\partial H}{\partial x}(x(t)) \mid \dot{x}(t) >= \\
&= e_R^T f_R(t) + e_P^T(t) f_P(t) \le e_P^T(t) f_P(t),
\end{aligned}
\tag{6.149}
$$

as was the case for input-state-output port-Hamiltonian systems. Thus port-Hamiltonian DAE systems are cyclo-passive with respect to the supply rate $e_P^T f_P$, and passive if H is bounded from below.

The *algebraic constraints* that are present in a port-Hamiltonian DAE system are determined by the distribution P_1 defined by $\mathcal{D}$ (cf. (6.142)), as well as by the Hamiltonian H. In fact, the condition

$$
\left(\frac{\partial H}{\partial x}(x), e_R, e_P \right) \in P_1(x), \quad x \in \mathcal{X},
\tag{6.150}
$$

may entail algebraic constraints on the state x.

On the other hand, the *Casimir functions* $C : \mathcal{X} \to \mathbb{R}$ of the port-Hamiltonian DAE system (6.148) are determined by the distribution G_1. Indeed, $\frac{dC}{dt} = \frac{\partial^T C}{\partial x}(x)\dot{x} = 0$ if and only if $\frac{\partial^T C}{\partial x}(x) f_S = 0$ for all f_S for which there exists f_R, f_P such that $(f_S, f_R, f_P) \in G_1(x)$. Furthermore, C is a Casimir for $f_P = 0$ if and only if $\frac{\partial^T C}{\partial x}(x) f_S = 0$ for all f_S for which there exists f_R such that $(f_S, f_R, 0) \in G_1(x)$.

Definition 6.7.2 is a geometric, coordinate-free, definition. *Equational representations* of port-Hamiltonian DAE systems are obtained by choosing a coordinate representation of the Dirac structure $\mathcal{D}$ as in (6.143). In case the Dirac structure $\mathcal{D}$ is given in *kernel representation*

$$
\begin{aligned}
\mathcal{D}(x) = \{ (f_S, f_R, f_P, e_S, e_R, e_P) \mid F_S(x)f_S + E_S(x)e_S + \\
F_R(x)f_R + E_R(x)e_R + F_P(x)f_P + E_P(x)e_P = 0 \}
\end{aligned}
\tag{6.151}
$$

for matrices $F_S(x), E_S(x), F_R(x), E_R(x), F_P(x), E_P(x)$ satisfying

$$
\begin{aligned}
&(i) \;\; E_S F_S^T + F_S E_S^T + E_R F_R^T + F_R E_R^T + E_P F_P^T + F_P E_P^T = 0 \\
&(ii) \; \text{rank} \left[F_S \vdots F_R \vdots F_P \vdots E_S \vdots E_R \vdots E_P \right] = \dim \mathcal{F}
\end{aligned}
\tag{6.152}
$$

this leads to the following specification of algebraic constraints and Casimirs. With respect to the algebraic constraints, we notice that

$$
e_S \in \text{im } F_S^T(x),
\tag{6.153}
$$

implying the algebraic constraints

$$\frac{\partial H}{\partial x}(x) \in \operatorname{im} F_S^T(x) \tag{6.154}$$

With respect to the Casimirs we notice that

$$f_S \in \operatorname{im} E_S^T(x), \tag{6.155}$$

implying that $C : \mathcal{X} \to \mathbb{R}$ is a Casimir function if and only if $\frac{dC}{dt}(x(t)) = \frac{\partial^T C}{\partial x}(x(t))$ $\dot{x}(t) = 0$ for all $\dot{x}(t) \in \operatorname{im} E_S^T(x(t))$. Hence, C is a Casimir of the port-Hamiltonian DAE system (6.148) if and only if it satisfies the set of pde's

$$E_S(x)\frac{\partial C}{\partial x}(x) = 0, \quad x \in \mathcal{X} \tag{6.156}$$

Finally, C is a Casimir function for $f_P = 0$ if and only if $\frac{\partial^T C}{\partial x}(x)E_S(x)\lambda = 0$ for all λ such that $E_P^T(x)\lambda = 0$. As a result, C is a Casimir function for $f_P = 0$ if and only if it satisfies the conditions

$$E_S(x)\frac{\partial C}{\partial x}(x) \in \operatorname{im} E_P(x), \quad x \in \mathcal{X} \tag{6.157}$$

Example 6.7.3 Consider the LC-circuit of Example 6.4.5 without voltage source $(V = 0)$, and where the two inductors are replaced by two capacitors with charges Q_1, Q_2, and dually the capacitor is replaced by an inductor with flux linkage φ. This does not change the Dirac structure (determined by Kirchhoff's current and voltage laws). However, while the original LC-circuit has a *Casimir* $\varphi_1 + \varphi_2$, in the present LC-circuit there is the *algebraic constraint*

$$\frac{\partial H}{\partial Q_1}(Q_1, Q_2, \varphi) + \frac{\partial H}{\partial Q_2}(Q_1, Q_2, \varphi) = 0 \tag{6.158}$$

constraining the state variables Q_1, Q_2.

Example 6.7.4 The constrained Hamiltonian equations (6.42) can be viewed as a port-Hamiltonian DAE system, with respect to the Dirac structure $\mathcal{D}$ given in constrained input–output representation (6.123) as

$$\mathcal{D} = \{(f_S, f_P, e_S, e_P) \mid 0 = A^T(q)e_S, \ e_P = B^T(q)e_S,$$

$$-f_S = \begin{bmatrix} 0 & I_n \\ -I_n & 0 \end{bmatrix} e_S + \begin{bmatrix} 0 \\ A(q) \end{bmatrix} \lambda + \begin{bmatrix} 0 \\ B(q) \end{bmatrix} f_P, \ \lambda \in \mathbb{R}^k \} \tag{6.159}$$

The kinematic constraints correspond to the following algebraic constraints on the state variables (q, p)

$$0 = A^T(q)\frac{\partial H}{\partial p}(q, p), \tag{6.160}$$

while the Casimir functions C for $u = 0$ are determined by the equations

$$\frac{\partial^T C}{\partial q}(q)\dot{q} = 0, \quad \text{for all } \dot{q} \text{ satisfying } A^T(q)\dot{q} = 0 \tag{6.161}$$

Hence, finding a Casimir function amounts to (partially) *integrating* the kinematic constraints $A^T(q)\dot{q} = 0$. In particular, if the kinematic constraints are *holonomic*, and thus can be expressed as in (6.38), then $\bar{q}_{n-k+1}, \ldots, \bar{q}_n$ generate all the Casimir functions.

The results concerning *composition* of Dirac structures as treated in the previous Sect. 6.6 imply that any power-conserving interconnection of port-Hamiltonian systems is *again a port-Hamiltonian system*. Indeed, let us consider k port-Hamiltonian DAE systems specified by Dirac structures

$$\mathcal{D}_i(x_i) \subset T_{x_i}\mathcal{X}_i \times T_{x_i}^*\mathcal{X}_i \times \mathcal{F}_R^i \times \mathcal{E}_R^i \times \mathcal{F}_P^i \times \mathcal{E}_P^i, \quad x_i \in \mathcal{X}_i, \quad i = 1, \ldots, k \tag{6.162}$$

together with Hamiltonians and energy-dissipating relations

$$H_i : \mathcal{X}_i \to \mathbb{R}, \quad \mathcal{R}_i \subset \mathcal{F}_R^i \times \mathcal{E}_R^i, \quad i = 1, \ldots, k \tag{6.163}$$

Furthermore, define an *interconnection Dirac structure*

$$\mathcal{D}_I \subset \mathcal{F}_P^1 \times \mathcal{E}_P^1 \times \cdots \times \mathcal{F}_P^k \times \mathcal{E}_P^k \times \mathcal{F}_P^e \times \mathcal{E}_P^e \tag{6.164}$$

with $\mathcal{F}_P^e, \mathcal{E}_P^e$ spaces of external flows and efforts. $\mathcal{D}_I$ specifies the way the flows and efforts f_P^i, e_P^i of the composing systems are connected to each other and to the new external flows and efforts f_P^e, e_P^e in a power-conserving manner. The *composition* through the shared flows and efforts in $\mathcal{F}_P^1 \times \mathcal{E}_P^1 \times \cdots \times \mathcal{F}_P^k \times \mathcal{E}_P^k$ defines a new Dirac structure

$$(\mathcal{D}_1(x_1) \times \cdots \times \mathcal{D}_k(x_k)) \circ \mathcal{D}_I \subset T_x\mathcal{X} \times T_x^*\mathcal{X} \times \mathcal{F}_R \times \mathcal{E}_R \times \mathcal{F}_P^e \times \mathcal{E}_P^e \tag{6.165}$$

(note that this amounts to the composition of *two* Dirac structures), where

$$x \in \mathcal{X} := \mathcal{X}_1 \times \cdots \times \mathcal{X}_k, \ \mathcal{F}_R := \mathcal{F}_R^1 \times \cdots \times \mathcal{F}_R^k, \ \mathcal{E}_R := \mathcal{E}_R^1 \times \cdots \times \mathcal{E}_R^k \tag{6.166}$$

As a result, the interconnected system is again a port-Hamiltonian DAE system on the product state space $\mathcal{X}$ with Hamiltonian $H : \mathcal{X} \to \mathbb{R}$ given as $H(x) = H_1(x_1) +$

$\cdots + H_k(x_k)$, and with energy-dissipating relation $\mathcal{R}$ given as the direct product of $\mathcal{R}_1, \ldots, \mathcal{R}_k$.

Example 6.7.5 (PID control) Consider the standard *Proportional–Integral–Derivative* (PID) controller

$$y_c = k_P u_c + k_I \int u_c dt + k_D \dot{u}_c \tag{6.167}$$

for certain positive constants k_P, k_I, k_D. Trivially rewriting (6.167) as

$$k_D \dot{u}_c = -k_P u_c - k_I \int u_c dt + y_c, \tag{6.168}$$

and defining $\xi = \int u_c dt$ (or equivalently $\dot{\xi} = u_c$) and $\eta = k_D u_c$, the PID-controller can be formulated as the linear input-state-output[9] port-Hamiltonian system

$$\begin{bmatrix} \dot{\xi} \\ \dot{\eta} \end{bmatrix} = \begin{bmatrix} 0 & 1 \\ -1 & -k_P \end{bmatrix} \begin{bmatrix} k_I \xi \\ \frac{\eta}{k_D} \end{bmatrix} + \begin{bmatrix} 0 \\ 1 \end{bmatrix} y_c$$
$$u_c = \begin{bmatrix} 0 & 1 \end{bmatrix} \begin{bmatrix} k_I \xi \\ \frac{\eta}{k_D} \end{bmatrix} \tag{6.169}$$

with Hamiltonian $H_c(\xi, \eta) = \frac{1}{2} k_I \xi^2 + \frac{1}{2k_D} \eta^2$.

Considering any plant input-state-output port-Hamiltonian system as in (6.1)

$$\dot{x} = [J(x) - R(x)] \frac{\partial H}{\partial x}(x) + g(x)u$$
$$y = g^T(x) \frac{\partial H}{\partial x}(x) \tag{6.170}$$

the closed-loop system arising from standard feedback $u = -y_c$, $u_c = y$ with the PID-controller is given by the port-Hamiltonian DAE system

$$\dot{x} = [J(x) - R(x)] \frac{\partial H}{\partial x}(x) + g(x)u$$
$$\begin{bmatrix} \dot{\xi} \\ \dot{\eta} \end{bmatrix} = \begin{bmatrix} 0 & 1 \\ -1 & -k_P \end{bmatrix} \begin{bmatrix} k_I \xi \\ \frac{\eta}{k_D} \end{bmatrix} - \begin{bmatrix} 0 \\ 1 \end{bmatrix} u$$
$$0 = g^T(x) \frac{\partial H}{\partial x}(x) - \begin{bmatrix} 0 & 1 \end{bmatrix} \begin{bmatrix} k_I \xi \\ \frac{\eta}{k_D} \end{bmatrix} \tag{6.171}$$

with total Hamiltonian $H(x) + \frac{1}{2} k_I \xi^2 + \frac{1}{2k_D} \eta^2$. This port-Hamiltonian system is in constrained input–output representation (6.123), with u acting as a vector of Lagrange multipliers.

---

[9]Note that y_c serves as an input to (6.169) and u_c as an output, contrary to the intuitive use of a PID-controller, where u_c equals the output of the *plant* system and $-y_c$ is the input applied to the plant system. This is of course caused by the fact that the D-action involves a differentiation.

We close this section by indicating two direct extensions from the input-state-output port-Hamiltonian case to the DAE case. The first concerns *shifted passivity*. Consider a port-Hamiltonian DAE system for which we can find coordinates in which the Dirac structure is *constant*. Furthermore, assume that the resistive structure $\mathcal{R}$ is *linear*. A *steady-state* $\bar{x}$, corresponding to steady-state values $\bar{f}_R, \bar{e}_R, \bar{f}_P, \bar{e}_P$, is characterized by

$$\left(0, \frac{\partial H}{\partial x}(\bar{x}), \bar{f}_R, \bar{e}_R, \bar{f}_P, \bar{e}_P\right) \in \mathcal{D}, \quad (\bar{f}_R, \bar{e}_R) \in \mathcal{R} \qquad (6.172)$$

Using now the linearity of $\mathcal{D}$ and $\mathcal{R}$, we can subtract (6.172) from (6.148), so as to obtain

$$\begin{aligned} &(-\dot{x}(t), \frac{\partial H}{\partial x}(x(t)) - \frac{\partial H}{\partial x}(\bar{x}), \\ &f_R(t) - \bar{f}_R, \, e_R(t) - \bar{e}_R, f_P(t) - \bar{f}_P, \, e_P(t) - \bar{e}_P) \in \mathcal{D} \\ &(f_R(t) - \bar{f}_R, \, e_R(t) - \bar{e}_R) \in \mathcal{R} \end{aligned} \qquad (6.173)$$

Similar to Proposition 6.5.1, this defines a *shifted* port-Hamiltonian system with respect to the same Dirac structure $\mathcal{D}$ and resistive structure $\mathcal{R}$, and with Hamiltonian given by the shifted Hamiltonian function $\widehat{H}_{\bar{x}}$, and shifted external port variables $f_P - \bar{f}_P, e_P - \bar{e}_P$.

The second extension concerns the notion of *steady-state input–output relation* (4.31). For a port-Hamiltonian DAE system, Σ this relation is given as

$$\begin{aligned} \Sigma_{ss} = \{(\bar{f}_P, \bar{e}_P) \mid \exists \bar{x}, \bar{f}_R, \bar{e}_R \text{ such that} \\ \left(0, \tfrac{\partial H}{\partial x}(\bar{x}), \bar{f}_R, \bar{e}_R, \bar{f}_P, \bar{e}_P\right) \in \mathcal{D}(\bar{x}), (\bar{f}_R, \bar{e}_R) \in \mathcal{R}\} \end{aligned} \qquad (6.174)$$

It directly follows that $\bar{e}_P^T \bar{f}_P \geq 0$ for all $(\bar{f}_P, \bar{e}_P) \in \Sigma_{ss}$.

6.8   Port-Hamiltonian Network Dynamics

Section 4.4 already presented a treatment of passive network systems. In this section we will go one step further, by identifying large classes of network systems as port-Hamiltonian systems, where the Dirac structure of the network system is determined by the *network interconnection structure*.

Let us start with some basic notions regarding graphs, extending the background already provided in Sect. 4.4. Like in Sect. 4.4 "graph" throughout means "directed graph." Given a graph, we define its *vertex space* Λ_0 as the vector space of all functions from $\mathcal{V}$ to some linear space $\mathcal{R}$. In the examples, $\mathcal{R}$ will be mostly $\mathcal{R} = \mathbb{R}$ in which case Λ_0 can be identified with $\mathbb{R}^N$. Furthermore, we define the *edge space* Λ_1 as the vector space of all functions from $\mathcal{E}$ to $\mathcal{R}$. Again, if $\mathcal{R} = \mathbb{R}$ then Λ_1 can be identified with $\mathbb{R}^M$. The dual spaces of Λ_0 and Λ_1 will be denoted by Λ^0, respectively, by Λ^1. The duality pairing between $f \in \Lambda_0$ and $e \in \Lambda^0$ is given as

$$< f \mid e >= \sum_{v \in \mathcal{V}} < f(v) \mid e(v) >, \qquad (6.175)$$

where $< \mid >$ on the right-hand side denotes the duality pairing between $\mathcal{R}$ and $\mathcal{R}^*$. A similar expression holds for $f \in \Lambda_1$ and $e \in \Lambda^1$ (with summation over the edges).

The incidence matrix D of the graph induces a linear map $\widehat{D}$ from the edge space to the vertex space as follows. Define $\widehat{D} : \Lambda_1 \to \Lambda_0$ as the linear map with matrix representation $D \otimes I$, where $I : \mathcal{R} \to \mathcal{R}$ is the identity map and $\otimes$ denotes the Kronecker product. $\widehat{D}$ will be called the *incidence operator*. For $\mathcal{R} = \mathbb{R}$ the incidence operator reduces to the linear map given by the matrix D itself, in which case we will throughout use D *both for the incidence matrix and for the incidence operator*. The adjoint map of $\widehat{D}$ is denoted as

$$\widehat{D}^* : \Lambda^0 \to \Lambda^1,$$

and is called the *coincidence* operator. For $\mathcal{R} = \mathbb{R}^3$ the coincidence operator is given by $D^T \otimes I_3$, while for $\mathcal{R} = \mathbb{R}$ the coincidence operator is simply given by the transposed matrix D^T, and we will throughout use D^T *both for the co-incidence matrix and for the coincidence operator.*

In order to define *open* network systems we will identify a subset $\mathcal{V}_b \subset \mathcal{V}$ of *boundary vertices*. The remaining subset $\mathcal{V}_i := \mathcal{V} - \mathcal{V}_b$ are called the *internal vertices* of the graph.

The splitting of the vertices into internal and boundary vertices induces a splitting of the vertex space and its dual, given as

$$\Lambda_0 = \Lambda_{0i} \oplus \Lambda_{0b}, \ \Lambda^0 = \Lambda^{0i} \oplus \Lambda^{0b}, \qquad (6.176)$$

where Λ_{0i} is the vertex space corresponding to the internal vertices and Λ_{0b} the vertex space corresponding to the boundary vertices. Consequently, the incidence operator $\widehat{D} : \Lambda_1 \to \Lambda_0$ splits as

$$\widehat{D} = \widehat{D}_i \oplus \widehat{D}_b, \qquad (6.177)$$

with $\widehat{D}_i : \Lambda_1 \to \Lambda_{0i}$ and $\widehat{D}_b : \Lambda_1 \to \Lambda_{0b}$. For $\mathcal{R} = \mathbb{R}$ we will simply write

$$D = \begin{bmatrix} D_i \\ D_b \end{bmatrix} \qquad (6.178)$$

Furthermore, we define the *boundary space* Λ_b as the linear space of all functions from the set of boundary vertices $\mathcal{V}_b$ to the linear space $\mathcal{R}$. Note that the boundary space Λ_b is equal to the linear space Λ_{0b}, and that the linear mapping $\widehat{D}_b$ can be also regarded as a mapping $\widehat{D}_b : \Lambda_1 \to \Lambda_b$. The dual space of Λ_b will be denoted as Λ^b. The elements $f_b \in \Lambda_b$ are called the *boundary flows* and the elements $e^b \in \Lambda^b$ the *boundary efforts*.

A paradigmatic example of a port-Hamiltonian network system is a *mass–spring–damper system*. Let us start with mass–spring systems, as already considered in Chap. 4, Example 4.4.5, as an example of a *passive* network system. Any mass–spring system is modeled by a graph $\mathcal{G}$ with N vertices corresponding to the masses and M edges corresponding to the springs, specified by an incidence matrix D. For ease of notation, consider first the situation that the mass–spring system is located in one-dimensional space $\mathcal{R} = \mathbb{R}$, and the springs are scalar. A vector in the vertex space Λ_0 then corresponds to the vector p of the scalar momenta of all N masses, i.e., $p \in \Lambda_0 = \mathbb{R}^N$. Furthermore, a vector in the dual edge space Λ^1 will correspond to the total vector q of extensions of all M springs, i.e., $q \in \Lambda^1 = \mathbb{R}^M$.

Next ingredient is the Hamiltonian $H : \Lambda^1 \times \Lambda_0 \to \mathbb{R}$, which is the sum of the kinetic and potential energies of each mass and spring. In the absence of boundary vertices the dynamics of the mass–spring system is described as

$$\begin{bmatrix} \dot{q} \\ \dot{p} \end{bmatrix} = \begin{bmatrix} 0 & D^T \\ -D & 0 \end{bmatrix} \begin{bmatrix} \frac{\partial H}{\partial q}(q,p) \\ \frac{\partial H}{\partial p}(q,p) \end{bmatrix}, \tag{6.179}$$

defined with respect to the constant Poisson structure on the linear state space $\Lambda^1 \times \Lambda_0$ given by the skew-symmetric matrix

$$J := \begin{bmatrix} 0 & D^T \\ -D & 0 \end{bmatrix} \tag{6.180}$$

implicitly already encountered in Sect. 4.4; see (4.83).

The inclusion of boundary vertices, and thereby of external interaction, can be done in different ways. The first option is to associate *boundary masses* to the boundary vertices. We are then led to the port-Hamiltonian input-state-output system

$$\dot{q} = D^T \frac{\partial H}{\partial p}(q,p)$$

$$\dot{p} = -D \frac{\partial H}{\partial q}(q,p) + E f_b \tag{6.181}$$

$$e^b = E^T \frac{\partial H}{\partial p}(q,p)$$

Here E is a matrix with as many columns as there are boundary vertices; each column consists of zeros except for exactly one 1 in the row corresponding to the associated boundary vertex. Furthermore $f_b \in \Lambda_b$ are the external *forces* exerted (by the environment) on the boundary masses, and $e_b \in \Lambda^b$ are the *velocities* of these boundary masses.

A second possibility is to regard the boundary vertices as being *massless*. In this case, we obtain the port-Hamiltonian input-state-output system (with p now denoting the vector of momenta of the masses associated to the *internal* vertices)

$$\dot{q} = D_i^T \frac{\partial H}{\partial p}(q, p) + D_b^T e^b$$

$$\dot{p} = -D_i \frac{\partial H}{\partial q}(q, p) \qquad\qquad (6.182)$$

$$f_b = D_b \frac{\partial H}{\partial q}(q, p)$$

with $e^b \in \Lambda^b$ the velocities of the massless boundary vertices, and $f_b \in \Lambda_b$ the forces at the boundary vertices as *experienced* by the environment. Note that in this second case the external velocities e^b of the boundary vertices can be considered to be *inputs* to the system and the forces f_b to be *outputs*; in contrast to the previously considered case (boundary vertices corresponding to boundary masses), where the forces f_b are inputs and the velocities e^b the outputs of the system.

For a *mass–spring–damper system* the edges will correspond partly to springs, and partly to dampers. This corresponds to an incidence matrix

$$D = \begin{bmatrix} D_s & D_d \end{bmatrix}, \qquad\qquad (6.183)$$

where the columns of D_s reflect the spring edges and the columns of D_d the damper edges. For the case *without* boundary vertices the dynamics of a mass–spring–damper system with linear dampers takes the form

$$\begin{bmatrix} \dot{q} \\ \dot{p} \end{bmatrix} = \begin{bmatrix} 0 & D_s^T \\ -D_s & -D_d R D_d^T \end{bmatrix} \begin{bmatrix} \dfrac{\partial H}{\partial q}(q, p) \\ \dfrac{\partial H}{\partial p}(q, p) \end{bmatrix} \qquad (6.184)$$

with R the diagonal matrix of damping coefficients. In the presence of boundary vertices, we may again distinguish between *massless* boundary vertices, with inputs being the boundary velocities and outputs the boundary (reaction) forces, and *boundary masses*, in which case the inputs are the external forces and the outputs the velocities of the boundary masses.

The formulation of mass–spring–damper systems in $\mathcal{R} = \mathbb{R}$ directly extends to $\mathcal{R} = \mathbb{R}^3$ using the incidence operator $\widehat{D} = D \otimes I_3$ as defined before. Furthermore, the set-up can be extended [298] to *multi-body systems* and *spatial mechanisms* (networks of rigid bodies in $\mathbb{R}^3$ related by joints) by considering the linear space $\mathcal{R} := \mathrm{se}^*(3)$, the dual of the Lie algebra of the Lie group $SE(3)$ describing the position of a rigid body in $\mathbb{R}^3$. Finally we note that other examples like hydraulic networks are analogous to mass–spring–damper system; see e.g., [287].

Remark 6.8.1 The example of a power network given in Example 4.4.4 defines a port-Hamiltonian system which is similar to a mass–spring–damper system, with the difference that in this case the *dampers* (corresponding to A) are associated to the *vertices* of the graph and the edges correspond to the transmission lines with potential energies $-\gamma_j \cos q_j, j = 1, \ldots, M$.

Remark 6.8.2 Note the slight discrepancy of the role of the flows f and efforts e with respect to the definition of a port-Hamiltonian DAE systems given in the previous Sect. 6.7. Indeed, in Sect. 6.7 flows and efforts are used interchangeably, with the exception of the flows and efforts f_S, e_S corresponding to the energy-storing elements, which (cf. (6.146)) are given by $f_S = -\dot{x}$ and $e_S = \frac{\partial H}{\partial x}(x)$. In the current network setting, the flows f are elements of the spaces Λ_0, Λ_1, and the efforts e of their *dual* spaces Λ^0, Λ^1. In particular, the flows in Λ_1 correspond to the classical [211] *through* variables, and the efforts in Λ^1 to the *across* variables.

The port-Hamiltonian formulation of the dynamics (6.184) leads to the following *stability analysis*. Without loss of generality[10], we throughout assume that the graph is connected, or equivalently, see Sect. 4.4, $\ker D_s^T \cap \ker D_d^T = \text{span } \mathbb{1}$, where $\mathbb{1}$ is the vector of all ones. We start with the following proposition regarding the *equilibria*.

Proposition 6.8.3 *Consider the dynamics (6.184). Its set of equilibria $\mathcal{E}$ is given as*

$$\mathcal{E} = \{(q, p) \in \Lambda^1 \times \Lambda_0 \mid \frac{\partial H}{\partial q}(q, p) \in \ker D_s, \frac{\partial H}{\partial p}(q, p) \in \text{span } \mathbb{1}\} \qquad (6.185)$$

Proof (q, p) is an equilibrium whenever

$$D_s^T \frac{\partial H}{\partial p}(q, p) = 0, \; D_s \frac{\partial H}{\partial q}(q, p) + D_d R D_d^T \frac{\partial H}{\partial p}(q, p) = 0 \qquad (6.186)$$

Premultiplication of the second equation by the row-vector $\frac{\partial^T H}{\partial p}(q, p)$, making use of the first equation, yields $\frac{\partial^T H}{\partial p}(q, p) B_d R B_d^T \frac{\partial H}{\partial p}(q, p) = 0$, or equivalently $D_d^T \frac{\partial H}{\partial p}$ $(q, p) = 0$, which implies $D_s \frac{\partial H}{\partial q}(q, p) = 0$. Hence, $\frac{\partial H}{\partial p}(q, p) \in \ker D_s^T \cap \ker D_d^T = \text{span } \mathbb{1}$. $\qquad \square$

In other words, for (q, p) to be an equilibrium, the elements of the vector of velocities $\frac{\partial H}{\partial p}(q, p)$ should be equal to each other, whereas $\frac{\partial H}{\partial q}(q, p)$ should be in the space $\ker D_s$ of *cycles* of the subgraph of masses and springs (resulting in zero net spring forces applied to the masses at the vertices).

Similarly, the *Casimirs* are computed as follows.

Proposition 6.8.4 *The Casimir functions of (6.184) are functions $C(q, p)$ satisfying*

$$\frac{\partial C}{\partial p}(q, p) \in \text{span } \mathbb{1}, \; \frac{\partial C}{\partial q}(q, p) \in \ker D_s \qquad (6.187)$$

Proof The function $C(q, p)$ is a Casimir if

$$\begin{bmatrix} \frac{\partial C}{\partial q}(q, p) & \frac{\partial C}{\partial p}(q, p) \end{bmatrix} \begin{bmatrix} 0 & D_s^T \\ -D_s & -D_d R D_d^T \end{bmatrix} = 0, \qquad (6.188)$$

or equivalently (see Proposition 6.4.2)

---

[10]Since otherwise the same analysis can be performed on each connected component of the graph.

$$\frac{\partial^T C}{\partial p}(q, p)D_s = 0, \quad \frac{\partial^T C}{\partial q}(q, p)D_s^T = 0, \quad \frac{\partial^T C}{\partial p}(q, p)D_d R D_d^T = 0 \qquad (6.189)$$

Post-multiplication of the third equation by $\frac{\partial C}{\partial p}(q, p)$, making use of the first equation, gives the result. ☐

Therefore, all Casimir functions can be expressed as functions of the *linear* Casimir functions

$$C(q, p) = \mathbb{1}^T p, \quad C(q, p) = k^T q, \ k \in \ker D_s \qquad (6.190)$$

This implies that starting from an arbitrary initial position $(q_0, p_0) \in \Lambda^1 \times \Lambda_0$ the solution of the mass–spring–damper system (6.184) will be contained in the affine space

$$\mathcal{A}_{(q_0, p_0)} := \begin{bmatrix} q_0 \\ p_0 \end{bmatrix} + \begin{bmatrix} 0 \\ \ker \mathbb{1}^T \end{bmatrix} + \begin{bmatrix} \text{im } D_s^T \\ 0 \end{bmatrix} \qquad (6.191)$$

i.e., for all t the difference $q(t) - q_0$ remains in the space im D_s^T of *co-cycles* of the mass–spring graph, while $\mathbb{1}^T p(t) = \mathbb{1}^T p_0$.

Under generically fulfilled conditions on the Hamiltonian $H(q, p)$, each affine space $\mathcal{A}_{(q_0, p_0)}$ will intersect the set of equilibria $\mathcal{E}$ in a *single* point (q_∞, p_∞), which qualifies as the point of asymptotic convergence starting from (q_0, p_0). For simplicity, consider *linear* mass–spring–damper systems, corresponding to a quadratic Hamiltonian function

$$H(q, p) = \frac{1}{2}q^T K q + \frac{1}{2}p^T G p, \qquad (6.192)$$

where K is the positive diagonal matrix of spring constants, and G is the positive diagonal matrix of reciprocals of the masses. In this case, the set of equilibria is given as $\mathcal{E} = \{(q, p) \in \Lambda^1 \times \Lambda_0 \mid Kq \in \ker B_s, Gp \in \text{span } \mathbb{1}\}$, while indeed it is easily seen that for each (q_0, p_0) there exists a *unique* point $(q_\infty, p_\infty) \in \mathcal{E} \cap \mathcal{A}_{(q_0, p_0)}$. In fact, q_∞ is given by the spring graph co-cycle/cycle decomposition

$$q_0 = v_0 + q_\infty, \quad v_0 \in \text{im } D_s^T \subset \Lambda^1, Kq_\infty \in \ker D_s \subset \Lambda_1 \qquad (6.193)$$

Furthermore, p_∞ is uniquely determined by

$$Gp_\infty \in \text{span } \mathbb{1}, \quad \mathbb{1}^T p_\infty = \mathbb{1}^T p_0 \qquad (6.194)$$

This leads to the following asymptotic stability theorem. First note that

$$\begin{aligned} \frac{d}{dt}H(q, p) &= -\frac{\partial^T H}{\partial p}(q, p)D_d R D_d^T \frac{\partial H}{\partial p}(q, p) \\ &= -p^T G D_d R D_d^T G p \leq 0 \end{aligned} \qquad (6.195)$$

Theorem 6.8.5 *Consider a linear mass–spring–damper system with $H(q, p) = \frac{1}{2}q^T Kq + \frac{1}{2}p^T Gp$, where K and G are diagonal positive matrices. Then for every (q_0, p_0), there exists a unique equilibrium point $(q_\infty, p_\infty) \in \mathcal{E} \cap \mathcal{A}_{(q_0,p_0)}$, determined by (6.193), (6.194). Define the spring Laplacian matrix $L_s := D_s K D_s^T$. Then for every (q_0, p_0) the following holds: the trajectory starting from (q_0, p_0) converges asymptotically to (q_∞, p_∞) if and only if the largest GL_s-invariant subspace contained in $\ker D_d^T$ is equal to span $\mathbb{1}$.*

The condition that the largest GL_s-invariant subspace contained in $\ker D_d^T$ is equal to span $\mathbb{1}$ amounts to *pervasive damping*: the influence of the dampers spreads through the whole system.

Another feature of the dynamics of the mass–spring–damper system (6.184) is its *robustness* with regard to constant external (disturbance) forces. Indeed, consider a mass–spring–damper system with boundary masses and general Hamiltonian $H(q, p)$, subject to *constant* forces $\bar{f}_b$

$$\begin{bmatrix} \dot{q} \\ \dot{p} \end{bmatrix} = \begin{bmatrix} 0 & D_s^T \\ -D_s & -D_d R D_d^T \end{bmatrix} \begin{bmatrix} \dfrac{\partial H}{\partial q}(q, p) \\ \dfrac{\partial H}{\partial p}(q, p) \end{bmatrix} + \begin{bmatrix} 0 \\ E \end{bmatrix} \bar{f}_b, \qquad (6.196)$$

where we *assume*[11] the existence of a $\bar{q}$ such that

$$D_s \frac{\partial H}{\partial q}(\bar{q}, 0) = E\bar{f}_b \qquad (6.197)$$

The shifted Hamiltonian $\widehat{H}_{(\bar{q},0)}(q, p) := H(q, p) - (q - \bar{q})^T \frac{\partial H}{\partial q}(\bar{q}, 0) - H(\bar{q}, 0)$ as defined before in (6.92) satisfies

$$\frac{d}{dt}\widehat{H}_{(\bar{q},0)}(q, p) = -\frac{\partial^T H}{\partial p}(q, p)D_d R D_d^T \frac{\partial H}{\partial p}(q, p) \leq 0 \qquad (6.198)$$

Specializing to a quadratic Hamiltonian $H(q, p) = \frac{1}{2}q^T Kq + \frac{1}{2}p^T Gp$ one obtains $\widehat{H}_{(\bar{q},0)}(q, p) = \frac{1}{2}(q - \bar{q})^T K(q - \bar{q}) + \frac{1}{2}p^T Gp$, leading to the following analog of Theorem 6.8.5.

Proposition 6.8.6 *Consider a linear mass–spring–damper system (6.196) with constant external disturbance $\bar{f}_b$ and Hamiltonian $H(q, p) = \frac{1}{2}q^T Kq + \frac{1}{2}p^T Gp$, where K and G are diagonal positive matrices. and with im $E \subset$ im D_s. The set of steady states is given by $\bar{\mathcal{E}} = \{(q, p) \in \Lambda^1 \times \Lambda_0 \mid D_s Kq = E\bar{f}_b, Gp \in \text{span } \mathbb{1}\}$. For every (q_0, p_0) there exists a unique equilibrium point $(\bar{q}_\infty, p_\infty) \in \bar{\mathcal{E}} \cap \mathcal{A}_{(q_0,p_0)}$. Here p_∞ is determined by (6.194), while $\bar{q}_\infty = \bar{q} + q_\infty$, with $\bar{q}$ such that $D_s K\bar{q} = E\bar{f}_b$ and q_∞ the unique solution of (6.193) with q_0 replaced by $q_0 - \bar{q}$. Furthermore, for each*

---

[11]If the mapping $q \mapsto \frac{\partial H}{\partial q}(q, 0)$ is surjective, then there exists for every $\bar{f}_b$ such a $\bar{q}$ if and only if im $E \subset$ im D_s.

(q_0, p_0) *the trajectory starting from* (q_0, p_0) *converges asymptotically to* $(\bar{q}_\infty, p_\infty)$ *if and only if the largest* GL_s-*invariant subspace contained in* $\ker D_d^T$ *is equal to* span $\mathbb{1}$.

Note that the above proposition has the classical interpretation in terms of robustness of *integral control* with regard to constant disturbances: the springs act as integral controllers which counteract the influence of the unknown external force $\bar{f}_b$ so that the vector of velocities $M^{-1}p$ will still converge to span $\mathbb{1}$.

An alternative to the above formulation of mass–spring–damper systems is to consider instead of the spring extensions q the configuration vector $q_c \in \Lambda^0 =: Q_c$ describing the *positions* of the masses. For ordinary springs, the relation between $q_c \in \Lambda^0$ and $q \in \Lambda^1$ describing the extensions of the springs is given as $q = D^T q_c$. Hence, the energy can be also expressed as the function H_c of (q_c, p) defined as

$$H_c(q_c, p) := H(D^T q_c, p) \tag{6.199}$$

It follows that the dynamics of the mass–spring–damper system is alternatively given by the following Hamiltonian equations in the state variables q_c, p

$$\dot{q}_c = \frac{\partial H_c}{\partial p}(q_c, p)$$

$$\dot{p} = -\frac{\partial H_c}{\partial q_c}(q_c, p) - D_d R D_d^T \frac{\partial H_c}{\partial p}(q_c, p) + E f_b \tag{6.200}$$

$$e^b = E^T \frac{\partial H_c}{\partial p}(q_c, p)$$

What is the relation with the formulation given before? It turns out that this relation is precisely given by the standard procedure of *symmetry reduction* of a Hamiltonian system. Indeed, since $\mathbb{1}^T D = 0$ the Hamiltonian function $H_c(q_c, p)$ given in (6.199) is *invariant* under the action of the group $\mathbb{R}$ acting on the phase space $\Lambda^0 \times \Lambda_0 \simeq \mathbb{R}^{2N}$ by the symplectic group action

$$(q_c, p) \mapsto (q_c + \alpha \mathbb{1}, p), \quad \alpha \in \mathbb{R} \tag{6.201}$$

From standard reduction theory of Hamiltonian dynamics with symmetries, see e.g., [179, 197], it thus follows that we may factor out the configuration space $Q_c := \Lambda^0$ to the *reduced configuration space*

$$Q := \Lambda^0 / \mathbb{R} \tag{6.202}$$

Using the identification $Q := \Lambda^0 / \mathbb{R} \simeq D^T \Lambda^0 \subset \Lambda^1$ the *reduced state space* of the mass–spring–damper system is given by im $D^T \times \Lambda_0$, with im $D^T \subset \Lambda^1$, and the Hamiltonian equations (6.200) on $\Lambda^0 \times \Lambda_0$ reduce to the port-Hamiltonian equations (6.184) on im $D^T \times \Lambda_0 \subset \Lambda^1 \times \Lambda_0$ as before.

The above example of a mass–spring–damper system on a graph can be generalized as follows. First note that a mass–spring–damper system with additive Hamiltonian H given by (6.192) can be also interpreted as the *interconnection* of port-Hamiltonian systems $\dot{p}_i = u_i^0$, $y_i^0 = \frac{\partial H_i^0}{\partial p_i}(p_i)$ corresponding to the masses (index i ranging over the vertices), port-Hamiltonian systems $\dot{q}_j = u_j^1$, $y_j^1 = \frac{\partial H_j^1}{\partial q_j}(q_j)$ corresponding to the springs (with j ranging over the spring edges), and static port-Hamiltonian systems $y_k^1 = r_j u_k^1$, $r_k > 0$, corresponding to dampers (with index k ranging over the damper edges), via the interconnection equations

$$u^v = -Dy^b, \quad u^b = D^T y^v \qquad (6.203)$$

Here the superscripts ^{v}, ^{b}, again refer to inputs and outputs of the port-Hamiltonian systems associated to, respectively, the vertices and edges (branches). In the same way we can therefore consider *arbitrary* port-Hamiltonian systems with scalar inputs and outputs associated with the vertices and the edges, interconnected by (6.203). Like in the general theory of interconnection of port-Hamiltonian systems this again defines a port-Hamiltonian DAE system, with Dirac structure determined by the Dirac structures of the port-Hamiltonian systems associated to the vertices and to the edges, and by the interconnection (6.203).

Remark 6.8.7 Similar to the second scenario considered for passive systems in Sect. 9.94, we may also consider the interconnection of single-input single-output port-Hamiltonian systems associated to the vertices of a graph by the interconnection $u = -Ly + e$, cf. (4.91), where L is a *balanced* Laplacian matrix. Decomposing L into its symmetric and skew-symmetric part we then obtain an interconnected port-Hamiltonian system with extra energy-dissipating terms corresponding to the symmetric part of L.

Remark 6.8.8 Another paradigmatic example of port-Hamiltonian systems on graphs are *RLC-electrical circuits*. In this case, all the energy-storing and energy-dissipating elements are associated to the *edges* of the circuit graph. This leads to the consideration of the *Kirchhoff–Dirac structure* defined as

$$\mathcal{D}_K := \{(f_1, e^1, f_b, e^b) \in \Lambda_1 \times \Lambda^1 \times \Lambda_b \times \Lambda^b \mid$$
$$D_i f_1 = 0, D_b f_1 = f_b, \exists e^{0i} \in \Lambda^{0i} \text{ s.t. } e^1 = -D_i^T e^{0i} - D_b^T e^b\} \qquad (6.204)$$

capturing Kirchhoff's current and voltage laws. The port-Hamiltonian formulation of the electrical circuit is obtained by supplementing the Kirchhoff–Dirac structure by energy-storage relations corresponding to either capacitors or inductors, and by energy-dissipating relations corresponding to the resistors [297].

6.9 Scattering of Port-Hamiltonian Systems

In Sects. 2.4 and 3.4, we already introduced the scattering transformation from *flow and effort* vectors f, e to *wave vectors* v, z. Thus, let $\mathcal{F}$ be an ℓ-dimensional linear space of flows, and consider the canonically defined symmetric bilinear form, cf. (6.114), on $\mathcal{F} \times \mathcal{E}$, with $\mathcal{E} = \mathcal{F}^*$, given as

$$\ll (f_1, e_1), (f_2, e_2) \gg := < e_1 \mid f_2 > + < e_2 \mid f_1 > \qquad (6.205)$$

for $f_i \in \mathcal{F}, e_i \in \mathcal{E}^*, i = 1, 2$. Furthermore, as in Sect. 2.4, let $\mathcal{V} \subset \mathcal{F} \times \mathcal{E}$ be any ℓ-dimensional *positive* space of $\ll, \gg$, and $\mathcal{Z} \subset \mathcal{F} \times \mathcal{E}$ an ℓ-dimensional *negative* space of $\ll, \gg$, which is *orthogonal* (in the sense of $\ll, \gg$) to $\mathcal{V}$. This means that

$$\mathcal{F} \times \mathcal{E} = \mathcal{V} \oplus \mathcal{Z} \qquad (6.206)$$

Now, consider a constant Dirac structure $\mathcal{D} \subset \mathcal{F} \times \mathcal{E}$, that is

$$\mathcal{D} = \mathcal{D}^{\perp\!\!\!\perp} \qquad (6.207)$$

with $\perp\!\!\!\perp$ denoting orthogonal companion with respect to $\ll, \gg$. It follows that $\ll, \gg$ is *zero* when restricted to $\mathcal{D}$, and thus

$$\mathcal{D} \cap \mathcal{V} = 0, \quad \mathcal{D} \cap \mathcal{Z} = 0 \qquad (6.208)$$

This implies that the Dirac structure $\mathcal{D}$ can be represented as the *graph* of an *invertible* linear map $\mathcal{O} : \mathcal{V} \to \mathcal{Z}$, that is,

$$\mathcal{D} = \{(f, e) = v + z \mid z = \mathcal{O}v\}, \qquad (6.209)$$

where $v + z \in \mathcal{V} \oplus \mathcal{Z}$ is the *scattering representation* of $(f, e) \in \mathcal{F} \times \mathcal{E}$ with respect to the scattering subspaces $\mathcal{V}, \mathcal{Z}$.

Furthermore, for any $(f_1, e_1), (f_2, e_2) \in \mathcal{D}$, with scattering representation $v_1 + z_1$, respectively, $v_2 + z_2$, we obtain by (2.35) and (6.207)

$$0 = < e_1 \mid f_2 > + < e_2 \mid f_1 > = < v_1, v_2 >_\mathcal{V} - < z_1, z_2 >_\mathcal{Z}, \qquad (6.210)$$

where $<, >_\mathcal{V}$ and $<, >_\mathcal{Z}$ are the inner-products on $\mathcal{V}$, respectively, $\mathcal{Z}$, induced from $\ll, \gg$; see Sect. 2.4, Eq. (2.35). This implies that

$$< z_1, z_2 >_\mathcal{Z} = < \mathcal{O}v_1, \mathcal{O}v_2 >_\mathcal{Z} = < v_1, v_2 >_\mathcal{V} \qquad (6.211)$$

for all $v_1, v_2 \in \mathcal{V}$. Hence, the linear map $\mathcal{O} : \mathcal{V} \to \mathcal{Z}$ is an *inner-product preserving* map from $\mathcal{V}$, with inner product $<, >_\mathcal{V}$, to $\mathcal{Z}$ with inner-product $<, >_\mathcal{Z}$. Conversely, let $\mathcal{O} : \mathcal{V} \to \mathcal{Z}$ be an inner-product preserving map. If we now *define* $\mathcal{D}$ by (6.209), then by (6.210) and (6.211)

$$0 =< v_1, v_2 >_\mathcal{V} - < z_1, z_2 >_\mathcal{Z}=< e_1 \mid f_2 > + < e_2 \mid f_1 > ,$$

and thus $\mathcal{D} \subset \mathcal{D}^{\perp\!\perp}$. Furthermore, because dim $\mathcal{D} = \ell$, we conclude $\mathcal{D} = \mathcal{D}^{\perp\!\perp}$, implying that $\mathcal{D}$ is a Dirac structure. *Hence constant Dirac structures $\mathcal{D} \subset \mathcal{F} \times \mathcal{E}$ are in
one-to-one correspondence with inner-product preserving linear maps $\mathcal{O} : \mathcal{V} \to \mathcal{Z}$.*
This leads to the following definition.

Definition 6.9.1 Let $\mathcal{D} \subset \mathcal{F} \times \mathcal{E}$ be a Dirac structure, and let $(\mathcal{V}, \mathcal{Z})$ be a pair of
scattering subspaces. The map $\mathcal{O} : \mathcal{V} \to \mathcal{Z}$ satisfying (6.209) is called the *scattering
representation* of $\mathcal{D}$.

A matrix representation of the scattering representation $\mathcal{O}$ of a Dirac structure $\mathcal{D}$ is
obtained as follows. Consider a basis $a_1, \ldots, a_\ell$ for $\mathcal{F}$ and dual basis $a_1^*, \ldots, a_\ell^*$ for
$\mathcal{E}$, together with the resulting scattering transformation as in (2.40). Furthermore,
corresponding to this basis let $\mathcal{D}$ be given in kernel representation as

$$\mathcal{D} = \{(f, e) \mid Ff + Ee = 0\}, \tag{6.212}$$

with F, E square $\ell \times \ell$ matrices satisfying

$$EF^T + FE^T = 0, \quad \text{rank}[F \vdots E] = \ell \tag{6.213}$$

Proposition 6.9.2 *Let the Dirac structure $\mathcal{D}$ be given by (6.212). The matrix representation of its scattering representation $\mathcal{O} : \mathcal{V} \to \mathcal{Z}$ is the orthonormal matrix*

$$\mathcal{O} = (F - E)^{-1}(F + E) \tag{6.214}$$

Proof $\mathcal{D}$ is equivalently given in image representation as $\mathcal{D} = \{(f, e) \mid f = E^T \lambda, e = F^T \lambda, \lambda \in \mathbb{R}^\ell\}$. The coordinate relation between $(f, e) \in \mathcal{F} \times \mathcal{E}$ and its scattering
representation $v + z$ is given as (cf. (2.41))

$$\begin{aligned} v &= \tfrac{1}{\sqrt{2}}(f + e) \\ z &= \tfrac{1}{\sqrt{2}}(-f + e) \end{aligned} \tag{6.215}$$

Thus in scattering representation $\mathcal{D}$ is given as

$$\mathcal{D} = \left\{ v + z \mid v = \frac{1}{\sqrt{2}}(E^T + F^T)\lambda, z = \frac{1}{\sqrt{2}}(-E^T + F^T)\lambda, \lambda \in \mathbb{R}^\ell \right\} \tag{6.216}$$

We claim that $E^T + F^T$ is invertible. Indeed, suppose $x \in \ker (E^T + F^T)$, that is,
$E^T x = -F^T x$. Since by (6.213) $EF^T x + FE^T x = 0$ for all x, this implies $EE^T x = -EF^T x = FE^T x = -FF^T x$, and thus

$$[EE^T + FF^T]x = 0, \tag{6.217}$$

which in view of rank $[F \vdots E] = \ell$ implies $x = 0$. Hence, $E^T + F^T$ and $F + E$ are invertible. Therefore

$$\mathcal{D} = \{(v, z) \mid z = (F^T - E^T)(F^T + E^T)^{-1}v\} \tag{6.218}$$

Similarly, it follows that $-E^T + F^T$ and thus $F - E$ are invertible. Comparing with (6.209) we conclude that $\mathcal{O} = (F^T - E^T)(F^T + E^T)^{-1}$. Finally, adding, respectively, subtracting, $EF^T + FE^T = 0$ to the expression $FF^T + EE^T$ yields the equality

$$(F + E)(F^T + E^T) = (F - E)(F^T - E^T) \tag{6.219}$$

and thus $\mathcal{O}$ is also expressed as in (6.214). Furthermore, (6.219) implies

$$\begin{aligned}
\mathcal{O}\mathcal{O}^T &= (F - E)^{-1}(F + E)(F^T + E^T)(F^T - E^T)^{-1} \\
&= (F - E)^{-1}(F - E)(F^T - E^T)(F^T - E^T)^{-1} = I_\ell,
\end{aligned}$$

showing that O is orthonormal. $\qquad\square$

Example 6.9.3 Let the Dirac structure $\mathcal{D}$ be given by a skew-symmetric matrix J, that is, $\mathcal{D} = \{(f, e) \mid f = Je, \ J = -J^T\}$. Then the scattering representation of $\mathcal{D}$ is the orthonormal matrix

$$\mathcal{O} = (I + J)^{-1}(I - J) \tag{6.220}$$

(known as the *Cayley transform* of J).

Remark 6.9.4 The same result holds for Dirac structures on a manifold $\mathcal{X}$. In this case, the Dirac structure is represented by an orthonormal matrix $\mathcal{O}(x)$ depending on $x \in \mathcal{X}$ (where also the scattering subspaces $\mathcal{V}$ and $\mathcal{Z}$ may depend on x). In particular, the scattering representation of the Dirac structure defined as the graph of $J(x) = -J^T(x)$ is $\mathcal{O}(x) = (I + J(x))^{-1}(I - J(x))$.

A special type of Dirac structures (called 0- and 1-*junctions*) are defined as follows

$$\begin{aligned}
\mathcal{D}_0 &= \{(f, e) \in \mathbb{R}^\ell \times \mathbb{R}^\ell \mid f_1 + \cdots + f_\ell = 0, \ e_1 = \cdots = e_\ell\} \\
\mathcal{D}_1 &= \{(f, e) \in \mathbb{R}^\ell \times \mathbb{R}^n \mid e_1 + \cdots + e_\ell = 0, \ f_1 = \cdots = f_\ell\}
\end{aligned} \tag{6.221}$$

Using scattering representations they can be characterized as follows.

Proposition 6.9.5 *Scattering representations $\mathcal{O}_0, \mathcal{O}_1$ of $\mathcal{D}_0, \mathcal{D}_1$ are given by*

$$\mathcal{O}_0 = \frac{2}{\ell}\mathbb{I}_\ell - I_\ell, \quad \mathcal{O}_1 = -\frac{2}{\ell}\mathbb{I}_\ell + I_\ell \tag{6.222}$$

where $\mathbb{I}_\ell$ denotes the $\ell \times \ell$ matrix filled with ones and I_ℓ is the $\ell \times \ell$ identity matrix. Moreover, $\mathcal{O}_0$ and $\mathcal{O}_1$ are the only orthonormal $\ell \times \ell$ matrices that have equal diagonal elements and equal off-diagonal elements.

Proof With respect to the last claim note that $\mathcal{O} = a\mathbb{I}_\ell + bI_\ell$ is orthonormal if and only $\ell a + 2b = 0$ and $b^2 = 1$. The case $b = 1$ gives $\mathcal{O}_1 = -\frac{2}{\ell}\mathbb{I}_\ell + I_\ell$, while $b = -1$ yields $\mathcal{O}_0 = \frac{2}{\ell}\mathbb{I}_\ell - I_\ell$. The rest follows by direct computation. $\square$

Similarly to Sect. 3.4, let us finally apply scattering to a standard input-state-output port-Hamiltonian form

$$
\begin{aligned}
\dot{x} &= [J(x) - R(x)]\frac{\partial H}{\partial x}(x) + g(x)u \\
y &= g^T(x)\frac{\partial H}{\partial x}(x)
\end{aligned}
\tag{6.223}
$$

Consider a scattering representation of $(f_P, e_P) = (u, y)$ (but *not* of (f_S, e_S)), defined as

$$
\begin{aligned}
v &= \tfrac{1}{\sqrt{2}}(u + y) \\
z &= \tfrac{1}{\sqrt{2}}(-u + y)
\end{aligned}
\tag{6.224}
$$

The inverse of this transformation is $u = \frac{1}{\sqrt{2}}(v - z), y = \frac{1}{\sqrt{2}}(v + z)$, which by substitution in (6.223) yields

$$
\begin{aligned}
\dot{x} &= \left[J(x) - R(x) - g(x)g^T(x)\right]\frac{\partial H}{\partial x}(x) + \sqrt{2}g(x)v \\
z &= \sqrt{2}g^T(x)\frac{\partial H}{\partial x}(x) - v
\end{aligned}
\tag{6.225}
$$

Note that, compared with (6.223), artificial *energy dissipation* has been inserted in two ways: (i) by an extra resistive structure matrix $g(x)g^T(x) \geq 0$, (ii) by a negative unity feedthrough from v to z.

Finally, *composition* of Dirac structures takes the following form in scattering formulation. Consider two Dirac structures $\mathcal{D}_A, \mathcal{D}_B$ as in Theorem 6.6.9, composed by setting

$$
f_A = -f_B \in \mathcal{F}, \quad e_A = e_B \in \mathcal{E}
\tag{6.226}
$$

Now consider scattering representations $(f_A, e_A) = v_A + z_A$ and $(f_B, e_B) = v_B + z_B$ with respect to the *same* scattering subspaces $\mathcal{V}, \mathcal{Z} \subset \mathcal{F} \times \mathcal{E}$. Then (6.226) becomes

$$
\begin{aligned}
z_A &= v_B \\
z_B &= v_A
\end{aligned}
\tag{6.227}
$$

expressing that the outgoing wave vector for $\mathcal{D}_A$ equals the incoming wave vector for $\mathcal{D}_B$, and conversely. Hence, the composition of $\mathcal{D}_A, \mathcal{D}_B$ is seen to correspond to the configuration depicted in Fig. 6.9, known as the *Redheffer star product* [259] of the orthonormal matrices $\mathcal{O}_A$ and $\mathcal{O}_B$. This is formulated in the next proposition.

Fig. 6.9 Redheffer star
product of $\mathcal{O}_A$ and $\mathcal{O}_B$

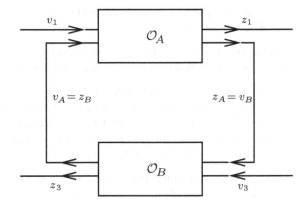

Proposition 6.9.6 *Let the orthonormal mappings $\mathcal{O}_A$ and $\mathcal{O}_B$ be scattering representations of $\mathcal{D}_A$ and $\mathcal{D}_B$ with respect to the same scattering subspaces. Then the scattering representation of $\mathcal{D}_A \circ \mathcal{D}_B$ is given by $\mathcal{O}_A \star \mathcal{O}_B$, with $\star$ denoting the Redheffer star product.*

Remark 6.9.7 Since $\mathcal{D}_A \circ \mathcal{D}_B$ is a Dirac structure it directly follows that the Redheffer star product of the orthonormal mappings $\mathcal{O}_A$ and $\mathcal{O}_B$ is again an orthonormal mapping.

6.10   Notes for Chapter 6

1. Port-Hamiltonian systems were originally introduced (under the slightly different name of *port-controlled Hamiltonian systems*) in Maschke & van der Schaft [201, 202], Maschke, van der Schaft & Breedveld [203] and van der Schaft & Maschke [294, 295].

2. A broad coverage of port-Hamiltonian systems and the background theory of port-based modeling, including application areas, can be found in [93] and the references quoted therein. The port-Hamiltonian formulation of bond-graph models is described in Golo, van der Schaft, Breedveld, Maschke [115]. A recent introductory survey of port-Hamiltonian systems theory, emphasizing new developments, is [291].

3. For port-Hamiltonian systems (6.1) two geometric structures play a role: (i) an (almost-)Poisson structure determined by the skew-symmetric matrix $J(x)$, (ii) the singular Riemannian metric determined by the symmetric positive semi-definite matrix $R(x)$. For some results and ideas on the interplay between these two structures, and its consequences for the resulting dynamics we refer to Morrison [223], and the references quoted therein. Similar structures have been used

in the description of thermodynamical systems, see e.g., Öttinger [246] and the references therein.

4. One can also define a bracket with respect to the *combined* structure $F(x) := J(x) - R(x)$, called the *Leibniz bracket*; see e.g., Ortega & Planas-Bielsa [236].

5. The formulation of detailed-balanced mass action kinetics chemical reaction networks as in Example 6.1.7 can be found in van der Schaft, Rao & Jayawardhana [299]; see also [258, 301] for the generalization to complex-balanced reaction networks. The port-Hamiltonian formulation was emphasized in van der Schaft, Rao & Jayawardhana [300].

6. A broader discussion about the obstruction to a port-Hamiltonian formulation indicated in Example 6.1.9 can be found in [213].

7. The formulation of autonomous Hamiltonian dynamics with regard to a Poisson structure which not necessarily has full rank, is standard in the literature on geometric mechanics, see e.g., Marsden & Ratiu [197], Olver [235].

8. The dual to any Lie algebra is endowed with a canonical Poisson structure, see e.g., Weinstein [347], Marsden & Ratiu [197]. For instance the Poisson structure given in Example 6.2.2 for the Hamiltonian formulation of Euler's equations is the Lie–Poisson structure on $so^*(3)$.

9. For an in-depth treatment of mechanical systems with kinematic constraints, including the constrained Euler–Lagrange equations, see e.g., Bloch [43], Bullo & Lewis [53], and the older reference Neimark & Fufaev [230]. For a classical survey on the kinematic model, regarding the admissible velocities as being directly controlled, we refer to Kolmanovsky & McClamroch [167].

10. The introduction of the new "momentum" variables $\tilde{p}$ in (6.45) is close to the classical use of quasi-coordinates (see e.g., Steigenberger [326] for a survey).

11. The description in Sect. 6.2 of mechanical systems with kinematic constraints as port-Hamiltonian systems defined with respect to the almost-Poisson structure J_c on the constrained state space $\mathcal{X}_c$ given by (6.48) and (6.51) is taken from van der Schaft & Maschke [293], where also the result can be found that J_c is a Poisson structure (i.e., satisfying the Jacobi-identity (6.33)) if and only if the kinematic constraints are holonomic. See also van der Schaft & Maschke [295] and Dalsmo & van der Schaft [78]. For a survey on almost-Poisson structures in nonholonomic mechanics see Cantrijn, de Leon & de Diego [60].

12. The port-Hamiltonian formulation of the classical eight-dimensional model of the synchronous generator, see e.g., Kundur [177], in Example 6.3.4 is taken from Shaik Fiaz, Zonetti, Ortega, Scherpen & van der Schaft [98]; see also van der Schaft & Stegink [303].

13. Hamiltonian functions involving two kinds of state variables in a non-separable way not only show up in multi-physics systems, as illustrated in Sect. 6.3, but also in "cyber-physical systems" such as *variable impedance control*. In its

most simple form a variable impedance controller is defined by a (virtual) linear spring with energy $H(q) = \frac{1}{2}kq^2$, where we regard, next to the extension q of the spring, also the spring "constant" k as a state variable whose value may change in time. This leads to the consideration of the port-Hamiltonian system with inputs u_1, u_2 and outputs y_1, y_2 given as

$$\begin{bmatrix} \dot{q} \\ \dot{k} \end{bmatrix} = \begin{bmatrix} u_1 \\ u_2 \end{bmatrix}, \quad \begin{bmatrix} y_1 \\ y_2 \end{bmatrix} = \begin{bmatrix} kq \\ \frac{1}{2}q^2 \end{bmatrix} \tag{6.228}$$

Here the port (u_1, y_1) corresponds to interaction with the environment (defining an impedance k), while the port (u_2, y_2) defines a control port, regulating the value of the impedance k based on the output $y_2 = \frac{1}{2}q^2$, possibly modulated by information about other variables in the total system. In robotics, this basic idea is referred to as *variable stiffness control*; see e.g., [354] for a survey.

14. An extensive treatment of Casimir functions for autonomous Hamiltonian dynamics as discussed in Sect. 6.4 can be found e.g., in Marsden & Ratiu [197] and Olver [235]. For the Energy-Casimir method see e.g., Marsden & Ratiu [197] and the references quoted in there. Here also the close connection with *symmetries* can be found. Solving the pde's (9.92) involves integrability conditions on the structure matrices $J(x)$ and $R(x)$. In particular, if $J(x)$ is a Poisson structure (i.e., satisfying the Jacobi-identity), then there always exist r independent solutions $C_1, \ldots, C_r$ of the pde's $\frac{\partial^T C}{\partial x}(x)J(x) = 0$, with $r = \dim \ker J(x)$.

15. System theoretic properties of the closely related class of *input–output Hamiltonian systems* introduced in Brockett [50] are investigated e.g., in van der Schaft [269], Crouch & van der Schaft [73], Nijmeijer & van der Schaft [233] (Chap. 12).

16. A subclass of port-Hamiltonian systems, called *reciprocal port-Hamiltonian systems* can be converted into a *gradient system* [71] formulation (with respect to an indefinite Hessian Riemannian metric); cf. van der Schaft [284] and van der Schaft & Jeltsema [291].

17. A systematic treatment of port-Hamiltonian systems with *switching* structure matrices (with applications to switching electrical circuits or mechanical systems) can be found e.g., in Escobar, van der Schaft & Ortega [95], van der Schaft & Camlibel [290], Valentin, Magos & Maschke [340]; see also van der Schaft & Jeltsema [291].

18. The property that a system is shifted passive with respect to *any* constant $\bar{u}$ and corresponding steady-state $\bar{x}$, cf. Sect. 6.5 and Proposition 6.5.1, was coined as *equilibrium independent passivity* in Arcak, Meissen & Packard [11].

19. Proposition 6.5.4 is due to Ferguson, Middleton & Donaire [97].

20. Example 6.5.3 is taken from Bürger & De Persis [54]; see also Arcak [10], van der Schaft & Stegink [303].

21. The construction of the modified Hamiltonian $\widetilde{H}$ in (6.106) can be found in [199].

22. The definition of Dirac structure was originally intended as a generalization of both Poisson and symplectic structures; cf. Courant [72], Dorfman [85]. The name apparently originates from the concept of the *Dirac bracket* as appearing for Hamiltonian systems with constraints in Dirac [81, 82]. The kernel, image and constrained input–output representations of Dirac structures can be found in Dalsmo & van der Schaft [78], see also Courant [72]. The hybrid input–output representation is due to Bloch & Crouch [47]. See also van der Schaft [282] for a survey.

23. The proof of Theorem 6.6.9, as well as of Proposition 6.6.10, can be found in Cervera, van der Schaft & Banos [63] using ideas from Narajanan [228]; see also van der Schaft [281].

24. The definition of port-Hamiltonian systems with respect to Dirac structures was first given in van der Schaft & Maschke [294], and further developed in van der Schaft & Maschke [189, 292]; see also Bloch & Crouch [47] for the use of Dirac structures in the modeling of general LC circuits. For a treatment of constrained mechanical systems in this context, see Maschke & van der Schaft [206].

25. Integrability of Dirac structures (generalizing the Jacobi-identity for Poisson structures) is treated in Courant [72], Dorfman [85]. See also Merker [218] and the references quoted therein for further developments. For applications of the integrability of Dirac structures to properties of port-Hamiltonian DAE systems, including the connection to integrability of kinematic constraints, we refer to Dalsmo & van der Schaft [78]; see also van der Schaft & Jeltsema [291] and the references quoted therein. Necessary and sufficient conditions for the integrability of composed Dirac structures are obtained in Blankenstein & van der Schaft [40].

26. A further treatment of port-Hamiltonian DAE systems and their equational representations can be found in van der Schaft [286].

27. Section 6.8 is largely based on [298]. The port-Hamiltonian modeling of general *LC* circuits can be found in Maschke, van der Schaft & Breedveld [205], Maschke & van der Schaft [207]. See also Blankenstein [39]. The formulation of *RLC*-circuits alluded to in Remark 6.8.8 can be found in van der Schaft & Maschke [297]; with the notion of Kirchhoff–Dirac structure in (6.204) given in van der Schaft & Maschke [298].

28. The scattering representation of Dirac structures as dealt with in Sect. 6.9 can be found in Cervera, van der Schaft & Banos [63]. The proof of Proposition 6.9.2 is based on ideas from Courant [72].

29. Proposition 6.9.5 is originally due to Hogan & Fasse [128].

30. Many of the definitions and results in this chapter can be extended to *distributed-parameter* port-Hamiltonian systems; see e.g., van der Schaft [296], Duindam, Macchelli, Stramigioli & Bruyninckx [93], van der Schaft & Jeltsema [291], and the references quoted therein.

31. For a theory of *symmetries* of port-Hamiltonian systems, and the resulting reduction and existence of Casimirs, see e.g., van der Schaft [280], Blankenstein & van der Schaft [41], Merker [218].

32. An extension of the port-Hamiltonian formalism to thermodynamical systems can be found in Eberard, Maschke & van der Schaft [94].

Chapter 7
Control of Port-Hamiltonian Systems

In this chapter we will exploit the port-Hamiltonian structure for control, going beyond passivity. We will mainly concentrate on the problem of *set-point stabilization*. Section 7.1 focusses on *control by interconnection*, by attaching a controller port-Hamiltonian system to the plant port-Hamiltonian system. Section 7.2 takes a different perspective by emphasizing direct shaping of the Hamiltonian and the structure matrices by state feedback. Other control opportunities will be indicated in Sect. 7.3; see also the Notes at the end of this chapter.

7.1 Stabilization by Interconnection

Consider an input-state-output port-Hamiltonian systems (6.1), for ease of exposition without feedthrough terms,

$$\Sigma : \begin{array}{ll} \dot{x} = [J(x) - R(x)]\frac{\partial H}{\partial x}(x) + g(x)u, & x \in \mathcal{X}, \ u \in \mathbb{R}^m \\ y = g^T(x)\frac{\partial H}{\partial x}(x), & y \in \mathbb{R}^m \end{array} \qquad (7.1)$$

The simplest situation in stabilization is when the set-point $x^* \in \mathcal{X}$ is a strict *minimum*[1] of the Hamiltonian H. In this case, the port-Hamiltonian system (6.1) is *passive*, and thus we may directly apply the asymptotic stabilization theory of passive systems provided in Chap. 4, Sect. 4.2, by employing output feedback $u = -Dy$, with $D = D^T$ a positive definite matrix, and using the Hamiltonian H as Lyapunov function. In particular we can directly apply Corollary 4.2.2 as illustrated by the next examples.

Example 7.1.1 Consider a fully actuated mechanical system (6.23), with H given by (6.24). Assume that the potential energy P is such that $P(q^*) = 0$ for the set-point configuration q^*, while $P(q) > 0$, $q \neq q^*$ and $\frac{\partial P}{\partial q}(q) \neq 0$, $q \neq q^*$. Then (6.23) with output $y = \dot{q}$ is zero-state detectable with respect to $(q^*, p^* = 0)$. Thus, the feedback

---

[1] "Local" or "global"; we leave this open for flexibility of exposition.

© Springer International Publishing AG 2017
A. van der Schaft, L_2-*Gain and Passivity Techniques in Nonlinear Control*,
Communications and Control Engineering, DOI 10.1007/978-3-319-49992-5_7

$u = -Dy$, $D = D^T > 0$, asymptotically stabilizes the equilibrium $(q^*, p = 0)$. In the case of an *underactuated* system (6.26) asymptotic stability results whenever the system is zero-state detectable with respect to the output $y = B^T(q)\dot{q}$.

Example 7.1.2 (Example 6.4.5 continued) H has a global minimum at $Q = 0$, $\varphi_1 = \varphi_2 = 0$, and the system is zero-state detectable. Therefore, the insertion of a resistor $u = -Ry$, $R > 0$, at the place of the voltage source will asymptotically stabilize this equilibrium.

Example 7.1.3 (Example 6.2.1 continued) The kinetic energy H has a global minimum at $p_x = p_y = p_z = 0$. Consider the feedback $u = -dy$, $d > 0$, resulting in $\frac{d}{dt}H = -dy^2$. The largest invariant set within $y = 0$ is determined as follows. First of all, consider the plane

$$P := \{(p_x, p_y, p_z) \mid y = 0\} = \left\{(p_x, p_y, p_z) \mid \frac{b_x}{I_x}p_x + \frac{b_y}{I_y}p_y + \frac{b_z}{I_z}p_z = 0\right\}$$

Second, we have the cone

$$C := \{(p_x, p_y, p_z) \mid \dot{y} = 0\} = \{(p_x, p_y, p_z) \mid \\ b_x(I_y - I_z)p_y p_z + b_y(I_z - I_x)p_z p_x + b_z(I_x - I_y)p_x p_y = 0\}$$

Furthermore, the trajectories of (6.31) for $u = 0$ remain on an energy level given by the ellipsoid $E_c = \{(p_x, p_y, p_z) \mid H(p_x, p_y, p_z) = c\}$ for a constant c. The intersection $P \cap C \cap E_c$ consists only of isolated points, which thus have to be equilibria of (6.31) for $u = 0$. The set of equilibria of (6.31) for $u = 0$ is given by the union of the p_x, p_y and p_z axis. Since these isolated points also have to be in P it follows that the largest invariant set within $y = 0$ equals the origin if and only if $b_x \neq 0$, $b_y \neq 0$ and $b_z \neq 0$. Hence, it follows from LaSalle's Invariance principle that the feedback $u = -dy$, $d > 0$, is asymptotically stabilizing to zero if and only if $b_x \neq 0$, $b_y \neq 0$, $b_z \neq 0$.

Up to now we only exploited the passivity of the port-Hamiltonian system. The situation becomes different when the set-point x^* is *not* a strict minimum of the Hamiltonian H. A new twist as compared to the stabilization theory of Sect. 4.2 based on passivity is then provided by the possible existence of *Casimir functions*; cf. Sect. 6.4. Indeed, assume one can find a Casimir function C such that the *modified Hamiltonian*

$$H_{\text{mod}}(x) := \Phi(H(x), C(x)) \tag{7.2}$$

for some map $\Phi : \mathbb{R}^2 \to \mathbb{R}$ has a strict minimum at the set-point x^*. Then the system can be rewritten as the *modified* input-state-output port-Hamiltonian system

$$\begin{aligned} \dot{x} &= [J(x) - R(x)]\frac{\partial H_{\text{mod}}}{\partial x}(x) + g(x)u \\ y_{\text{mod}} &:= g^T(x)\frac{\partial H_{\text{mod}}}{\partial x}(x), \end{aligned} \tag{7.3}$$

which is passive with respect to the *modified* output y_{mod}. Hence, as above, asymptotic stabilization of x^* may be achieved by feedback $u = -Dy_{\text{mod}}$, $D > 0$, employing H_{mod} as Lyapunov function, provided a detectability condition is satisfied.

Example 7.1.4 Consider the LC-circuit of Example 6.4.5. We may consider as modified Hamiltonian any quadratic function

$$H_{\text{mod}}(x) = \frac{1}{2L_1}\varphi_1^2 + \frac{1}{2L_2}\varphi_2^2 + \gamma(\varphi_1 + \varphi_2), \quad \gamma \in \mathbb{R} \tag{7.4}$$

serving as candidate Lyapunov function for non-zero set-points $(0, \varphi^*, \varphi^*)$. This leads to the modified output $y_{\text{mod}} = \frac{1}{L_1}\varphi_1 + \gamma$. We note that this can be naturally extended to arbitrary *nonlinear* inductors.

The stability analysis method of finding Casimirs $C_1, \ldots, C_r$ such that a suitable function $\Phi(H, C_1, \ldots, C_r)$ of the energy H and the Casimirs has a strict minimum at the equilibrium under consideration is called the *Energy-Casimir method*. We will use the same terminology for the corresponding asymptotic stabilization method.

Remark 7.1.5 Note that the Energy-Casimir method can be extended to the asymptotic stabilization of steady states $\bar{x}$ corresponding to a non-zero input $\bar{u}$; at least in the case that the matrices J, R, g are all constant. Indeed, in this case the input-state-output port-Hamiltonian system may be rewritten as a shifted port-Hamiltonian system with shifted external variables $u - \bar{u}$, $y - \bar{y}$ and shifted Hamiltonian $\hat{H}$; see Sect. 6.5. Candidate Lyapunov functions will now be sought as combinations of $\hat{H}$ and the Casimirs.

Next, suppose that H does not have a strict minimum at x^*, and that no useful Casimirs C can be found for direct application of the Energy-Casimir method. In this case, we will take recourse to *dynamical* controller systems that are also in port-Hamiltonian form, in order to generate Casimirs for the *closed-loop system*. As known from Sect. 6.7 any power-conserving interconnection of port-Hamiltonian systems is again port-Hamiltonian. In particular, the interconnection of the *plant* port-Hamiltonian system Σ given by (7.1) with a *controller* port-Hamiltonian system

$$\Sigma_c : \begin{array}{l} \dot{\xi} = [J_c(\xi) - R_c(\xi)]\frac{\partial H_c}{\partial \xi}(\xi) + g_c(\xi)u_c, \; \xi \in \mathcal{X}_c, \; u_c \in \mathbb{R}^m \\ y_c = g_c^T(\xi)\frac{\partial H_c}{\partial \xi}(\xi), \qquad\qquad\qquad\quad y_c \in \mathbb{R}^m \end{array} \tag{7.5}$$

with $J_c(\xi) = -J_c^T(\xi)$, $R_c(\xi) = R_c^T(\xi) \geq 0$, via the standard negative feedback interconnection

$$\begin{array}{l} u = -y_c + v \\ u_c = y + v_c \end{array} \tag{7.6}$$

where v, v_c are external input signals, results in the closed-loop system

$$
\begin{bmatrix} \dot{x} \\ \dot{\xi} \end{bmatrix} = \left(\underbrace{\begin{bmatrix} J(x) & -g(x)g_c^T(\xi) \\ g_c(\xi)g^T(x) & J_c(\xi) \end{bmatrix}}_{J_{cl}(x,\xi)} - \underbrace{\begin{bmatrix} R(x) & 0 \\ 0 & R_c(\xi) \end{bmatrix}}_{R_{cl}(x,\xi)} \right) \begin{bmatrix} \frac{\partial H}{\partial x}(x) \\ \frac{\partial H_c}{\partial \xi}(\xi) \end{bmatrix}
$$
$$
+ \begin{bmatrix} g(x) & 0 \\ 0 & g_c(\xi) \end{bmatrix} \begin{bmatrix} v \\ v_c \end{bmatrix}
$$
$$
\begin{bmatrix} y \\ y_c \end{bmatrix} = \begin{bmatrix} g^T(x) & 0 \\ 0 & g_c^T(\xi) \end{bmatrix} \begin{bmatrix} \frac{\partial H}{\partial x}(x) \\ \frac{\partial H_c}{\partial \xi}(\xi) \end{bmatrix}
$$
(7.7)

This is again an input-state-output port-Hamiltonian system (7.1), with state space $\mathcal{X} \times \mathcal{X}_c$, Hamiltonian $H(x) + H_c(\xi)$, interconnection structure matrix $J_{cl}(x, \xi)$, resistive structure matrix $R_{cl}(x, \xi)$, inputs (v, v_c) and outputs (y, y_c).

Now let us investigate the Casimir functions of this closed-loop system, especially those relating the state variables ξ of the controller port-Hamiltonian system to the state variables x of the plant port-Hamiltonian system. Indeed, if we can find Casimir functions $C_i(x, \xi)$, $i = 1, \ldots, r$, relating ξ to x, then by the Energy-Casimir method the Hamiltonian $H(x) + H_c(\xi)$ of the closed-loop system may be replaced by a candidate Lyapunov function $\Phi(H + H_c, C_1, \ldots, C_r)$ for any choice of $\Phi : \mathbb{R}^{r+1} \to \mathbb{R}$ and H_c, thus creating the possibility of shaping the dependence on x in a suitable way.

The following basic example illustrates the main idea.

Example 7.1.6 Consider as plant system Σ an actuated mass m, written in port-Hamiltonian form as

$$
\Sigma : \quad \begin{bmatrix} \dot{q} \\ \dot{p} \end{bmatrix} = \begin{bmatrix} 0 & 1 \\ -1 & 0 \end{bmatrix} \begin{bmatrix} \frac{\partial H}{\partial q} \\ \frac{\partial H}{\partial p} \end{bmatrix} + \begin{bmatrix} 0 \\ 1 \end{bmatrix} u
$$
$$
y = \begin{bmatrix} 0 & 1 \end{bmatrix} \begin{bmatrix} \frac{\partial H}{\partial q} \\ \frac{\partial H}{\partial p} \end{bmatrix}
$$
(7.8)

with q the position and p the momentum of the mass, with plant Hamiltonian $H(q, p) = \frac{1}{2m} p^2$ (kinetic energy). Suppose we want to asymptotically stabilize the mass to a set-point $(q^*, p^* = 0)$. Clearly H does not have a strict minimum at $(q^*, p^* = 0)$, while there are no nontrivial Casimirs. Now interconnect as in Fig. 7.1

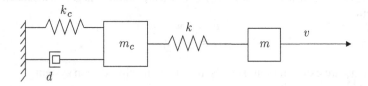

Fig. 7.1 Controlled mass

Σ given by (7.8) via feedback $u = -y_c + v$, $u_c = y$ to a port-Hamiltonian controller system Σ_c consisting of another mass m_c and two springs k_c, k, given by

$$\Sigma_c: \begin{bmatrix} \dot{q}_c \\ \dot{p}_c \\ \dot{\Delta q} \end{bmatrix} = \begin{bmatrix} 0 & 1 & 0 \\ -1 & -d & 1 \\ 0 & -1 & 0 \end{bmatrix} \begin{bmatrix} \frac{\partial H_c}{\partial q_c} \\ \frac{\partial H_c}{\partial p_c} \\ \frac{\partial H_c}{\partial \Delta q} \end{bmatrix} + \begin{bmatrix} 0 \\ 0 \\ 1 \end{bmatrix} u_c \tag{7.9}$$

$$y_c = \frac{\partial H_c}{\partial \Delta q}$$

where q_c is the extension of the spring k_c, Δq the extension of the spring k, p_c the momentum of the mass m_c, $d \geq 0$ is a damping constant, and v is an external force. In particular, the controller Hamiltonian is

$$H_c(q_c, p_c, \Delta q) = \frac{1}{2}\frac{p_c^2}{m_c} + \frac{1}{2}k(\Delta q)^2 + \frac{1}{2}k_c q_c^2 \tag{7.10}$$

The closed-loop system with $v = 0$ is seen to possess the Casimir functions

$$C(q, \Delta q_c, \Delta q) = q - \Delta q - q_c - \delta \tag{7.11}$$

for constant δ. Hence we may consider for the closed-loop system a candidate Lyapunov function

$$V(q, \Delta q, q_c, p, p_c) := \frac{1}{2m}p^2 + \frac{1}{2m_c}p_c^2 + \frac{1}{2}k(\Delta q)^2 + \frac{1}{2}k_c q_c^2 + \gamma(q - \Delta q - q_c - \delta)^2 \tag{7.12}$$

with the positive constants k, k_c, m_c, as well as δ, γ yet to be designed. It is clear that for any set-point q^* we may select these constants in such a way that V has a minimum at $p = 0$, $q = q^*$, for some accompanying values $(\Delta q)^*$, q_c^*, p_c^* of the controller states. By a direct application of LaSalle's Invariance principle it follows that this equilibrium of the closed-loop system is asymptotically stable whenever $d > 0$. Furthermore, the procedure can be extended to a plant Hamiltonian $H(q, p) = \frac{1}{2m}p^2 + P(q)$ for any plant potential energy function $P(q)$. In this case the constants need to be chosen such that

$$P(q) + \frac{1}{2}k(\Delta q)^2 + \frac{1}{2}k_c q_c^2 + \gamma(q - \Delta q - q_c - \delta)^2 \tag{7.13}$$

has a strict minimum at $(q^*, q_c^*, (\Delta q)^*)$ for some value s $q_c^*, (\Delta q)^*$.

Remark 7.1.7 In the above example, the dimension of the controller port-Hamiltonian system is *larger* than the dimension of the plant port-Hamiltonian system. This is mainly due to the fact that we chose to insert the *damping into the controller system*, instead of directly adding damping to the plant system by a feedback $-dy$. Following this latter strategy, one could have replaced the controller port-Hamiltonian system by a one-dimensional system corresponding to a single spring k with extension Δq. On the other hand, the control strategy described above has the advantage of being relatively insensitive to the presence of noise in the measurement of the velocity y. Furthermore, the availability of extra degrees of freedom makes it easier to control the transient behavior, and to attenuate disturbances.

The general theory proceeds as follows. Consider a plant port-Hamiltonian system Σ given by (7.1) in feedback interconnection to a port-Hamiltonian controller system Σ_c given by (7.5). However, for the moment we will *not* assume that $R_c(\xi) \geq 0$ (thus allowing for internal energy creation in the controller).

Without much loss of generality[2] we consider Casimir functions for the closed-loop of the form

$$\xi_i - F_i(x), \qquad i = 1, \dots, r \leq \dim n_c \tag{7.14}$$

This means, see (6.72), that we are looking for solutions of the pde's (with e_i denoting the i-th basis vector)

$$\left[-\frac{\partial^T F_i}{\partial x}(x) \; e_i^T \right] \left[\begin{array}{cc} J(x) - R(x) & -g(x)g_c^T(\xi) \\ g_c(\xi)g^T(x) & J_c(\xi) - R_c(\xi) \end{array} \right] = 0,$$

or written out

$$\frac{\partial^T F_i}{\partial x}(x) \left[J(x) - R(x) \right] - g_c^i(\xi)g^T(x) = 0$$

$$\frac{\partial^T F_i}{\partial x}(x)g(x)g_c^T(\xi) + J_c^i(\xi) - R_c^i(\xi) = 0 \tag{7.15}$$

with $\frac{\partial^T F_i}{\partial x}$ denoting as before the gradient vector $\left(\frac{\partial F_i}{\partial x_1}, \dots, \frac{\partial F_i}{\partial x_n} \right)$, and g_c^i, J_c^i, R_c^i denoting the i-th row of g_c, J_c, respectively, R_c. Defining $F := (F_1, \dots, F_r)^T$, we write (7.15) for $i = 1, \dots, r$ in compact form as

$$\frac{\partial^T F}{\partial x}(x) \left[J(x) - R(x) \right] - \tilde{g}_c(\xi)g^T(x) = 0$$

$$\frac{\partial^T F}{\partial x}(x)g(x)g_c^T(\xi) + \tilde{J}(\xi) - \tilde{R}_c(\xi) = 0 \tag{7.16}$$

with $\tilde{g}_c$ denoting the submatrix of g_c composed of the first r rows, and $\tilde{J}_c$, $\tilde{R}_c$ the submatrix of J_c, respectively, R_c, composed of its first r rows.

By post-multiplication of the first equation of (7.16) by $\frac{\partial F}{\partial x}(x)$, and using the second equation, one obtains

$$\frac{\partial^T F}{\partial x}(x) \left[J(x) - R(x) \right] \frac{\partial F}{\partial x}(x) = \bar{J}_c(\xi) + \bar{R}_c(\xi) \tag{7.17}$$

---

[2]Indeed, suppose we have r Casimirs $C_i(x, \xi), i = 1, \dots, r$, where the partial Jacobian matrix $\frac{\partial C}{\partial \xi}(x, \xi)$ of the map $C : \mathcal{X} \times \mathcal{X}_c \to \mathbb{R}^r$ with components C_i has full rank r. Then by an application of the Implicit Function theorem the level sets $C_1(x, \xi) = c_1, \dots, C_r(x, \xi) = c_r$ for constants $c_1, \dots, c_r$, can be equivalently described by level sets of functions of the form (7.14).

with $\bar{J}_c(\xi)$, $\bar{R}_c(\xi)$ the $r \times r$ left-upper submatrices of J_c, respectively, R_c. Collecting on both sides of (7.17) the skew-symmetric and the symmetric parts we conclude that (7.17) is equivalent to

$$\frac{\partial^T F}{\partial x}(x) J(x) \frac{\partial F}{\partial x}(x) = \bar{J}_c(\xi) \tag{7.18}$$

$$-\frac{\partial^T F}{\partial x}(x) R(x) \frac{\partial F}{\partial x}(x) = \bar{R}_c(\xi) \tag{7.19}$$

From the second equation, it follows that if $\bar{R}_c(\xi) \geq 0$ (which is the case if the controller system is a true port-Hamiltonian system), then necessarily

$$\frac{\partial^T F}{\partial x}(x) R(x) \frac{\partial F}{\partial x}(x) = 0, \quad \bar{R}_c(\xi) = 0 \tag{7.20}$$

Furthermore since $R(x) \geq 0$ Eq. (7.20) implies

$$R(x) \frac{\partial F}{\partial x}(x) = 0 \tag{7.21}$$

Proposition 7.1.8 *Consider the plant port-Hamiltonian system Σ given by (7.1) in negative feedback interconnection $u = -y_c + v$, $u_c = y + v_c$ with the controller port-Hamiltonian system Σ_c given by (7.1). The functions $\xi_i - F_i(x)$, $i = 1, \ldots, r \leq n_c$, satisfy (7.15), and thus are Casimirs of the closed-loop system (7.7) for $v = 0$, $v_c = 0$, if and only if $F = (F_1, \ldots, F_r)^T$ satisfies*

$$\begin{aligned}
&\frac{\partial^T F}{\partial x}(x) J(x) \frac{\partial F}{\partial x}(x) = \bar{J}_c(\xi) \\
&\frac{\partial^T F}{\partial x}(x) J(x) = \bar{g}_c(\xi) g^T(x) \\
&R(x) \frac{\partial F}{\partial x}(x) = 0 \\
&\bar{R}_c(\xi) = 0
\end{aligned} \tag{7.22}$$

Furthermore, if $\Phi : \mathbb{R}^{r+1} \to \mathbb{R}$ and $H_c(\xi)$ can be found such that the candidate Lyapunov function

$$V(x, \xi) = \Phi(H(x) + H_c(\xi), \xi_1 - F_1(x), \ldots, \xi_r - F_r(x)) \tag{7.23}$$

has a strict minimum at (x^*, ξ^*) for a certain choice of ξ^*, then (x^*, ξ^*) is a stable equilibrium. Moreover, the additional control action

$$v = -Dg^T(x) \frac{\partial V}{\partial x}(x, \xi), \quad v_c = -D_c g_c^T(\xi) \frac{\partial V}{\partial \xi}(x, \xi) \tag{7.24}$$

for certain matrices $D = D^T \geq 0$, $D_c = D_c^T \geq 0$, results in

$$\frac{d}{dt} V(x, \xi) \leq -\frac{\partial^T V}{\partial x} (x, \xi) g(x) D g^T (x) \frac{\partial V}{\partial x} (x, \xi)$$
$$-\frac{\partial^T V}{\partial \xi} (x, \xi) g_c(\xi) D_c g_c^T (\xi) \frac{\partial V}{\partial \xi} (x, \xi) \tag{7.25}$$

and asymptotic stability of (x^*, ξ^*) can be investigated by application of LaSalle's Invariance principle.

Proof Only the second equality of (7.22) remains to be shown. This follows directly from the first line of (7.16) together with (7.21). □

Remark 7.1.9 As can be seen from (7.1) (and in accordance with Proposition 6.4.2), the presence of energy dissipation imposes constraints on the availability of Casimir functions, which can be only mitigated by allowing for energy creation ($R_c(\xi) \not\equiv 0$) in the controller system. This is referred to as the *dissipation obstacle*.

In the special case $r = n_c$ the Eqs. (7.22) for $F = (F_1, \ldots, F_{n_c})^T$ amount to

$$\frac{\partial^T F}{\partial x} (x) J(x) \frac{\partial F}{\partial x} (x) = J_c (\xi)$$
$$\frac{\partial^T F}{\partial x} (x) J(x) = g_c (\xi) g^T (x) \tag{7.26}$$
$$R(x) \frac{\partial F}{\partial x} (x) = 0 = R_c(\xi)$$

Example 7.1.10 A mechanical system with damping and actuated by external forces u is described by the port-Hamiltonian system (see (6.26) for the undamped case)

$$\begin{bmatrix} \dot{q} \\ \dot{p} \end{bmatrix} = \left(\begin{bmatrix} 0 & I_k \\ -I_k & 0 \end{bmatrix} - \begin{bmatrix} 0 & 0 \\ 0 & D(q) \end{bmatrix} \right) \begin{bmatrix} \frac{\partial H}{\partial q} (q, p) \\ \frac{\partial H}{\partial p} (q, p) \end{bmatrix} + \begin{bmatrix} 0 \\ B(q) \end{bmatrix} u \tag{7.27}$$
$$y = B^T (q) \frac{\partial H}{\partial p} (q, p)$$

with $q \in \mathbb{R}^k$ the vector of generalized position coordinates, $p \in \mathbb{R}^k$ the vector of generalized momenta, and $D(q) = D^T (q) \geq 0$ the damping matrix. The outputs $y \in \mathbb{R}^m$ are the generalized velocities corresponding to the generalized external forces $u \in \mathbb{R}^m$.

Now consider a port-Hamiltonian controller system (7.5) with $n_c = m$. Then (7.26) with $F = (F_1(q, p), \ldots, F_m(q, p))^T$ reduces to

$$J_c = 0, \qquad \frac{\partial F}{\partial q} (q, p) = g_c^T (\xi) B(q), \qquad \frac{\partial F}{\partial p} (q, p) = 0 \tag{7.28}$$

Hence with $g_c(\xi)$ equal to the $m \times m$ identity matrix, there exists a solution F to (7.28) if and only if the columns of the input force matrix $B(q)$ satisfy the integrability conditions

$$\frac{\partial B_{il}}{\partial q_j} (q) = \frac{\partial B_{jl}}{\partial q_i} (q), \qquad i, j = 1, \ldots k, \quad l = 1, \ldots m \tag{7.29}$$

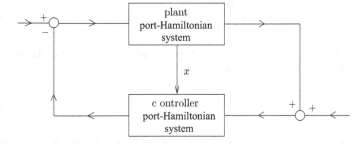

Fig. 7.2 Modulated control by interconnection

A useful *relaxation* of the conditions (7.26) for set-point stabilization is to allow the port-Hamiltonian controller system Σ_c given by (7.5) to be *modulated* by the plant state variables x. This means that the matrices J_c, R_c and g_c are allowed to depend on x; see Fig. 7.2.

In this case the closed-loop system (7.7) continues to be a port-Hamiltonian system, while the conditions (7.26) take the more amenable form

$$\frac{\partial^T F}{\partial x}(x)J(x)\frac{\partial F}{\partial x}(x) = J_c(\xi, x)$$

$$\frac{\partial^T F}{\partial x}(x)J(x) = g_c(\xi, x)g^T(x) \qquad (7.30)$$

$$R(x)\frac{\partial F}{\partial x}(x) = 0 = R_c(\xi, x)$$

In particular, the integrability conditions (7.29) on the input force matrix $B(q)$ may be avoided by taking $g_c^T(\xi, x) = B^{-1}(q)$ (assuming B to be invertible).

Remark 7.1.11 Allowing only g_c to depend on x may equivalently be formulated as modifying the feedback $u = -y_c$, $u_c = y$ to the *state modulated* feedback

$$u = -\beta(x)y_c, \quad u_c = \beta^T(x)y \qquad (7.31)$$

for some matrix $\beta(x)$, which can be considered as an "*integrating factor*" for the pde's (7.26).

In many cases of interest, including the above Example 7.1.10, the conditions (7.26) for $r = n_c$ simplify to

$$\frac{\partial^T F}{\partial x}(x)J(x)\frac{\partial F}{\partial x}(x) = 0 = J_c(\xi)$$

$$\frac{\partial^T F}{\partial x}(x)J(x) = g_c(\xi)g^T(x) \qquad (7.32)$$

$$R(x)\frac{\partial F}{\partial x}(x) = 0 = R_c(\xi)$$

If additionally $g_c(\xi)$ equals the $m \times m$ identity matrix, then the action of the controller port-Hamiltonian system amounts to (nonlinear) *integral action* on the output y of the plant port-Hamiltonian system, i.e., the controller system is given as

$$u = -\frac{\partial H_c}{\partial \xi}(\xi) + v, \quad \dot{\xi} = y + v_c \tag{7.33}$$

Example 7.1.12 Consider the mathematical pendulum with angle q, having Hamiltonian (with all parameters set equal to 1)

$$H(q, p) = \frac{1}{2}p^2 + (1 - \cos q) \tag{7.34}$$

and actuated by a torque u with output $y = p$ (angular velocity). Suppose we wish to stabilize the pendulum at a *non-zero* angle q^* and $p^* = 0$. Apply the nonlinear integral control

$$u = -\frac{\partial H_c}{\partial \xi}(\xi) + v, \quad \dot{\xi} = y + v_c \tag{7.35}$$

The Casimirs $C(q, p, \xi)$ of the closed-loop system for $v = 0$, $v_c = 0$ are found by solving

$$\begin{bmatrix} \frac{\partial C}{\partial q} & \frac{\partial C}{\partial p} & \frac{\partial C}{\partial \xi} \end{bmatrix} \begin{bmatrix} 0 & 1 & 0 \\ -1 & 0 & -1 \\ 0 & 1 & 0 \end{bmatrix} = 0 \tag{7.36}$$

leading to solutions $C(q, p, \xi) = K(q - \xi)$ for arbitrary $K : \mathbb{R} \to \mathbb{R}$. Consider a candidate Lyapunov function

$$V(q, p, \xi) = \frac{1}{2}p^2 + (1 - \cos q) + H_c(\xi) + K(q - \xi) \tag{7.37}$$

In order to let V have a local minimum at $(q^*, p^* = 0, \xi^*)$ for some ξ^*, determine K, H_c, and ξ^* such that

• *Equilibrium assignment*

$$\begin{array}{c} \sin q^* + \frac{dK}{dz}(q^* - \xi^*) = 0 \\ -\frac{dK}{dz}(q^* - \xi^*) + \frac{dH_c}{d\xi}(\xi^*) = 0 \end{array} \tag{7.38}$$

• *Minimum condition*

$$\begin{bmatrix} \cos q^* + \frac{d^2 K}{dz^2}(q^* - \xi^*) & 0 & -\frac{d^2 K}{dz^2}(q^* - \xi^*) \\ 0 & 1 & 0 \\ -\frac{d^2 K}{dz^2}(q^* - \xi^*) & 0 & \frac{d^2 K}{dz^2}(q^* - \xi^*) + \frac{d^2 H_c}{d\xi^2}(\xi^*) \end{bmatrix} > 0 \tag{7.39}$$

This has many solutions K, H_c, ξ^*. Additional damping feedback $v = -d\frac{\partial V}{\partial p}$ $(q, p, \xi) = -dp$, $v_c = -d_c \frac{dH_c}{d\xi}(q, p, \xi) + d_c \frac{dK}{dz}(q - \xi)$ with $d, d_c > 0$ will asymptotically stabilize this equilibrium.

The integral action perspective also motivates the following extension of the method. First note that given the plant port-Hamiltonian system (7.1) we could consider instead of the given output $y = g^T(x)\frac{\partial H}{\partial x}(x)$ any other output

$$y_A := [G(x) + P(x)]^T \frac{\partial H}{\partial x}(x) + [M(x) + S(x)]u \qquad (7.40)$$

for G, P, M, S satisfying

$$g(x) = G(x) - P(x), \quad M(x) = -M^T(x), \quad \begin{bmatrix} R(x) & P(x) \\ P^T(x) & S(x) \end{bmatrix} \geq 0 \qquad (7.41)$$

Indeed, by (6.5) any such output satisfies

$$\frac{d}{dt}H(x) \leq u^T y_A \qquad (7.42)$$

We call an output y_A satisfying (7.42) an *alternate passive output*.

A special choice of alternate passive output is obtained by rewriting the dynamics $\dot{x} = [J(x) - R(x)]\frac{\partial H}{\partial x}(x) + g(x)u$ as (assuming $J(x) - R(x)$ to be invertible)

$$\dot{x}^T[J(x) - R(x)]^{-1}\dot{x} = \dot{x}^T\frac{\partial H}{\partial x}(x) + \dot{x}^T[J(x) - R(x)]^{-1}g(x)u \qquad (7.43)$$

Since $\dot{x}^T[J(x) - R(x)]^{-1}\dot{x} \leq 0$ and $\dot{x}^T\frac{\partial H}{\partial x}(x) = \frac{d}{dt}H(x)$ this leads to (7.42) with respect to the alternate passive output defined as

$$\begin{aligned} y_A := &\, g^T(x)[J(x) + R(x)]^{-1}[J(x) - R(x)]\frac{\partial H}{\partial x}(x) + \\ &\, g^T(x)[J(x) + R(x)]^{-1}g(x)u \end{aligned} \qquad (7.44)$$

called the "*swapping the damping*" alternate passive output. If $R = 0$ this particular alternate passive output is just the original output with the addition of the feedthrough term $g^T(x)J(x)^{-1}g(x)u$, while in the special case $J = 0$ it equals

$$y_A = -g^T(x)\frac{\partial H}{\partial x}(x) + g^T(x)R(x)^{-1}g(x)u, \qquad (7.45)$$

corresponding to[3] $G(x) = 0, P(x) = -g(x), S(x) = g^T(x)R(x)^{-1}g(x)$.

Now in order to generate novel Casimirs based upon integral action of alternate passive outputs we consider the following alternate passive outputs. Assuming that im $g(x) \subset$ im $[J(x) - R(x)]$, $x \in \mathcal{X}$, there exists an $n \times m$ matrix $\Gamma(x)$ (unique if $[J(x) - R(x)]$ has full rank) such that

---

[3]Note that in this case the extended resistive structure matrix $\begin{bmatrix} R(x) & P(x) \\ P^T(x) & S(x) \end{bmatrix}$ is a minimal rank extension of $R(x)$, since its rank is *equal* to the rank of $R(x)$ ($= n$).

$$[J(x) - R(x)]\Gamma(x) = g(x) \tag{7.46}$$

Then define the alternate output

$$
\begin{aligned}
y_A := & [J(x)\Gamma(x) + R(x)\Gamma(x)]^T \frac{\partial H}{\partial x}(x) \\
& + [-\Gamma^T(x)J(x)\Gamma(x) + \Gamma^T(x)R(x)\Gamma(x)]u
\end{aligned} \tag{7.47}
$$

It is directly checked that this is an alternate passive output satisfying (7.41). Integral action $\dot{\xi} = y_A$ for arbitrary H_c leads to the following closed-loop system for $v = 0, v_c = 0$ (leaving out arguments x)

$$
\begin{bmatrix} \dot{x} \\ \dot{\xi} \end{bmatrix} = \begin{bmatrix} J - R & -J\Gamma + R\Gamma \\ -\Gamma^T J + \Gamma^T R & \Gamma^T J\Gamma - \Gamma^T R\Gamma \end{bmatrix} \begin{bmatrix} \frac{\partial H}{\partial x}(x) \\ \frac{\partial H_c}{\partial \xi}(\xi) \end{bmatrix} \tag{7.48}
$$

(Recall that $g(x) = J(x)\Gamma(x) - R(x)\Gamma(x)$.) It is immediately verified that at any $x \in \mathcal{X}$

$$
\begin{bmatrix} J - R & -J\Gamma + R\Gamma \\ -\Gamma^T J + \Gamma^T R & \Gamma^T J\Gamma - \Gamma^T R\Gamma \end{bmatrix} \begin{bmatrix} \Gamma \\ I_m \end{bmatrix} = 0, \tag{7.49}
$$

where I_m is the $m \times m$ identity matrix. Hence if there exist[4] functions $F_1, \dots, F_m$ such that the columns of $\Gamma(x)$ satisfy

$$
\Gamma_j(x) = -\frac{\partial F_j}{\partial x}(x), \quad j = 1, \dots, m, \tag{7.50}
$$

then the functions $\xi_j - F_j(x)$, $j = 1, \dots, m$, are Casimirs of the closed-loop system, yielding new possibilities for the construction of a suitable Lyapunov function $V(x.\xi)$. Finally, by Poincaré's lemma local existence of solutions $F_1, \dots, F_m$ to (7.50) is equivalent to $\Gamma(x)$ satisfying the integrability conditions

$$
\frac{\partial \Gamma_{ji}}{\partial x_k}(x) = \frac{\partial \Gamma_{jk}}{\partial x_i}(x), \quad i, j, k = 1, \dots, n \tag{7.51}
$$

Remark 7.1.13 Obviously, also partially integrable situations (corresponding to some of the columns of $\Gamma(x)$ being integrable) may be considered, leading to a smaller number of Casimirs for the closed-loop system.

Example 7.1.14 Consider an RLC-circuit with voltage source u, where the capacitor is in parallel with the resistor. The dynamics is given as

---

[4]In view of (7.46) this means that the input vector fields g_j are "Leibniz vector field" with respect to the Leibniz structure $J(x) - R(x)$ and potential functions $-F_j$. Note that the same assumption was made in (6.106) in the context of generating Lyapunov functions for port-Hamiltonian systems driven by constant inputs.

$$\begin{bmatrix} \dot{Q} \\ \dot{\phi} \end{bmatrix} = \begin{bmatrix} -\frac{1}{R} & 1 \\ -1 & 0 \end{bmatrix} \begin{bmatrix} \frac{Q}{C} \\ \frac{\phi}{L} \end{bmatrix} + \begin{bmatrix} 0 \\ 1 \end{bmatrix} u \tag{7.52}$$

with R, L, C, respectively, the resistance, inductance, and capacitance. Suppose we want to stabilize the system at some feasible non-zero set-point value $(Q^*, \phi^*) = (C\bar{u}, \frac{L}{R}\bar{u})$ for $\bar{u} \neq 0$. Integral action of the natural passive output $y = \frac{\phi}{L}$ (the current through the voltage source) does not help in creating Casimirs. Instead consideration of (7.46) yields the solution $\Gamma^T = \begin{bmatrix} 1 & \frac{1}{R} \end{bmatrix}$, and the resulting alternate passive output

$$y_A = \frac{2}{R}\frac{Q}{C} - \frac{\phi}{L} + \frac{1}{R}u \tag{7.53}$$

Integral action of this alternate passive output yields the Casimir $Q + \frac{1}{R}\phi - \xi$ for the closed-loop system, resulting in a candidate Lyapunov function $V(Q, \phi, \xi) = \frac{1}{2C}Q^2 + \frac{1}{2L}\phi^2 + H_c(\xi) + \Phi(Q + \frac{1}{R}\phi - \xi)$. It can be verified that H_c and Φ can be found such that V has a minimum at (Q^*, ϕ^*, ξ^*) for some ξ^*.

Note that in alternative case of a *series* RLC-circuit integral action of the natural output suffices, resulting in a controller system that emulates an extra capacitor, thereby shifting the zero equilibrium to any feasible equilibrium $(Q^*, \phi^*) = (C\bar{u}, 0)$. From a physical point of view a main difference between these two cases is that in the parallel RLC-circuit there is energy dissipation at equilibrium whenever $\bar{u} \neq 0$, in contrast with the series RLC-circuit.

7.2 Passivity-Based Control

Suppose there exists a solution F to the Casimir equation (7.26) with $r = n_c$, in which case all controller states ξ are related to the plant states x. It follows that for any choice of the vector of constants $\lambda = (\lambda_1, \ldots, \lambda_{n_c})$ the multi-level set

$$L_\lambda := \{(x, \xi) \mid \xi_i = F_i(x) + \lambda_i, \ i = 1, \ldots, n_c\} \tag{7.54}$$

is an *invariant manifold* of the closed-loop system for $v = 0$, $v_c = 0$. Furthermore, the dynamics restricted to L_λ is given as

$$\dot{x} = [J(x) - R(x)]\frac{\partial H}{\partial x}(x) - g(x)g_c^T(F(x) + \lambda)\frac{\partial H_c}{\partial \xi}(F(x) + \lambda) \tag{7.55}$$

Using the second and third equality of (7.26) and the chain-rule for differentiation

$$\frac{\partial H_c(F(x) + \lambda)}{\partial x} = \frac{\partial F}{\partial x}(x)\frac{\partial H_c}{\partial \xi}(F(x) + \lambda) \tag{7.56}$$

Equation (7.55) can be rewritten as

$$\dot{x} = [J(x) - R(x)]\frac{\partial H_s}{\partial x}(x) \tag{7.57}$$

with

$$H_s(x) := H(x) + H_\lambda(x), \quad H_\lambda(x) := H_c(F(x) + \lambda) \tag{7.58}$$

This constitutes a port-Hamiltonian system with the *same* interconnection structure matrix $J(x)$ and resistive structure matrix $R(x)$ as the original plant port-Hamiltonian system, but with *shaped* Hamiltonian H_s.

An energy-balancing interpretation of the extra term H_λ in H_s is the following. Since by (7.26) $R_c(\xi) = 0$, the controller Hamiltonian H_c satisfies

$$\frac{dH_c}{dt} = u_c^T y_c = -u^T y \tag{7.59}$$

in view of $u = -y_c$ and $u_c = y$. Therefore on any multi-level set L_λ, up to a constant,

$$H_\lambda(x(t)) = -\int_0^t u^T(\tau)y(\tau)d\tau, \tag{7.60}$$

which is *minus the energy* supplied to the plant system (7.1) by the controller system (7.5).

Alternatively, the dynamics (7.57) could have been obtained *directly* by applying to the plant port-Hamiltonian system (7.1) a *state feedback* $u = \alpha_\lambda(x)$ such that

$$g(x)\alpha_\lambda(x) = [J(x) - R(x)]\frac{\partial H_\lambda}{\partial x}(x) \tag{7.61}$$

In fact, by (7.55) such an $\alpha_\lambda(x)$ is given as

$$\alpha_\lambda(x) = -g_c^T(F(x) + \lambda)\frac{\partial H_c}{\partial \xi}(F(x) + \lambda) \tag{7.62}$$

The state feedback $u = \alpha_\lambda(x)$ is customarily called a *passivity-based state feedback*, since it is based on the passivity properties of the original plant system (7.1), and transforms (7.1) into *another* passive system with *shaped* storage function; in this case H_s.

Seen from this perspective the passivity-based state feedback $u = \alpha_\lambda(x)$ satisfying (7.61) *can be derived* from the feedback interconnection of the port-Hamiltonian plant system (7.1) with a controller port-Hamiltonian system (7.5), and thus inherits its robustness properties.

Finally, we note that since the Casimirs are anyway defined up to a constant we can also leave out the dependence on λ and simply consider the passivity-based state feedbacks

$$\alpha(x) := -g_c^T(F(x))\frac{\partial H_c}{\partial \xi}(F(x)) \tag{7.63}$$

for any solution F to (7.26).

Example 7.2.1 (Example 7.1.10 continued) Assuming g_c to be the $m \times m$ identity matrix and $B(q)$ to satisfy (7.29) there exist Casimirs $\xi_i - F_i(q), i = 1, \ldots, m$. The resulting passivity-based state feedback is given by

$$\alpha(q) = -\frac{\partial H_c}{\partial \xi}(F_1(q), \ldots, F_m(q)) \tag{7.64}$$

with shaped Hamiltonian $H_s(q, p) = H(q, p) + H_c(F_1(q), \ldots, F_m(q))$. In particular, if H is in the common "kinetic plus potential energy" form $\frac{1}{2}p^T M^{-1}(q)p + P(q)$, then this corresponds to shaping of the potential energy $P(q)$ into

$$P(q) + H_c(F_1(q), \ldots, F_m(q)) \tag{7.65}$$

A comparison of the actual *implementation* of the passivity-based state feedback law (7.63) with respect to the implementation of the port-Hamiltonian controller system (7.5) depends on the application. For example, in a robotics context the measurements of the generalized velocities $y = B^T(q)\dot{q}$ are more noisy than the measurements of the generalized positions q, since typically the first are obtained by differentiation of the latter.

Example 7.2.2 (Example 7.1.12 continued) Consider, as a special case of Example 7.2.1, the equations of a normalized pendulum as in Example 7.1.12. The solution of (7.28) for $g_c(\xi) = 1$ is $F(q) = q$. Let q^* be the set-point angle of the pendulum, and let us try to shape the potential energy $P(q) = 1 - \cos q$ in such a way that it has a minimum at $q = q^*$. The simplest choice for the controller Hamiltonian $H_c(\xi)$ in order to achieve this is to take

$$H_c(\xi) = \cos \xi + \frac{1}{2}(\xi - q^*)^2 \tag{7.66}$$

The shaped potential energy (cf. (7.65)) is then given as

$$(1 - \cos q) + \cos(q + \lambda) + \frac{1}{2}(q + \lambda - q^*)^2 \tag{7.67}$$

which has a minimum at $q = q^*$ for $\lambda = 0$. Hence, the resulting passivity-based state feedback is simply

$$\alpha_0(x) = -\frac{\partial H_c}{\partial \xi}(q) = \sin q - (q - q^*) \tag{7.68}$$

which is the well-known "proportional plus gravity compensation control."

In the rest of this section we develop a *general* theory of passivity-based state feedbacks $u = \alpha(x)$ for the asymptotic stabilization of a set-point x^* of the plant port-Hamiltonian system (7.1). This proceeds in two steps: (I) shape by state feedback the

Hamiltonian H in such a way that it has a strict minimum at $x = x^*$, implying that x^* is a (marginally) stable equilibrium of the controlled system, (II) add damping to the system such that x^* becomes an *asymptotically* stable equilibrium of the controlled system. Note that Step II has been treated before, see in particular Sects. 7.1 and 4.2, so that we may concentrate on Step I.

Proposition 7.2.3 *Consider the port-Hamiltonian system (7.1). Assume we can find a feedback $u = \alpha(x)$ and a vector function $h(x)$ satisfying*

$$[J(x) - R(x)]\, h(x) = g(x)\alpha(x) \tag{7.69}$$

such that

$$(i)\ \ \tfrac{\partial h_i}{\partial x_j}(x) = \tfrac{\partial h_j}{\partial x_i}(x), \quad i, j = 1, \ldots, n$$

$$(ii)\ \ h(x^*) = -\tfrac{\partial H}{\partial x}(x^*) \tag{7.70}$$

$$(iii)\ \ \tfrac{\partial h}{\partial x}(x^*) > -\tfrac{\partial^2 H}{\partial x^2}(x^*)$$

with $\frac{\partial h}{\partial x}(x)$ the $n \times n$ matrix with i-th column given by $\frac{\partial h_i}{\partial x}(x)$, and $\frac{\partial^2 H}{\partial x^2}(x^*)$ denoting the Hessian matrix of H at x^*. Then x^* is a stable equilibrium of the closed-loop system

$$\dot{x} = [J(x) - R(x)] \frac{\partial H_d}{\partial x}(x) \tag{7.71}$$

where $H_d(x) := H(x) + H_a(x)$, with H_a defined locally around x^* by

$$h(x) = \frac{\partial H_a}{\partial x}(x) \tag{7.72}$$

Proof From (7.70)(i) it follows by Poincaré's lemma that locally around x^* there exists H_a satisfying (7.72). Then by (7.69) the closed-loop system equals (7.71). Furthermore by (7.70)(ii) x^* is an equilibrium of (7.71), which is stable by (iii) since $\frac{\partial^2 H_d}{\partial x^2}(x^*) > 0$, implying that H_d has a strict minimum at x^*.   □

Example 7.2.4 Consider a mechanical system (7.27), with Hamiltonian $H(q, p) = \frac{1}{2}p^T M^{-1}(q)p + P(q)$. Equation (7.69) reduces to

$$h(q, p) = \begin{bmatrix} 0 \\ h^2(q, p) \end{bmatrix}, \quad h^2(q, p) = B(q)\alpha(q, p) \tag{7.73}$$

Furthermore, from (7.70)(i) it follows that h^2 (and thus α) only depends on q, while

$$\frac{\partial h_i^2}{\partial q_j}(q) = \frac{\partial h_j^2}{\partial q_i}(q) \quad i, j = 1, \ldots k \tag{7.74}$$

Finally, (7.70)(ii) and (iii) reduce to, respectively,

$$h^2(q^*) = -\frac{\partial P}{\partial q}(q^*)$$

$$\frac{\partial h^2}{\partial q}(q^*) > -\frac{\partial^2 P}{\partial q^2}(q^*)$$

(7.75)

This implies that the shaped potential energy $P_d(q) = P(q) + P_a(q)$, with P_a such that $h^2(q) = \frac{\partial P_a}{\partial q}(q)$, has a strict minimum at $q = q^*$, and thus $H_d(q, p) = \frac{1}{2}p^T M^{-1}(q)p + P_d(q)$ has a strict minimum at $(q^*, p = 0)$.

Example 7.2.5 (Example 6.2.3 continued) The Hamiltonian H of the rolling coin does not have a strict minimum at the desired equilibrium $x = y = \theta = \phi = 0$, $p_1 = p_2 = 0$, since the potential energy is zero. Equations (7.69) and (7.70) reduce to

$$\begin{bmatrix} 0 & 0 & 0 & -1 \\ -\cos\phi & -\sin\phi & -1 & 0 \end{bmatrix} \begin{bmatrix} \frac{\partial P_a}{\partial x} \\ \frac{\partial P_a}{\partial y} \\ \frac{\partial P_a}{\partial \theta} \\ \frac{\partial P_a}{\partial \phi} \end{bmatrix} = \begin{bmatrix} 0 & 1 \\ 1 & 0 \end{bmatrix} \begin{bmatrix} \alpha_1 \\ \alpha_2 \end{bmatrix}$$

(7.76)

with P_a and α_1, α_2 functions of x, y, θ, ϕ. Taking $P_a(x, y, \theta, \phi) = \frac{1}{2}(x^2 + y^2 + \theta^2 + \phi^2)$ leads to the state feedback

$$u_1 = -x \cos\phi - y \sin\phi - \theta + v_1$$

$$u_2 = -\phi + v_2$$

(7.77)

By adding damping $v_1 = -y_1 = -p_2$, $v_2 = -p_1$, the trajectories of the closed-loop system will converge to the set of equilibria

$$\{(x, y, \theta, \phi) \mid p_1 = p_2 = 0, \quad \phi = 0, \quad x + \theta = 0\}$$

(7.78)

The use of different potential functions P_a leads to different sets of equilibria which, however, always contain an infinite number of points, in accordance with the fact that mechanical systems with nonholonomic kinematic constraints cannot be asymptotically stabilized using continuous state feedback; see Corollary 6.4.7.

Remark 7.2.6 The conditions (7.70) (i), (iii) can be stated in a more explicit way if there exist functions $F_1, \ldots, F_m$ such that

$$[J(x) - R(x)]\frac{\partial F_j}{\partial x}(x) = -g_j(x), \quad j = 1, \ldots, m$$

(7.79)

(Note that these are exactly the conditions involved in the generation of Casimirs for the closed-loop system resulting from integral action on alternate passive outputs; cf. (7.46) and (7.50).) In this case, any feedback of the form

$$u = \alpha(x) := -\frac{\partial G}{\partial \xi}(F_1(x), \ldots, F_m(x))$$

(7.80)

for some $G : \mathbb{R}^m \to \mathbb{R}$, satisfies (7.69), and (7.70)(i) with $h(x) = \frac{\partial H_a}{\partial x}(x)$, and $H_a(x) = G(F_1(x), \ldots, F_m(x))$. Furthermore, there exists a feedback (7.80) such that additionally (7.70)(iii) is satisfied if and only if the Hessian $\frac{\partial^2 H}{\partial x^2}(x^*)$ is *positive definite* when restricted to the subspace ker $F_x(x^*)$, where $F_x(x^*)$ is the $m \times n$ matrix of gradient row vectors of $F_1, \ldots, F_m$ [143, 270].

The conditions for existence of solutions $h(x)$ and $\alpha(x)$ to (7.69) can be simplified as follows.

Proposition 7.2.7 *Consider the plant port-Hamiltonian system (7.1), where it is assumed that $g(x)$ has full column rank for every $x \in \mathcal{X}$. Denote by $g^\perp(x)$ a matrix of maximal rank such that $g^\perp(x)g(x) = 0$. Let $h(x), \alpha(x)$ be a solution to (7.69). Then $h(x)$ is a solution to*

$$g^\perp(x)[J(x) - R(x)]h(x) = 0 \qquad (7.81)$$

Conversely, if $h(x)$ is a solution to (7.81) then there exists $\alpha(x)$ such that $h(x), \alpha(x)$ is a solution to (7.69). In fact, $\alpha(x)$ is uniquely given as

$$\alpha(x) = \left(g^T(x)g(x)\right)^{-1} g^T(x) [J(x) - R(x)] h(x) \qquad (7.82)$$

Proof The first claim is immediate. The expression (7.82) follows by noting that $\left(g^T(x)g(x)\right)^{-1} g^T(x)$ is the Moore–Penrose pseudoinverse of $g(x)$.   □

The Eq. (7.81) is called the *energy shaping matching equation*, and characterizes the possible vectors $h(x)$, and therefore H_a, which may be used in order to shape H to $H_d = H + H_a$. Note that this matching equation is closely related to the Eq. (7.46) for constructing alternate passive outputs for Casimir generation. Indeed, by allowing for modulated feedback (7.31) as in Remark 7.1.11 the alternate passive output Eq. (7.46) generalizes to $[J(x) - R(x)]\Gamma(x) = g(x)\beta(x)$ for some matrix $\beta(x)$, which is equivalent to

$$g^\perp(x)[J(x) - R(x)]\Gamma(x) = 0 \qquad (7.83)$$

Hence, the set of solutions $h(x)$ to (7.81) is contained in im $\Gamma(x)$, and in this sense energy shaping by state feedback is *equivalent* to finding alternate passive outputs for Casimir generation.

An important extension to the above method for energy shaping, which considerably enlarges the set of possible H_a and therefore of H_d, is to additionally allow for shaping of the structure matrices $J(x)$ and $R(x)$ in (7.1). Thus, the *interconnection and damping assignment passivity-based control* (IDA-PBC) design objective is to obtain by state feedback $u = \alpha(x)$ a closed-loop system of the shaped port-Hamiltonian form

$$\dot{x} = [J_d(x) - R_d(x)] \frac{\partial H_d}{\partial x}(x), \qquad (7.84)$$

where $J_d(x) = -J_d^T(x)$ and $R_d(x) = R_d^T(x) \geq 0$ are *desired* interconnection and resistive structure matrices, and, as before, H_d is a desired (shaped) Hamiltonian having a minimum at x^*. Transforming (7.1) into (7.84) by application of state feedback $u = \alpha(x)$ amounts to the equation

$$[J(x) - R(x)]\frac{\partial H}{\partial x}(x) + g(x)\alpha(x) = [J_d(x) - R_d(x)]\frac{\partial H_d}{\partial x}(x) \qquad (7.85)$$

Proposition 7.2.8 *Consider the port-Hamiltonian system (7.1) and a desired equilibrium x^* to be stabilized. Assume that we can find $\alpha(x)$ and $H_d(x) = H(x) + H_a(x)$, and matrices $J_d(x)$ and $R_d(x)$ satisfying (7.85) and such that the following conditions are met:*
• *Equilibrium assignment: at x^* the gradient of $H_a(x)$ satisfies*

$$\frac{\partial H_a}{\partial x}(x^*) + \frac{\partial H}{\partial x}(x^*) = 0 \qquad (7.86)$$

• *Minimum condition: the Hessian of $H_a(x)$ at x^* satisfies*

$$\frac{\partial^2 H_a}{\partial x^2}(x^*) + \frac{\partial^2 H}{\partial x^2}(x^*) > 0 \qquad (7.87)$$

Then, x^* is a stable equilibrium of the closed-loop system (7.84). It will be asymptotically stable if, in addition, the largest invariant set under the closed-loop dynamics contained in

$$\left\{ x \in \mathcal{X} \left| \frac{\partial^T H_d}{\partial x}(x) R_d(x) \frac{\partial H_d}{\partial x}(x) = 0 \right. \right\}$$

equals $\{x^*\}$.

As before, see (7.81) in Proposition 7.2.7, the dependence of (7.85) on the unknown $\alpha(x)$ can be eliminated. Indeed, assume that $g(x)$ has maximal column rank for all $x \in \mathcal{X}$. Denoting by $g^{\perp}(x)$ a maximal rank annihilator of $g(x)$, premultiplication of both sides of (7.85) by $g^{\perp}(x)$ results in

$$g^{\perp}(x)[J(x) - R(x)]\frac{\partial H}{\partial x}(x) = g^{\perp}(x)[J_d(x) - R_d(x)]\frac{\partial H_d}{\partial x}(x) \qquad (7.88)$$

This is called the *IDA-PBC matching equation*. Interconnection and damping assignment passivity-based control is thus concerned with finding J_d, R_d and H_a satisfying (7.88) such that H_d has its minimum at the desired equilibrium x^*. Once (7.88) has been solved, the required state feedback $u = \alpha(x)$ *follows*, and is uniquely given as

$$
\begin{aligned}
\alpha(x) = {}& \left(g^T(x)g(x)\right)^{-1} g^T(x) \\
& \times \left([J_d(x) - R_d(x)]\frac{\partial H_d}{\partial x}(x) - [J(x) - R(x)]\frac{\partial H}{\partial x}(x)\right)
\end{aligned}
\qquad (7.89)
$$

Solving the IDA-PBC matching Eq. (7.88) can be a tedious process; basically because there are many degrees of freedom J_d, R_d in order to construct H_d having its minimum at x^*.

Example 7.2.9 (Example 6.3.2 continued) Suppose that we first try to stabilize the levitated ball at a desired height q^* by only shaping the Hamiltonian without altering the interconnection and resistive structure matrices $J(x)$, $R(x)$ given in (6.5.9). Then (7.69) for $h(x) = \frac{\partial H_a}{\partial x}(x)$ amounts to

$$\frac{\partial H_a}{\partial q} = 0, \quad \frac{\partial H_a}{\partial p} = 0, \quad -R\frac{\partial H_a}{\partial \varphi} = \alpha(x) \tag{7.90}$$

The first two equalities mean that H_a can only depend on φ, while the third defines the state feedback $\alpha(x)$. Thus, the closed-loop Hamiltonian $H_d = H + H_a$, with H the original Hamiltonian given by (6.60), takes the form

$$H_d(q, p, \varphi) = mgq + \frac{p^2}{2m} + \frac{1}{2L(q)}\varphi^2 + H_a(\varphi) \tag{7.91}$$

Even though, with a suitable selection of H_a, we can satisfy the equilibrium assignment condition, H_d will never be positive definite at this equilibrium. This is due to the lack of effective coupling between the electrical and mechanical subsystems. Indeed, the interconnection matrix J only couples position with momentum. To overcome this problem, a coupling between the flux-linkage φ and the momentum p is enforced by modifying J to

$$J_d = \begin{bmatrix} 0 & 1 & 0 \\ -1 & 0 & 1 \\ 0 & -1 & 0 \end{bmatrix} \tag{7.92}$$

Furthermore, we take $R_d = R$. Then $H_d = H + H_a$ solves (7.85) if

$$\frac{\partial H_a}{\partial p} = 0, \quad -\frac{\partial H_a}{\partial q} + \frac{\partial H_a}{\partial \varphi} + \frac{\varphi}{L(q)} = 0 \tag{7.93}$$

The first equation implies that H_a only depends on q, φ. For $L(q)$ given by $L(q) = \frac{k}{1-q}$ the second equation has the following set of solutions

$$H_a(q, \varphi) = \frac{\varphi^3}{6k} - \frac{1}{2k}(1-q)\varphi^2 + \Phi(q + \varphi), \tag{7.94}$$

with $\Phi : \mathbb{R}^2 \to \mathbb{R}$ arbitrary. This results in the class of shaped Hamiltonians

$$H_d(q, p, \varphi) = mgq + \frac{p^2}{2m} + \frac{\varphi^3}{6k} + \Phi(q + \varphi) \tag{7.95}$$

By a suitable choice of Φ the shaped Hamiltonian H_d indeed has a minimum at $x^* = (q^*, 0, \sqrt{2kmg})$.

An interesting case of IDA-PBC matching equations is obtained for mechanical systems, as treated in Sect. 6.2; see also Example 7.1.10. Consider the mechanical system

$$\Sigma : \quad \begin{bmatrix} \dot{q} \\ \dot{p} \end{bmatrix} = \begin{bmatrix} 0 & I_n \\ -I_n & 0 \end{bmatrix} \begin{bmatrix} \frac{\partial H}{\partial q}(q, p) \\ \frac{\partial H}{\partial p}(q, p) \end{bmatrix} + \begin{bmatrix} 0 \\ G(q) \end{bmatrix} u, \tag{7.96}$$

where $q \in \mathbb{R}^n$ are the generalized position coordinates, $p \in \mathbb{R}^n$ are the momenta, given as $p = M\dot{q}$ with M the inertia matrix, and the Hamiltonian is given by

$$H(q, p) = \frac{1}{2} p^\top M^{-1}(q) p + V(q) \tag{7.97}$$

We assume that $G(q)$ has constant rank $m \le n$. Hence a matrix $G^\perp(q)$ of rank $n - m$ exists such that $G^\perp(q) G(q) = 0$. Mechanical IDA-PBC is to design a state feedback $u = \alpha(q, p) + v$ that transforms Σ into a closed-loop system, which is *again* a mechanical system

$$\Sigma_d : \quad \begin{bmatrix} \dot{q} \\ \dot{p} \end{bmatrix} = (J_d(q, p) - R_d(q, p)) \begin{bmatrix} \frac{\partial H_d}{\partial q}(q, p) \\ \frac{\partial H_d}{\partial p}(q, p) \end{bmatrix} + \begin{bmatrix} 0 \\ G(q) \end{bmatrix} v \tag{7.98}$$

with $H_d(q, p) = \frac{1}{2} p^\top M_d^{-1}(q) p + V_d(q)$ the desired total energy, where $M_d > 0$ is the desired inertia matrix and $V_d(q)$ the desired potential energy function. The upper-half of the IDA-PBC matching equations immediately implies that J_d is restricted to the form

$$J_d = \begin{bmatrix} 0 & M^{-1}(q) M_d(q) \\ -M_d(q) M^{-1}(q) & J_2(q, p) \end{bmatrix} \tag{7.99}$$

for some $J_2(q, p) = -J_2^\top(q, p)$. The resulting lower-half of the IDA-PBC matching equations is given as

$$G^\perp(q) M_d(q) M^{-1}(q) \tfrac{\partial}{\partial q} \left(\tfrac{1}{2} p^\top M_d^{-1}(q) p \right) - G^\perp(q) J_2(q, p) M_d^{-1}(q) p$$
$$+ G^\perp(q) M_d(q) M^{-1}(q) \tfrac{\partial}{\partial q} V_d(q) \tag{7.100}$$
$$= G^\perp(q) \tfrac{\partial}{\partial q} \left(\tfrac{1}{2} p^\top M^{-1}(q) p \right) + G^\perp(q) \tfrac{\partial}{\partial q} V(q)$$

These equations split into an p-dependent and p-independent part, given by

$$G^\perp(q) M_d(q) M^{-1}(q) \tfrac{\partial}{\partial q} \left(\tfrac{1}{2} p^\top M_d^{-1}(q) p \right) - G^\perp(q) J_2(q, p) M_d^{-1}(q) p =$$
$$G^\perp(q) \tfrac{\partial}{\partial q} \left(\tfrac{1}{2} p^\top M^{-1}(q) p \right) \tag{7.101}$$
$$G^\perp(q) M_d(q) M^{-1}(q) \tfrac{\partial}{\partial q} V_d(q) = G^\perp(q) \tfrac{\partial}{\partial q} V(q)$$

It follows that $J_2(q, p)$ is necessarily *linear* in p. The aim is to find a shaped potential energy function V_d that has a minimum at the set-point q^*. In case $M_d = M$ (energy shaping passivity-based control) the second equation confines shaping of the potential energy to take place in the *actuated* degrees of freedom corresponding to im $G(q)$ (as we have seen before). By allowing shaping of M to M_d (and thereby of J to J_d, with J_2 an additional degree of freedom), the options for shaping of V into V_d are considerably enlarged.

7.3   Control by Energy-Routing

In the previous sections, we addressed the problem of asymptotic set-point stabilization of port-Hamiltonian systems. The main idea was to shape the Hamiltonian into a suitable Lyapunov function, either by interconnection with a dynamical controller system in port-Hamiltonian form or by direct state feedback, and then to add damping to the system by feeding back the resulting passive outputs. In this final section, we indicate other uses of the port-Hamiltonian structure for control. In particular, we explore the option of controlling the Poisson or Dirac structure in order to regulate the power flow through the system.

As a first example of this, we consider the problem of *energy transfer*. Consider two port-Hamiltonian systems Σ_i (without internal dissipation) in input-state-output form

$$\Sigma_i : \quad \begin{aligned} \dot{x}_i &= J_i(x_i)\frac{\partial H_i}{\partial x_i}(x_i) + g_i(x_i)u_i, & u_i &\in \mathbb{R}^m \\ y_i &= g_i^T(x_i)\frac{\partial H_i}{\partial x_i}(x_i), & y_i &\in \mathbb{R}^m \end{aligned} \qquad i = 1, 2, \qquad (7.102)$$

both satisfying the power-balance $\dot{H}_i(x_i) = y_i^T u_i$. Suppose now that we want to transfer the energy from system Σ_1 to system Σ_2, while keeping the total energy $H_1 + H_2$ constant. This can be done by using the following output feedback

$$\begin{bmatrix} u_1 \\ u_2 \end{bmatrix} = \begin{bmatrix} 0 & -y_1 y_2^T \\ y_2 y_1^T & 0 \end{bmatrix} \begin{bmatrix} y_1 \\ y_2 \end{bmatrix}, \qquad (7.103)$$

which, due to its skew-symmetry property, is power preserving. Hence, the closed-loop system composed of Σ_1 and Σ_2 is energy preserving, that is $\dot{H}_1 + \dot{H}_2 = 0$. However, if we consider the individual energies then we notice that

$$\dot{H}_1(x) = -y_1^T y_1 y_2^T y_2 = -||y_1||^2 ||y_2||^2 \le 0, \qquad (7.104)$$

implying that H_1 is decreasing as long as $||y_1||$ and $||y_2||$ are different from 0. Conversely, as expected since the total energy is constant,

$$\dot{H}_2(x) = y_2^T y_2 y_1^T y_1 = ||y_2||^2 ||y_1||^2 \geq 0, \qquad (7.105)$$

implying that H_2 is increasing at the same rate. In particular, if H_1 has a minimum at the zero equilibrium, and Σ_1 is zero-state detectable, then all the energy H_1 of Σ_1 will be transferred to Σ_2, as long as $||y_2||$ is not zero. If there is internal energy dissipation in Σ_1 and/or Σ_2, then this energy transfer mechanism still works. However, the fact that H_2 grows or not will depend on the balance between the energy delivered by Σ_1 to Σ_2 and the internal loss of energy in Σ_2 due to dissipation. We conclude that this particular scheme of power-conserving energy transfer is accomplished by a skew-symmetric output feedback, which is *modulated* by the values of the output vectors of both systems. Note that this implies that the interconnection matrix of the closed-loop system is, strictly speaking, *not* an (almost-)Poisson structure, since it is modulated by the co-tangent vectors $\frac{\partial H_1}{\partial x_1}$ and $\frac{\partial H_2}{\partial x_2}$, instead of by the state variables x_1, x_2.

A related scenario for energy-routing is the case where the interconnection matrix J (or, more generally, the Dirac structure) is depending on the control u. Such a case, for example, occurs in the control of power converters, where different switch positions lead to different circuit topologies (and thus to different J matrices), and where the duty ratio's may be taken as control variables, leading to $J(u)$, see, e.g., [95, 200]. Especially in case external sources and sinks are present, this allows for a control of the power flow through the system in such a manner that certain crucial variables are kept close to their desired values. In a robotics context continuously variable transmission components, corresponding to a modulated Dirac structure, may be designed for energy-efficient control [89]. Finally, control dependence may also appear in the energy storage and energy dissipation. For the first case, we refer to Note 13 in Sect. 6.10 (variable stiffness control).

7.4 Notes for Chapter 7

1. Example 7.1.3 is based on Aeyels & Szafranski [3].

2. The idea of stabilizing mechanical systems by shaping the potential energy via feedback and by adding damping can be traced back at least to Takegaki & Arimoto [334], and is one of the starting points of passivity-based control; see Ortega & Spong [243]), as well as Ortega, Loria, Nicklasson & Sira-Ramirez [239]. For the closely related topic of stabilization of Hamiltonian input-output systems, see van der Schaft [270], Nijmeijer & van der Schaft [233].

3. Instead of controlling the system to the minimal level of its energy by damping injection as in Sect. 7.1, one may also use similar strategies to bring the system to any desired level of energy $\bar{H}$ by controlling $(H - \bar{H})^2$ to zero. Furthermore, this can be extended to stabilizing the system to any (multi-)level set of its *Casimir* functions. This idea is explored in Fradkov, Makarov, Shiriaev & Tomchina [102] and the references quoted therein.

4. The Energy-Casimir method for investigating stability of Hamiltonian dynamics can be found in e.g., Marsden & Ratiu [197] and the references quoted therein. A classical example is the stability analysis of non-zero equilibria of Euler's equations 6.31 for the dynamics of the angular momenta of a rigid body. See Aeyels [2] for the stabilization application. The method of generating Casimirs for the closed-loop system resulting from interconnection with a controller port-Hamiltonian systems is initiated in Stramigioli, Maschke & van der Schaft [330], and further developed in Ortega, van der Schaft, Mareels & Maschke [241] and Ortega, van der Schaft, Maschke & Escobar [242].

5. Example 7.1.6 is taken from Stramigioli, Maschke & van der Schaft [330]; see also Stramigioli [328].

6. Closed-loop Casimir generation with controller systems *not* satisfying $R_c(\xi) \geq 0$ is addressed in [168].

7. The "swapping the damping" alternate passive output in (7.44) is introduced in Jeltsema, Ortega & Scherpen [147].

8. The construction to generate Casimirs based on integral action of alternate passive outputs and as described at the end of Sect. 7.1, see (7.46), is initiated (in a different context and form) in Maschke, Ortega & van der Schaft [198, 199]. For a general approach to alternate passive outputs for port-Hamiltonian DAE systems see Venkatraman & van der Schaft [342].

9. Section 7.2 is largely based on Ortega, van der Schaft, Mareels & Maschke [241] and Ortega, van der Schaft, Maschke & Escobar [242]. A survey on IDA-PBC is Ortega & Garcia–Canseco [238]. See Ortega, van der Schaft, Castanos & Astolfi [240] and Castanos, Ortega, van der Schaft & Astolfi [62] for further developments, in particular about the close relations between "control by interconnection" and IDA-PBC. Example 7.2.9 is from Ortega, van der Schaft, Mareels & Maschke [241]. Example 7.2.5 is based on Maschke & van der Schaft [204].

10. IDA-PBC control of linear port-Hamiltonian systems is investigated in Prajna, van der Schaft & Meinsma [256].

11. The IDA-PBC matching equations for mechanical systems (7.100) and (7.101) are due to Ortega, Spong, Gomez–Estern & Blankenstein [244]; see also Blankenstein, Ortega & van der Schaft [42]. The extension to matching equations for mechanical systems with damping is treated in Gomez-Estern & van der Schaft [116]. In parallel to IDA-PBC of mechanical systems the theory of *controlled Lagrangians* is developed in Bloch, Leonard & Marsden [44, 45], Bloch, Chang, Leonard & Marsden [46], Chang, Bloch, Leonard, Marsden & Woolsey [64]), with eventually equivalent results; see also Auckly, Kapitanski & White [17] and Hamberg [118].

12. Relatively little attention has been paid to the extension of the control theory of this chapter to port-Hamiltonian DAE systems; see however Macchelli [191]. The characterization of achievable Dirac structures and achievable Casimirs

through control by interconnection is addressed in Pasumarthy & van der Schaft [251].

13. The energy transfer scheme (7.103) is taken from Duindam & Stramigioli [91]; see also Duindam, Blankenstein & Stramigioli [90], Duindam, Stramigioli & Scherpen [92], where it is used for asymptotic path-following of mechanical systems (converting kinetic energy orthogonal to the tangent of the desired path into kinetic energy along it). The fact that the interconnection matrix of the closed-loop system is not a (almost-)Poisson structure matrix anymore (since it is depending on the gradient vector of the Hamiltonian) is similar to the situation encountered in thermodynamical systems, see Eberard, Maschke & van derSchaft [94].

14. Related to the ideas exposed in Sect. 7.3 is the theory of *impedance control* as formulated in Hogan [127]. See furthermore Stramigioli [329], and the references quoted in there, for various ideas on "energy-aware control" in robotics.

Chapter 8
L_2-Gain and the Small-Gain Theorem

In this chapter, we elaborate on the characterization of finite L_2-gain for state space systems, continuing on the general theory of dissipative systems in Chap. 3. Within this framework we revisit the Small-gain theorem and its implications for robustness (Sect. 8.2), and extend the small-gain condition to network systems (Sect. 8.3). Furthermore, we provide an alternative characterization of L_2-gain in terms of response to periodic input functions (Sect. 8.4), and in Sect. 8.5 we end by sketching the close relationships to the theory of (integral-)input-to-state stability.

8.1 L_2-Gain of State Space Systems

Recall from Chap. 3 the following basic definitions regarding L_2-gain of a general state space system Σ

$$\Sigma : \begin{array}{l} \dot{x} = f(x, u), \ x \in \mathcal{X}, \ u \in \mathbb{R}^m \\ y = h(x, u), \ y \in \mathbb{R}^p \end{array} \tag{8.1}$$

with n-dimensional state space manifold $\mathcal{X}$.

Definition 8.1.1 A *state space system* Σ given by (8.1) has *L_2-gain* $\leq \gamma$ if it is dissipative with respect to the supply rate $s(u, y) = \frac{1}{2}\gamma^2||u||^2 - \frac{1}{2}||y||^2$; that is, there exists a storage function $S : \mathcal{X} \to \mathbb{R}^+$ such that

$$S(x(t_1)) - S(x(t_0)) \leq \frac{1}{2} \int_{t_0}^{t_1} (\gamma^2||u(t)||^2 - ||y(t)||^2)dt \tag{8.2}$$

along all input functions $u(\cdot)$ and resulting state trajectories $x(\cdot)$ and output functions $y(\cdot)$, for all $t_0 \leq t_1$. Equivalently, if S is C^1

© Springer International Publishing AG 2017

A. van der Schaft, *L_2-Gain and Passivity Techniques in Nonlinear Control*,
Communications and Control Engineering, DOI 10.1007/978-3-319-49992-5_8

$$S_x(x)f(x,u) \leq \frac{1}{2}\gamma^2||u||^2 - \frac{1}{2}||h(x,u)||^2 , \qquad \forall\, x, u \qquad\qquad (8.3)$$

The L_2-gain of Σ is defined as $\gamma(\Sigma) = \inf\{\gamma \mid \Sigma \text{ has } L_2\text{-gain} \leq \gamma\}$. Σ is said to have L_2-gain $< \gamma$ if there exists $\tilde{\gamma} < \gamma$ such that Σ has L_2-gain $\leq \tilde{\gamma}$. Finally, Σ is called *inner* if it is conservative with respect to $s(u,y) = \frac{1}{2}||u||^2 - \frac{1}{2}||y||^2$.

A more explicit form of the differential dissipation inequality (8.3) can be obtained for systems that are *affine* in the input u and *without feedthrough* term

$$\Sigma_a : \quad \begin{aligned} \dot{x} &= f(x) + g(x)u \\ y &= h(x), \end{aligned} \qquad\qquad (8.4)$$

with $g(x)$ an $n \times m$ matrix. In this case, the differential dissipation inequality (8.3) for Σ_a amounts to

$$S_x(x)[f(x) + g(x)u] - \frac{1}{2}\gamma^2||u||^2 + \frac{1}{2}||h(x)||^2 \leq 0, \qquad \forall\, x, u \qquad (8.5)$$

This can be simplified by computing the maximizing u^* (as a function of x) for the left-hand side, i.e.,

$$u^* = \frac{1}{\gamma^2}g^T(x)S_x^T(x), \qquad\qquad (8.6)$$

and substituting (8.6) into (8.5) to obtain the *Hamilton–Jacobi inequality*

$$S_x(x)f(x) + \frac{1}{2}\frac{1}{\gamma^2}S_x(x)g(x)g^T(x)S_x^T(x) + \frac{1}{2}h^T(x)h(x) \leq 0, \qquad \forall\, x \in \mathcal{X} \qquad (8.7)$$

Thus, Σ_a has L_2-gain $\leq \gamma$ with a C^1 storage function if and only if there exists a C^1 solution $S \geq 0$ to (8.7). Furthermore, it follows from the theory of *dynamic programming* that if the available storage S_a and required supply S_r (assuming existence) are C^1, they are actually solutions of the Hamilton–Jacobi (-Bellman) *equality*

$$S_x(x)f(x) + \frac{1}{2}\frac{1}{\gamma^2}S_x(x)g(x)g^T(x)S_x^T(x) + \frac{1}{2}h^T(x)h(x) = 0 \qquad (8.8)$$

More information on the structure of the solution set of the Hamilton–Jacobi inequality (8.7) and equality (8.8) will be provided in Chap. 11.

An alternative view on L_2-gain, directly relating to Proposition 1.2.9 in Chap. 1, is provided by the following definition and proposition.

Definition 8.1.2 Given the state space system Σ given by (8.1) define the *associated Hamiltonian input–output system* Σ^H as

$$\Sigma^H : \begin{aligned} \dot{x} &= \frac{\partial H}{\partial p}(x, p, u) \\ \dot{p} &= -\frac{\partial H}{\partial x}(x, p, u) \\ y_a &= \frac{\partial H}{\partial u}(x, p, u) \end{aligned} \tag{8.9}$$

with state (x, p), inputs u and outputs y_a, where the Hamiltonian $H(x, p, u)$ is defined as

$$H(x, p, u) := p^T f(x, u) + \frac{1}{2} h^T(x, u) h(x, u) \tag{8.10}$$

Here $p \in T_x^* \mathcal{X} \simeq \mathbb{R}^n$ denotes the co-state vector.

Remark 8.1.3 Geometrically, the state space of Σ^H is given by the cotangent bundle $T^* \mathcal{X}$, with $p \in T_x^* \mathcal{X}$. See [24, 73] for further information. In case of a linear system, $\dot{x} = Ax + Bu$, $y = Cx + Du$ with transfer matrix $K(s) = C(sI - A)^{-1}B + D$, the transfer matrix of the associated input–output Hamiltonian system is given as $K^T(-s)K(s)$.

Proposition 8.1.4 *Consider Σ satisfying $f(0, 0) = 0$, $h(0, 0) = 0$, with input–output map $G_0 : L_{2e}(\mathbb{R}^m) \to L_{2e}(\mathbb{R}^p)$, which is assumed to be L_2-stable. Then the input–output map $G_{(0,0)}^H$ of Σ^H for initial state $(x, p) = (0, 0)$ is equal to the composed map*

$$G_{(0,0)}^H(u) = (DG_0(u))^* \circ G_0(u) \tag{8.11}$$

with $DG_0(u)$ the Fréchet derivative of the map G_0. In particular, it follows that the input–output map G_0 is inner if and only if $G_{(0,0)}^H$ is the identity-map.

Proof Follows directly from the definition of the Fréchet derivative $DG_0(u)$ of G_0 at u. The last statement follows from Proposition 1.2.9 in Chap. 1. $\qquad\square$

8.2 The Small-Gain Theorem Revisited

In this section, we provide a state space interpretation of the Small-gain Theorem 2.1.1 for input–output maps in Chap. 2.

Let us consider, as in Chap. 4, the standard feedback configuration $\Sigma_1 \|_f \Sigma_2$ of two input-state-output systems

$$\Sigma_i : \begin{aligned} \dot{x}_i &= f_i(x_i, u_i), \quad x_i \in \mathcal{X}_i, \; u_i \in U_i \\ y_i &= h_i(x_i, u_i), \qquad\qquad\quad y_i \in Y_i \end{aligned} \quad i = 1, 2 \tag{8.12}$$

with $U_1 = Y_2$, $U_2 = Y_1$, cf. Fig. 4.1. Suppose Σ_1 and Σ_2 in (8.12) have L_2-gain $\leq \gamma_1$, respectively $\leq \gamma_2$. Denote the storage functions of Σ_1, Σ_2 by S_1, S_2, with corresponding dissipation inequalities

$$S_1(x_1(t_1)) - S_1(x_1(t_0)) \leq \frac{1}{2} \int_{t_0}^{t_1} (\gamma_1^2 ||u_1(t)||^2 - ||y_1(t)||^2) dt$$
$$S_2(x_2(t_1)) - S_2(x_2(t_0)) \leq \frac{1}{2} \int_{t_0}^{t_1} (\gamma_2^2 ||u_2(t)||^2 - ||y_2(t)||^2) dt \qquad (8.13)$$

Consider now the feedback interconnection with $e_1 = e_2 = 0$

$$u_1 = -y_2, \quad u_2 = y_1, \qquad (8.14)$$

and assume $\gamma_1 \cdot \gamma_2 < 1$. Then we can find an α such that

$$\gamma_1 < \alpha < \frac{1}{\gamma_2} \qquad (8.15)$$

Substitute (8.14) into (8.13), multiply the second inequality of (8.13) by α^2, and add both resulting inequalities, to obtain

$$S(x_1(t_1), x_2(t_1)) - S(x_1(t_0), x_2(t_0)) \leq$$
$$\frac{1}{2} \int_{t_0}^{t_1} [(\alpha^2 \gamma_2^2 - 1)||y_1(t)||^2 + (\gamma_1^2 - \alpha^2)||y_2(t)||^2] dt \qquad (8.16)$$

where $S(x_1, x_2) := S_1(x_1) + \alpha^2 S_2(x_2)$. Since α satisfies (8.15) it immediately follows that

$$S(x_1(t_1), x_2(t_1)) - S(x_1(t_0), x_2(t_0)) \leq -\varepsilon \int_{t_0}^{t_1} ||y_1(t)||^2 + ||y_2(t)||^2 dt \qquad (8.17)$$

for a certain $\varepsilon > 0$. Thus S is a candidate Lyapunov function for the closed-loop system. In fact, we may immediately apply the reasoning of Lemma 3.2.16, resulting in

Theorem 8.2.1 (Small-gain theorem; state space version) *Suppose Σ_1 and Σ_2 have L_2-gain $\leq \gamma_1$ and $\leq \gamma_2$, with $\gamma_1 \cdot \gamma_2 < 1$. Suppose $S_1, S_2 \geq 0$ satisfying (8.13) are C^1 and have strict local minima at $x_1^* = 0, x_2^* = 0$, and Σ_1 and Σ_2 are zero-state detectable. Then $x^* = (x_1^*, x_2^*)$ is an asymptotically stable equilibrium of the closed-loop system $\Sigma_1 \|_f \Sigma_2$ with $e_1 = e_2 = 0$, which is globally asymptotically stable if additionally S_1, S_2 have global minima at x_1^*, x_2^* and are proper.*

We leave the refinement of Theorem 8.2.1 to positive semidefinite S_1 and S_2 based on Theorem 3.2.19 to the reader (see also [312]). Instead we formulate the following version based on Proposition 3.2.22. For simplicity, assume that Σ_i, $i = 1, 2$, are *affine* systems

$$\Sigma_{ai} : \begin{array}{l} \dot{x}_i = f_i(x_i) + g_i(x_i)u_i \\ y_i = h_i(x_i) \end{array} \qquad (8.18)$$

Proposition 8.2.2 *Suppose the affine systems Σ_{a1} and Σ_{a2} as in (8.18) have L_2-gain $\leq \gamma_1$ and $\leq \gamma_2$, with $\gamma_1 \cdot \gamma_2 < 1$. Suppose $S_1, S_2 \geq 0$ satisfying (4.52) are C^1 and*

$S_1(x_1^*) = S_2(x_2^*) = 0$ *(that is, S_1 and S_2 are positive semidefinite at x_1^*, respectively,* *x_2^*). Furthermore, assume that x_i^* is an asymptotically stable equilibrium of $\dot{x}_i = f_i(x_i)$, $i = 1, 2$. Then (x_1^*, x_2^*) is an asymptotically stable equilibrium of the closed-loop system $\Sigma_1 \|_f \Sigma_2$ with $e_1 = 0$, $e_2 = 0$.*

Proof The closed-loop system $\Sigma_1 \|_f \Sigma_2$ with $e_1 = 0$, $e_2 = 0$ can be written as

$$\dot{x} = f(x) + g(x)k(x)$$

with $x = (x_1, x_2)$, and

$$f = \begin{bmatrix} f_1 \\ f_2 \end{bmatrix}, \quad g = \begin{bmatrix} g_1 & 0 \\ 0 & g_2 \end{bmatrix}, \quad k = \begin{bmatrix} -h_2 \\ h_1 \end{bmatrix} \tag{8.19}$$

By (8.17)

$$S_x(x_1, x_2)[f(x) + g(x)k(x)] \le -\varepsilon \|k(x)\|^2, \tag{8.20}$$

while by assumption (x_1^*, x_2^*) is an asymptotically stable equilibrium of $\dot{x} = f(x)$. The statement now follows from Proposition 3.2.22.   □

Remark 8.2.3 Contrary to the Small-gain Theorem 2.1.1 for input–output maps in Chap. 2 we can relax the small-gain condition $\gamma_1 \cdot \gamma_2 < 1$ to $\gamma_1 \cdot \gamma_2 \le 1$, in the sense that for $\gamma_1 \cdot \gamma_2 = 1$ the inequalities (8.17) and (8.20) remain to hold with $\varepsilon = 0$. Hence under appropriate conditions on S_1, S_2 stability continues to hold if $\gamma_1 \cdot \gamma_2 = 1$.

Remark 8.2.4 Note that the small-gain theorem is equally valid for the *positive* feedback interconnection $u_1 = y_2$, $u_2 = y_1$.

Theorem 8.2.1 and Proposition 8.2.2 have immediate applications to robustness analysis. A simple corollary of Theorem 8.2.1 is the following.

Corollary 8.2.5 *Consider a nominal set of differential equations $\dot{x} = f(x)$, $f(0) = 0$, with perturbation model*

$$\dot{x} = f(x) + \bar{g}(x)\Delta \bar{h}(x), \tag{8.21}$$

where $\bar{g}(x)$ is a known $n \times \bar{m}$ matrix, $\bar{h} : \mathcal{X} \to \mathbb{R}^{\bar{p}}$, $\bar{h}(0) = 0$, is a known mapping, and Δ is an unknown $\bar{m} \times \bar{p}$ matrix representing the uncertainty. Suppose the system

$$\bar{\Sigma} : \begin{array}{ll} \dot{x} = f(x) + \bar{g}(x)\bar{u}, & \bar{u} \in \mathbb{R}^m \\ \bar{y} = \bar{h}(x), & \bar{y} \in \mathbb{R}^p \end{array} \tag{8.22}$$

has L_2-gain $\le \gamma$ (from $\bar{u}$ to $\bar{y}$), with C^1 storage function having a strict local minimum at 0, or having a local minimum at 0 while 0 is an asymptotically stable equilibrium of $\dot{x} = f(x)$. Then 0 is an asymptotically stable equilibrium of the perturbed system (8.21) for all perturbations Δ having largest singular value less than $\frac{1}{\gamma}$.

Proof Take in Propositions 8.2.1 or 8.2.2 $\Sigma_1 = \bar{\Sigma}$, and Σ_2 equal to the static system corresponding to multiplication by Δ. The L_2-gain of Σ_2 is the largest singular value of Δ. □

Another direct consequence of the preceding theory concerns *robustness* of finite L_2-gain.

Corollary 8.2.6 *Consider the perturbed state space system*

$$\Sigma_p : \begin{array}{l} \dot{x} = [f(x) + \bar{g}(x)\Delta\bar{h}(x)] + g(x)u, \ x \in \mathcal{X}, \ u \in \mathbb{R}^m, \\ y = h(x), \qquad\qquad\qquad\qquad\qquad y \in \mathbb{R}^p \end{array} \tag{8.23}$$

where $\bar{g}, \bar{h}$, and Δ are as in Corollary 8.2.5, with $\Delta = 0$ representing the nominal state space system. Suppose there exists a solution $S \geq 0$ to the parametrized Hamilton–Jacobi inequality

$$S_x(x)f(x) + \tfrac{1}{2}\tfrac{1}{\gamma^2}S_x(x)g(x)g^T(x)S_x^T(x) + \tfrac{1}{2}h^T(x)h(x) + $$
$$\tfrac{1}{2}\tfrac{1}{\gamma^2}\tfrac{1}{\varepsilon^2}S_x(x)\bar{g}(x)\bar{g}^T(x)S_x^T(x) + \tfrac{1}{2}\varepsilon^2\bar{h}^T(x)\bar{h}(x) \leq 0 \tag{8.24}$$

(with ε a fixed but arbitrary scaling parameter), meaning that the extended system

$$\begin{array}{l} \dot{x} = f(x) + g(x)u + \tfrac{1}{\varepsilon}\bar{g}(x)\bar{u} \\ y = h(x) \\ \bar{y} = \varepsilon\bar{h}(x) \end{array} \tag{8.25}$$

has L_2-gain $\leq \gamma$ from $(u, \bar{u})$ to $(y, \bar{y})$. Then the perturbed system Σ_p has L_2-gain $\leq \gamma$ for all perturbations Δ having largest singular value $\leq \frac{1}{\gamma}$.

Proof For all Δ with largest singular value $\leq \frac{1}{\gamma}$

$$S_x(x)\bar{g}(x)\Delta\bar{h}(x) \leq \frac{1}{2}\frac{1}{\gamma^2}\frac{1}{\varepsilon^2}S_x(x)\bar{g}(x)\bar{g}^T(x)S_x^T(x) + \frac{1}{2}\varepsilon^2\bar{h}^T(x)\bar{h}(x)$$

and thus the expression

$$S_x(x)[f(x) + \bar{g}(x)\Delta\bar{h}(x)] + \frac{1}{2}\frac{1}{\gamma^2}S_x(x)g(x)g^T(x)S_x^T(x) + \frac{1}{2}h^T(x)h(x)$$

is bounded from above by the left-hand side of (8.24), implying that Σ_p has L_2-gain $\leq \gamma$. □

This last corollary can be extended to *dynamic* perturbations Δ in the following way.

Proposition 8.2.7 *Consider the extended system (8.25), and assume that there exists a solution $S \geq 0$ to (8.24) (implying that (8.25) has L_2-gain $\leq \gamma$ from $(u, \bar{u})$ to*

$(y, \bar{y}))$. *Consider another dynamical system* Δ *with state* ξ, *inputs* $\bar{y}$ *and outputs* $\bar{u}$, *and having* L_2-*gain* $\leq \frac{1}{\gamma}$ *with* C^1 *storage function* $S_\Delta(\xi) \geq 0$. *Then the closed-loop system has* L_2-*gain* $\leq \gamma$ *from* u *to* y, *with storage function* $S(x) + \gamma^2 S_\Delta(\xi)$.

Proof By (8.24)

$$\dot{S} \leq \frac{1}{2}\gamma^2 \|u\|^2 - \frac{1}{2}\|y\|^2 + \frac{1}{2}\gamma^2 \|\bar{u}\|^2 - \frac{1}{2}\|\bar{y}\|^2 \qquad (8.26)$$

Furthermore, since Δ has L_2-gain $\leq \frac{1}{\gamma}$ with storage function S_Δ, i.e.,

$$\dot{S}_\Delta \leq \frac{1}{2}\frac{1}{\gamma^2}\|\bar{y}\|^2 - \frac{1}{2}\|\bar{u}\|^2 \qquad (8.27)$$

Premultiplying (8.27) by γ^2, and adding to (8.26) yields

$$\dot{S} + \gamma^2 \dot{S}_\Delta \leq \frac{1}{2}\gamma^2\|u\|^2 - \frac{1}{2}\|y\|^2 \qquad (8.28)$$

$\square$

Similar to the developments in Chap. 4, cf. Definition 4.7.1, we can formulate the following state space version of *incremental* L_2-*gain*, as already defined for input–output maps in Definition 2.1.5.

Definition 8.2.8 Consider a system (8.1). The system Σ has *incremental* L_2-*gain* $\leq \gamma$ if there exists a function, called the *incremental storage function*,

$$S : \mathcal{X} \times \mathcal{X} \rightarrow \mathbb{R}^+ \qquad (8.29)$$

such that

$$\begin{aligned} S(x_1(T), x_2(T)) &\leq S(x_1(0), x_2(0)) \\ &+ \tfrac{1}{2}\int_0^T \gamma^2\|u_1(t) - u_2(t)\|^2 + \|y_1(t) - y_2(t)\|^2 \, dt, \end{aligned} \qquad (8.30)$$

for all $T \geq 0$, and for all pairs of input functions $u_1, u_2 : [0, T] \rightarrow \mathbb{R}^m$ and all pairs of initial conditions $x_1(0), x_2(0)$, with resulting pairs of state and output trajectories $x_1, x_2 : [0, T] \rightarrow \mathcal{X}$, $y_1, y_2 : [0, T] \rightarrow \mathbb{R}^p$.

As before in the context of incremental passivity, cf. Remark 4.7.2, it can be assumed without loss of generality that the storage function $S(x_1, x_2)$ satisfies the symmetry property $S(x_1, x_2) = S(x_2, x_1)$.

The *differential* version of the incremental dissipation inequality (8.29) takes the form

$$\begin{aligned} S_{x_1}(x_1, x_2) f(x_1, u_1) &+ S_{x_2}(x_1, x_2) f(x_2, u_2) \\ &\leq \tfrac{1}{2}\gamma^2\|u_1 - u_2\|^2 + \tfrac{1}{2}\|y_1 - y_2\|^2 \end{aligned} \qquad (8.31)$$

for all $x_1, x_2, u_1, u_2, y_1 = h(x_1, u_1)$, $y_2 = h(x_2, u_2)$, where $S_{x_1}(x_1, x_2)$ and S_{x_2} (x_1, x_2) denote row vectors of partial derivatives with respect to x_1, respectively x_2.

Let us, as above, specialize to systems (8.4), in which case the incremental dissipation inequality (8.31) becomes

$$
\begin{aligned}
& S_{x_1}(x_1, x_2)f(x_1) + S_{x_1}(x_1, x_2)g(x_1)u_1 + \\
& S_{x_2}(x_1, x_2)f(x_2) + S_{x_1}(x_1, x_2)g(x_2)u_2 \\
& \quad \leq \tfrac{1}{2}\gamma^2 \|u_1 - u_2\|^2 + \tfrac{1}{2}\|h(x_1) - h(x_2)\|^2
\end{aligned}
\tag{8.32}
$$

In general, this inequality is hard to solve for the unknown incremental storage function $S(x_1, x_2)$. Assuming[1] that $g(x)$ is independent of x and restricting attention to incremental storage functions of the form $S(x_1, x_2) = \bar{S}(x_1 - x_2)$, implying that $S_{x_1}(x_1, x_2) = -S_{x_2}(x_1, x_2) = \bar{S}_x(x_1 - x_2)$, the inequality (8.32) reduces to

$$
\begin{aligned}
& \bar{S}_x(x_1 - x_2)[f(x_1) - f(x_2)] + \bar{S}_x(x_1 - x_2)g[u_1 - u_2] \\
& \quad \leq \tfrac{1}{2}\gamma^2\|u_1 - u_2\|^2 + \tfrac{1}{2}\|h(x_1) - h(x_2)\|^2
\end{aligned}
\tag{8.33}
$$

By "completion of the squares" in the difference $u_1 - u_2$, this can be seen to further reduce to the Hamilton–Jacobi inequality

$$
\begin{aligned}
& \bar{S}_x(x_1 - x_2)[f(x_1) - f(x_2)] + \tfrac{1}{\gamma^2}\bar{S}_x(x_1 - x_2)gg^T\bar{S}_x^T(x_1 - x_2) \\
& \quad + \tfrac{1}{2}\|h(x_1) - h(x_2)\|^2 \leq 0
\end{aligned}
\tag{8.34}
$$

for all x_1, x_2.

Remark 8.2.9 A trivial example is provided by a *linear* system having L_2-gain $\leq \gamma$ with quadratic storage function $\tfrac{1}{2}x^T Q x$. In this case, $S(x_1, x_2) := \tfrac{1}{2}(x_1 - x_2)^T Q(x_1 - x_2)$ defines an incremental storage function, and hence the system also has incremental L_2-gain $\leq \gamma$.

8.3 Network Version of the Small-Gain Theorem

The small-gain theorem concerns the stability of the interconnection of two systems in negative or positive feedback interconnection. A *network* version of this can be formulated as follows.

Consider a multiagent system, corresponding to a directed graph $\mathcal{G}$ with N vertices and input-state-output systems Σ_i, $i = 1, \ldots, N$, associated to these vertices. Furthermore, assume that the edges of the graph are specified by an $N \times N$ *adjacency*

---

[1]Or assuming the existence of coordinates in which $g(x)$ is constant, which is, under the assumption that rank$g(x) = m$, equivalent to the Lie brackets of the vector fields $g_1, \ldots, g_m$ defined by the columns of $g(x)$ to be zero [233].

matrix [48, 114] $\mathcal{A}$ with elements 0, 1, corresponding to interconnections[2]

$$u_i = y_j \tag{8.35}$$

if and only if the (i, j)-th element of $\mathcal{A}$ is equal to 1.

Now assume that the systems Σ_i have L_2-gain $\leq \gamma_i$, $i = 1, \ldots, N$. This means that there exist storage functions $S_i : \mathcal{X}_i \to \mathbb{R}^+$ such that[3]

$$\dot{S}_i \leq \frac{1}{2}\gamma_i^2 \|u_i\|^2 - \frac{1}{2}\|y_i\|^2, \quad i = 1, \ldots, N \tag{8.36}$$

Then define the following $N \times N$ matrix with nonnegative elements

$$\Gamma := \mathrm{diag}(\gamma_1^2, \ldots, \gamma_N^2)\mathcal{A} \tag{8.37}$$

The Perron–Frobenius theorem yields the following lemma.

Lemma 8.3.1 ([79]) *Denote by $z > 0$ a (column or row) vector z with all elements positive, and by $z \geq 0$ a vector with nonnegative elements. Furthermore, let I_N denote the $N \times N$ identity matrix.*

Consider an $N \times N$ matrix Γ with all nonnegative elements. Then there exists a vector $\mu > 0$ such that

$$\mu^T(\Gamma - I_N) < 0 \tag{8.38}$$

if and only if

$$r(\Gamma) < 1, \tag{8.39}$$

where $r(\Gamma)$ denotes the spectral radius of Γ.

Proof (If). If $r(\Gamma) < 1$ then by the Perron–Frobenius theorem there exists $\mu > 0$ such that $\mu^T\Gamma < \mu^T$, or equivalently $\mu^T(\Gamma - I_N) < 0$.
(Only if). Conversely, if $r(\Gamma) \geq 1$ then there exists $\nu \geq 0$, $\nu \neq 0$, such that $(\Gamma - I_N)\nu \geq 0$, whence for any $\mu > 0$ we have $\mu^T(\Gamma - I_N)\nu \geq 0$, contradicting $\mu^T(\Gamma - I_N) < 0$. $\square$

Based on this lemma we obtain the following theorem.

Theorem 8.3.2 (Small-gain network theorem) *Consider a directed graph $\mathcal{G}$ with systems Σ_i associated to its vertices, which have L_2-gains $\leq \gamma_i$ with C^1 storage functions S_i, $i = 1, \ldots, N$, and which are interconnected through the adjacency matrix $\mathcal{A}$ defined by (8.35). Consider the matrix Γ given by (8.37). If the spectral*

---

[2]The typical situation being that each row in the $\mathcal{A}$ matrix contains only one 1. Multiple occurrence of ones in a row is allowed but will imply an *equality constraint* on the corresponding outputs, leading to algebraic constraints between the state variables.

[3]The subsequent argumentation directly extends to the case of non-differentiable storage functions, replacing the differential dissipation inequalities by their integral counterparts.

radius $r(\Gamma) < 1$, then there exists $\mu > 0$ such that $\mu^T(\Gamma - I_N) < 0$ and the nonnegative function

$$S(x_1, \ldots, x_N) := \sum_{i=1}^{N} \mu_i S_i(x_i) \tag{8.40}$$

satisfies along trajectories of the interconnected system

$$\dot{S} \le -\varepsilon_1 \|y_1\|^2 - \varepsilon_2 \|y_2\|^2 \cdots - \varepsilon_N \|y_N\|^2 \tag{8.41}$$

for certain positive constants $\varepsilon_1, \ldots, \varepsilon_N$. Hence, if S_i has a strict minimum at x_i^*, $i = 1, \ldots, N$, then $x^* = (x_1^*, \ldots, x_N^*)$ is a stable equilibrium of the interconnected system, which is asymptotically stable provided the interconnected system is zero-state detectable.

Proof Denote the vector with components $\|y_i\|^2$, $i = 1, \ldots, N$, by $\widehat{y}$. Then

$$\begin{aligned}\dot{S} = \sum_{i=1}^{N} \mu_i \dot{S}_i(x_i) \le \tfrac{1}{2} \sum_{i=1}^{N} \mu_i(\gamma_i^2 \|u_i\|^2 - \|y_i\|^2) = \tfrac{1}{2}\mu^T(\Gamma - I_N)\widehat{y} \\ \le -\varepsilon_1 \|y_1\|^2 - \varepsilon_2 \|y_2\|^2 \cdots - \varepsilon_N \|y_N\|^2\end{aligned} \tag{8.42}$$

for certain positive constants $\varepsilon_1, \ldots, \varepsilon_N$.   □

Example 8.3.3 In case of the feedback interconnection $u_1 = y_2$, $u_2 = y_1$ of two systems Σ_1, Σ_2 with L_2-gain $\le \gamma_1$, respectively, $\le \gamma_2$, application of Theorem 8.3.2 leads to the consideration of the matrix

$$\Gamma = \begin{bmatrix} 0 & \gamma_1^2 \\ \gamma_2^2 & 0 \end{bmatrix}, \tag{8.43}$$

which has spectral radius < 1 if and only if $\gamma_1 \cdot \gamma_2 < 1$; thus recovering the small-gain condition of Theorem 8.2.1.

8.4 L_2-Gain as Response to Periodic Input Functions

An interesting interpretation of the L_2-gain of a nonlinear state space system Σ given by (8.1) in terms of the response to *periodic* input functions was obtained in [140].

Assume that $f(0, 0) = 0$, and that the matrix $A := \frac{\partial f}{\partial x}(0, 0)$ has all its eigenvalues in the open left half plane. Let Σ have L_2-gain $\le \gamma$, and consider a *periodic* input function $u_p(\cdot)$ (of period $T > 0$), which is generated by a dynamical system ("*exosystem*")

$$\Sigma_e : \dot{u} = s(u), \quad s(0) = 0, \tag{8.44}$$

whose linearization has all its eigenvalues on the imaginary axis. Then it follows from center manifold theory (see, e.g., [61]) that the series interconnection of the exo-system Σ_e with Σ, given as

$$\begin{aligned}\dot{x} &= f(x, u)\\ \dot{u} &= s(u)\end{aligned} \tag{8.45}$$

and having augmented state (x, u), has a *center manifold*, which is the graph $\mathcal{C} = \{(x, u) \mid x = c(u)\}$ of a mapping $x = c(u)$. Furthermore, for initial conditions close enough to $\mathcal{C}$ the solutions of the composed system (8.45) tend exponentially to $\mathcal{C}$, and thus, since $u_p(\cdot)$ is periodic of period T, the solution $x(\cdot)$ converges to a *steady state* solution $x_p(\cdot)$, with $x_p(t) = c(u_p(t))$, also having period T (see [141]).

It follows that $y_p(t) = h(x_p(t), u_p(t))$ has period T as well, and furthermore by the L_2-gain dissipation inequality (8.2)

$$\int_{t_0}^{t_0+T} \|y_p(t)\|^2 dt \ \leq\ \gamma^2 \int_{t_0}^{t_0+T} \|u_p(t)\|^2 dt \tag{8.46}$$

(since $x_p(t_0 + T) = x_p(t_0)$), for all t_0.

Defining the *rms* values of any periodic signal z with period T as

$$\|z\|_{rms} := \frac{1}{T}\left(\int_{t_0}^{t_0+T} \|z(t)\|^2 dt\right)^{1/2}, \tag{8.47}$$

the property that Σ has L_2-gain $\leq \gamma$ therefore implies that

$$\|y_p\|_{rms} \ \leq \gamma \ \|u_p\|_{rms}\ , \tag{8.48}$$

for all periodic input functions u_p which are generated by the exo-system $\dot{u} = s(u)$, $s(0) = 0$, and all initial conditions close enough to $\mathcal{C}$.

8.5 Relationships with IIS- and iIIS-Stability

An alternative approach to extending Lyapunov stability theory of autonomous systems $\dot{x} = f(x)$ to systems with inputs $\dot{x} = f(x, u)$ is offered by the theory of (integral-)Input-to-State Stability ((i)ISS). In this section, we will briefly indicate the close relationships of (i)ISS with dissipative systems theory and (a generalized) version of L_2-gain. We will do so by *not* giving the original definitions of IIS and iIIS, but instead by providing their equivalent characterizations in terms of dissipative systems theory.

By itself the close relation between dissipative systems theory on the one hand and stability theory based on the IIS or iIIS property on the other is not surprising. Indeed, we have already seen in Chap. 3 how dissipativity implies Lyapunov stability properties of the system for $u = 0$. Furthermore, for $u \neq 0$ the dissipation inequality may be used for deriving properties of the relationships between input, state and output functions.

Recall that a function $\alpha : \mathbb{R}^+ \to \mathbb{R}^+$ is of class $\mathcal{K}_\infty$, denoted $\alpha \in \mathcal{K}_\infty$, if it is continuous, strictly increasing, unbounded, and satisfies $\alpha(0) = 0$.

Turning the characterization of input-to-state stability (IIS) as given in [323] into a definition we formulate

Definition 8.5.1 ([323]) A system

$$\dot{x} = f(x, u), \quad f(0, 0) = 0, \qquad x \in \mathbb{R}^n, u \in \mathbb{R}^m \tag{8.49}$$

is ISS if there exist functions $\alpha, \beta \in \mathcal{K}_\infty$ such that the input-state-output system $\dot{x} = f(x, u), y = x$, is dissipative with respect to the supply rate

$$s(u, y) = \beta(\|u\|) - \alpha(\|y\|) \tag{8.50}$$

with a C^1 and radially unbounded storage function S satisfying $S(x) > 0, x \neq 0, S(0) = 0$.

Note the conceptual difference with dissipative systems theory (apart from minor technical differences like the a priori assumption of strict positivity and radial unboundedness of S). In dissipative systems theory, one starts with a *given* supply rate on the space of inputs and outputs, and derives properties of the system based on this. On the other hand, ISS theory aims at providing stability results for systems with inputs $\dot{x} = f(x, u), f(0, 0) = 0$, and seeks criteria, which can be translated into the *existence* of a supply rate (8.50), where the functions α, β may not be easy to determine explicitly.

We remark that the supply rate (8.50) can be regarded as a generalization of the L_2-gain supply rate. In fact, by taking α and β to be the quadratic functions $\alpha(r) = \frac{1}{2}r^2$ and $\beta(r) = \frac{1}{2}\gamma^2 r^2$ for some constant γ one recovers the L_2-gain $\leq \gamma$ case for $y = x$. Conversely, by allowing for arbitrary nonlinear coordinate transformations on the space $\mathcal{U} = \mathbb{R}^m$ of inputs and $\mathcal{X} = \mathbb{R}^n$ of states, the dissipation inequality for L_2-gain (from u to $y = x$) can be seen to transform into (8.50) for certain $\alpha, \beta \in \mathcal{K}^\infty$. See also the discussion in Note 2 in Sect. 2.5 on the related notion of nonlinear L_2-gain.

With regard to integral input-to-state stability (iIIS) case we have the following characterization in terms of dissipative systems theory.

Definition 8.5.2 ([9]) A system (8.49) is iIIS if there exists a function $\gamma \in \mathcal{K}_\infty$ and a function $\alpha : \mathbb{R}^+ \to \mathbb{R}^+$ with $\alpha(0) = 0, \alpha(r) > 0, r \neq 0$, such that the input-state-output system $\dot{x} = f(x, u), y = x$, is dissipative with respect to the supply rate (8.50) with a C^1 and radially unbounded storage function S satisfying $S(x) > 0, x \neq 0, S(0) = 0$.

From the above characterization, it is clear that IIS implies iIIS, while the converse does not hold. That is, there are systems which are iIIS but not IIS. An example is provided by the system

$$\dot{x} = -x + ux \qquad (8.51)$$

See, e.g., [9, 319] for a further discussion regarding the differences between ISS and iISS.

Another version of the relationship between iIIS and dissipative systems theory is provided in the following proposition.

Proposition 8.5.3 ([9]) *A system 8.49 is iIIS if and only if there exists a continuous output function $y = h(x)$ with $h(0) = 0$ for which the resulting system $\dot{x} = f(x, u)$, $y = h(x)$ is zero-detectable and there exists $\beta \in \mathcal{K}_\infty$ and a function $\alpha : \mathbb{R}^+ \to \mathbb{R}^+$ with $\alpha(0) = 0, \alpha(r) > 0, r \neq 0$, such that the system is dissipative with respect to the supply rate (8.50) with a C^1 and radially unbounded storage function S satisfying $S(x) > 0, x \neq 0, S(0) = 0$.*

Note that the "if" part of the last proposition with $\alpha(r) = \frac{1}{2}r^2$ and $\beta(r) = \frac{1}{2}\gamma^2 r^2$ implies that a zero-detectable input-state-output system with finite L_2-gain having a C^1 and radially unbounded storage function S satisfying $S(x) > 0, x \neq 0, S(0) = 0$, is iIIS.

8.6 Notes for Chapter 8

1. Corollary 8.2.6 is a nonlinear generalization of a result given for linear systems in Xie & De Souza [358], and may be found in Shen & Tamura [313].

2. See Angeli [8] for further information regarding incremental L_2-gain as discussed in Sect. 8.2.

3. Applications of incremental L_2-gain to model reduction of nonlinear systems can be found in Besselink, van de Wouw & Nijmeijer [38].

4. Section 8.3, in particular Lemma 8.3.1, is largely based on Dashkovskiy, Ito & Wirth [79]. There is a wealth of literature on other network versions of the small-gain theorem; see, e.g., Jiang & Wang [150], Liu, Hill & Jiang [182], Rüffer [261].

5. See Pavlov, van de Wouw & Nijmeijer [254] and the references quoted therein for theory related to Sect. 8.4.

6. The notions of (integral-)Input-to-State Stability were introduced and explored by Sontag and co-workers, see, e.g., Sontag & Wang [323], Sontag [318, 319, 324], Angeli, Sontag & Wang [9] and the references quoted in there. See also the survey Sontag [321], and the account of the relation with dissipative systems theory provided in Isidori [139].

7. The characterization of the set of pairs of functions (β, α) such that the system is dissipative with respect to the supply rate (8.50) is addressed in Sontag & Teel [322].

Chapter 9
Factorizations of Nonlinear Systems

In this chapter, we apply the L_2-gain concepts and techniques from Chaps. 3 and 8 to obtain some useful types of representations of nonlinear systems, different from the standard input-state-output representation. In Sect. 9.1 we will derive *stable kernel* and *stable image* representations of nonlinear systems, and we will use them in order to formulate *nonlinear perturbation* models (with L_2-gain bounded uncertainties) in Sect. 9.2. In Sect. 9.3 we will employ stable kernel representations in order to derive a parametrization of stabilizing controllers, analogous to the Youla–Kucera parametrization in the linear case. Section 9.4 deals with the *factorization* of nonlinear systems into a series interconnection of a minimum phase system and a system which preserves the L_2-norm. This allows us to control the system based on its minimum phase factor system, using a nonlinear version of the Smith predictor.

9.1 Stable Kernel and Image Representations

A cornerstone of linear robust control theory is the theory of *stable factorizations* of transfer matrices. Let $G(s)$ be a $p \times m$ rational proper transfer matrix. A stable *left*, respectively *right*, factorization of $G(s)$ is $G(s) = D^{-1}(s)N(s)$, respectively, $G(s) = \tilde{N}(s)\tilde{D}^{-1}(s)$, where $D(s)$, $N(s)$, $\tilde{D}(s)$ and $\tilde{N}(s)$ are all *stable* proper rational matrices. These two factorizations can be alternatively interpreted as follows. The relation $y = D^{-1}(s)N(s)u$ for a stable left factorization (with u and y denoting the inputs and outputs in the frequency domain) can be equivalently rewritten as

$$0 = z = [D(s) \vdots -N(s)] \begin{bmatrix} y \\ u \end{bmatrix}, \tag{9.1}$$

© Springer International Publishing AG 2017

A. van der Schaft, *L_2-Gain and Passivity Techniques in Nonlinear Control*,
Communications and Control Engineering, DOI 10.1007/978-3-319-49992-5_9

and thus can be considered as a *stable kernel representation* of the system corresponding to $G(s)$. Indeed, $[D(s) \vdots -N(s)]$ is a stable transfer matrix with "inputs" y and u, and "outputs" z, while the input–output behavior of the system corresponding to $G(s)$ are all pairs (y, u) which are mapped by $[D(s) \vdots -N(s)]$ onto $z = 0$.

On the other hand, the relation $y = \tilde{N}(s)\tilde{D}^{-1}(s)u$ for a stable right factorization can be rewritten as

$$\begin{bmatrix} y \\ u \end{bmatrix} = \begin{bmatrix} \tilde{N}(s) \\ \tilde{D}(s) \end{bmatrix} l, \tag{9.2}$$

where l is an arbitrary m-vector of *auxiliary* variables. Thus the input–output behavior of the system corresponding to $G(s)$ are all pairs (y, u) (in the frequency domain) which are in the image of the stable transfer matrix $\begin{bmatrix} \tilde{N}(s) \\ \tilde{D}(s) \end{bmatrix}$. Hence a stable right factorization can be regarded as a *stable image representation*.

Let us generalize this to nonlinear input-state-output systems. For simplicity of exposition we only consider *affine* systems Σ_a, which throughout Sects. 9.1 and 9.2 of this chapter will be simply denoted as

$$\Sigma \ : \ \begin{array}{ll} \dot{x} = f(x) + g(x)u \,, & x \in \mathcal{X}, \ u \in \mathbb{R}^m \\ y = h(x) \,, & y \in \mathbb{R}^p \end{array} \tag{9.3}$$

In analogy with Chap. 1 we will say that the state space system Σ is L_2-*stable* if all its input–output maps G_{x_0}, $x_0 \in \mathcal{X}$, are L_2-stable. Recall from Chap. 1 that if Σ has L_2-gain $\leq \gamma$ for some γ then it is L_2-stable. Since we will only consider L_q-stability for $q = 2$ the terminology "L_2-stable" will be sometimes abbreviated to "stable".

Definition 9.1.1 A *kernel representation* of Σ given by (6.8) is any system

$$K_\Sigma \ : \ \begin{array}{ll} \dot{x} = F(x, u, y) \,, & x \in \mathcal{X}, \ u \in \mathbb{R}^m, \ y \in \mathbb{R}^p \\ z = G(x, u, y) \,, & z \in \mathbb{R}^q \end{array} \tag{9.4}$$

such that for every initial condition $x(0) = x_0 \in \mathcal{X}$ and every function $u(\cdot)$ there exists a unique solution $y(\cdot)$ to (9.4) with $z = 0$, which equals the output of Σ for the same initial condition and same input $u(\cdot)$. K_Σ is called a L_2-*stable kernel representation* of Σ if moreover K_Σ is L_2-stable (from (u, y) to z).

An *image representation* of Σ is any system

$$I_\Sigma \ : \ \begin{array}{ll} \dot{x} \ = F(x, l) \,, & x \in \mathcal{X}, \ l \in \mathbb{R}^k \\ \begin{bmatrix} y \\ u \end{bmatrix} = G(x, l) \end{array} \tag{9.5}$$

such that for every initial condition $x(0) = x_0 \in \mathcal{X}$ and every input function $u(\cdot)$ and resulting output function $y(\cdot)$ of Σ there exists a function $l(\cdot)$ such that the pair (u, y)

produced by I_Σ for the same initial condition x_0 coincides with the input–output pair (u, y) of Σ. I_Σ is an L_2-*stable image representation* if I_Σ is L_2-stable from l to (u, y).

In case Σ is already itself an L_2-*stable* input-state-output system, a stable kernel representation of Σ is simply

$$K_\Sigma \; : \; \begin{array}{l} \dot{x} = f(x) + g(x)u \\ z = y - h(x) \end{array} \tag{9.6}$$

while a stable image representation of Σ is given by

$$I_\Sigma \; : \; \begin{array}{l} \dot{x} = f(x) + g(x)l \\ y = h(x) \\ u = l \end{array} \tag{9.7}$$

If Σ is not L_2-stable then we may proceed as follows. Consider the following two versions of the same Hamilton–Jacobi equation

$$V_x(x)f(x) - \frac{1}{2}V_x(x)g(x)g^T(x)V_x^T(x) + \frac{1}{2}h^T(x)h(x) = 0 \tag{9.8}$$

$$W_x(x)f(x) + \frac{1}{2}W_x(x)g(x)g^T(x)W_x^T(x) - \frac{1}{2}h^T(x)h(x) = 0 \tag{9.9}$$

Suppose there exists a C^1 solution $W \geq 0$ to (9.9). Additionally, assume there exists an $n \times p$ matrix $k(x)$ satisfying

$$W_x(x)k(x) = h^T(x) \tag{9.10}$$

Then the system with "inputs" u and y and "outputs" z given by

$$K_\Sigma \; : \; \begin{array}{l} \dot{x} = [f(x) - k(x)h(x)] + g(x)u + k(x)y \\ z = y - h(x) \end{array} \tag{9.11}$$

is a *stable kernel representation* of Σ. Indeed, by setting $z = 0$ in (9.11) we recover the original system Σ. Second, from (9.9), (9.10) it follows that

$$W_x(x)([f(x) - k(x)h(x)] + g(x)u + k(x)y) =$$
$$-\frac{1}{2}||u - g^T(x)W_x^T(x)||^2 - \frac{1}{2}||z||^2 + \frac{1}{2}||u||^2 + \frac{1}{2}||y||^2 \leq$$
$$-\frac{1}{2}||z||^2 + \frac{1}{2}||u||^2 + \frac{1}{2}||y||^2 \tag{9.12}$$

showing that K_Σ has L_2-gain ≤ 1 from (u, y) to z. Moreover, from $z = y - h(x)$ it is easily seen that there does not exist $\gamma < 1$ such that the L_2-gain of K_Σ is

$\leq \gamma$. Therefore, the L_2-gain of K_Σ is actually *equal* to 1. We summarize this in the following proposition.

Proposition 9.1.2 *Assume there exist solutions $W \geq 0$ to (9.9) and $k(x)$ to (9.10), then K_Σ given by (9.11) is a stable kernel representation of Σ which has L_2-gain equal to* 1.

Remark 9.1.3 Equation (9.9) is the Hamilton–Jacobi–Bellman equation corresponding to the "optimal control problem in reversed time" or the "optimal filtering problem"

$$\min_u \int_{-\infty}^{0} (||u(t)||^2 + ||y(t)||^2)dt \quad , \quad x(-\infty) = 0, \quad x(0) = x, \tag{9.13}$$

and W is the corresponding value function. Conditions for local solvability of (9.9) will be derived in Chap. 11 (see Remark 11.2.4).

Remark 9.1.4 Suppose W has a minimum (or, more generally, a stationary point) at some x_0 with $h(x_0) = 0$. For simplicity set $x_0 = 0$. Then $W_x(0) = 0$ and $h(0) = 0$. It follows by standard arguments (see, e.g., [136, 233]) that we may write, locally about 0,

$$W_x(x) = x^T M(x) \quad , \quad h(x) = C(x)x \tag{9.14}$$

for certain matrices $M(x)$, $C(x)$, with entries smoothly depending on x. If $M(x)$ is invertible then the unique solution $k(x)$ to (9.10) is given as

$$k(x) = M^{-1}(x)C^T(x) \tag{9.15}$$

In Chap. 11, Remark 11.2.4, it will be shown that under minimality assumptions on the system linearized at $x = 0$, the matrix $M(0)$, and thus $M(x)$ for x near 0, is always invertible.

Remark 9.1.5 Substitution of $u = 0$, $y = 0$ in (9.12) yields

$$W_x(x)[f(x) - k(x)h(x)] \leq -\frac{1}{2}h^T(x)h(x) \tag{9.16}$$

Assume that x_0 is an equilibrium, i.e., $f(x_0) = 0$ and $h(x_0) = 0$, and that W has a strict local minimum at x_0. It follows that the zero-output constrained dynamics of Σ (see Sect. 5.1) is at least stable around x_0. This property can be regarded as a weak form of zero-state detectability (see Definition 3.2.15). Conversely, if Σ is zero-state detectable then (cf. Lemma 3.2.16) it follows that x_0 is an asymptotically stable equilibrium of $\dot{x} = f(x) - k(x)h(x)$.

Remark 9.1.6 For a *linear* system Σ, K_Σ equals the normalized left coprime factorization, see, e.g., [212].

Let us now move to *image* representations. In this case, we consider the Hamilton–Jacobi equation (9.8), and the basic proposition is as follows.

Proposition 9.1.7 *Assume there exists a C^1 solution $V \geq 0$ to (9.8), then the system*

$$
I_\Sigma : \begin{array}{l} \dot{x} = [f(x) - g(x)g^T(x)V_x^T(x)] + g(x)s \\ y = h(x) \\ u = s - g^T(x)V_x^T(x) \end{array}
\tag{9.17}
$$

with auxiliary variables s is a stable image representation of Σ, which has L_2-gain (from s to (y, u)) equal to 1.

Proof Clearly I_Σ is an image representation. (Eliminate from the last equations the auxiliary variables as $s = u + g^T(x)V_x^T(x)$.) In view of (9.8)

$$
V_x(x)(f(x) - g(x)g^T(x)V_x^T(x)) + V_x(x)g(x)s =
$$
$$
-\frac{1}{2}V_x(x)g(x)g^T(x)V_x^T(x) - \frac{1}{2}h^T(x)h(x) + V_x(x)g(x)s =
$$
$$
-\frac{1}{2}||s - g^T(x)V_x^T(x)||^2 + \frac{1}{2}||s||^2 - \frac{1}{2}||y||^2 =
\tag{9.18}
$$
$$
\frac{1}{2}||s||^2 - \frac{1}{2}||u||^2 - \frac{1}{2}||y||^2
$$

implying that I_Σ has L_2-gain = 1. □

Remark 9.1.8 In fact I_Σ is *conservative* (see Definition 3.1.2) with regard to the supply rate $\frac{1}{2}||s||^2 - \frac{1}{2}||u||^2 - \frac{1}{2}||y||^2$.

Remark 9.1.9 (Compare with Remark 9.1.3 and (3.64)) Equation (9.8) is the Hamilton–Jacobi–Bellman equation for the optimal control problem

$$
\min_u \int_0^\infty (||u(t)||^2 + ||y(t)||^2)dt, \quad x(0) = x, \quad x(\infty) = 0,
\tag{9.19}
$$

and conditions for local solvability of (9.8) will be derived in Chap. 10, Sect. 10.2. Clearly, the existence of a solution V requires some kind of stabilizability of Σ.

Remark 9.1.10 For a *linear* system Σ, I_Σ equals the normalized right coprime factorization (cf. [212]).

Finally assume that C^1 solutions $V \geq 0$, $W \geq 0$ and k to, respectively, (9.8)–(9.10) all exist, and thus that both K_Σ and I_Σ given by (9.11) and (9.17) are well-defined. In this case, we note that a *right inverse* to K_Σ is given by

$$
K_\Sigma^{-1} : \begin{array}{l} \dot{p} = f(p) - g(p)g^T(p)V_p^T(p) + k(p)\xi \\ u = -g^T(p)V_p^T(p) \\ y = \xi + h(p) \end{array}
\tag{9.20}
$$

Indeed, if $p(0) = x(0)$, then the input–output map (from ξ to z) of $K_\Sigma \circ K_\Sigma^{-1}$ is the identity map. Furthermore, a *left inverse* to I_Σ is given by

$$I_\Sigma^{-1} \; : \quad \begin{aligned} \dot{p} &= [f(p) - k(p)h(p)] + g(p)u + k(p)y \\ \xi &= u + g^T(p)V_p^T(p) \end{aligned} \tag{9.21}$$

since the input–output map of $I_\Sigma^{-1} \circ I_\Sigma$ (from s to ξ) for $p(0) = x(0)$ is again the identity map.

Remark 9.1.11 Note that I_Σ^{-1} is itself a *kernel* representation, and K_Σ^{-1} is itself an *image* representation of the *same* system

$$\widetilde{\Sigma} \; : \quad \begin{aligned} \dot{p} &= f(p) - g(p)g^T(p)V_p^T(p) + k(p)[y - h(p)] \\ u &= -g^T(p)V_p^T(p) \end{aligned} \tag{9.22}$$

The system $\widetilde{\Sigma}$ is a "*certainty equivalence*" controller: in the *optimal state feedback* $u = -g^T(x)V_x^T(x)$ corresponding to the optimal control problem (9.19) the actual state x is replaced by its *estimate* p, with the dynamics of p being driven by the *observation error* $y - h(p)$. In fact, if Σ is a linear system then $\widetilde{\Sigma}$ is exactly the optimal LQG controller (interpreted in a deterministic manner).

As a direct application of the preceding developments, let us consider the control figuration depicted in Fig. 9.1 (customarily called an *observer–controller* configuration), with Σ the plant system, having a stable image representation I_Σ with inverse I_Σ^{-1} given by (9.21). Furthermore, let M denote an input-state-output system with state variables z, defining input–output maps M_{z_0} for every initial state z_0.

Taking $p(0) = x(0) = x_0$ we obtain the input–output relations

$$u = v - M_{z_0}(I_\Sigma^{-1}(I_\Sigma(s))) = v - M_{z_0}(s) \tag{9.23}$$

Fig. 9.1 Observer–controller configuration

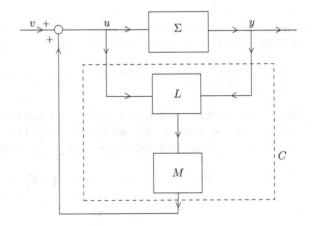

Furthermore, since I_Σ is an image representation of Σ, the input signal u is generated by (cf. (9.17))

$$\dot{x} = \left[f(x) - g(x)g^T(x)V_x^T(x)\right] + g(x)s, \quad x(0) = x_0$$
$$u = s - g^T(x)V_x^T(x)$$
(9.24)

Denoting the input–output map (from s to u) of (9.24) by D_{x_0}, we obtain from (9.23)

$$D_{x_0}(s) = v - M_{z_0}(s)$$
(9.25)

Hence, whenever M_{z_0} is such that the relation R_{vs} defined by (9.25) is L_2-stable, we conclude that for $v \in L_2$ also s will be in L_2, and thus also u, $y \in L_2$. Thus we have obtained a *parametrization* of the stabilizing controllers C in observer–controller configuration. More explicitly, all systems M with the property that $D_{x_0} + M_{z_0}$ has a L_2-stable inverse define a stabilizing controller C.

9.2 L_2-Gain Perturbation Models

As in the linear case [212], the stable kernel representation K_Σ and the stable image representation I_Σ both give rise to nonparametric *perturbation models* of Σ.

Recall that given a stable left factorization $G(s) = D^{-1}(s)N(s)$ a *perturbed* transfer matrix is $G_\Delta(s) := [D(s) + \Delta_D(s)]^{-1}[N(s) + \Delta_N(s)]$, with Δ_D, Δ_N stable transfer matrices. This corresponds to a *perturbed stable kernel representation*

$$0 = [D(s) + \Delta_D(s) \vdots - N(s) - \Delta_N(s)]\begin{bmatrix} y \\ u \end{bmatrix}$$
$$= [D(s) \vdots - N(s)]\begin{bmatrix} y \\ u \end{bmatrix} + [\Delta_D(s) \vdots - \Delta_N(s)]\begin{bmatrix} y \\ u \end{bmatrix}$$
(9.26)

Analogously, based on the stable kernel representation K_Σ of Σ given by (9.11) we consider the *perturbed stable kernel representation*

$$K_{\Sigma,\Delta} : \quad \begin{aligned} \dot{x} &= (f(x) - k(x)h(x)) + [g(x) \vdots k(x)]\begin{bmatrix} u \\ y \end{bmatrix} \\ e_p &:= e + w = y - h(x) + w, \end{aligned}$$
(9.27)

where w is the output of a nonlinear system Δ with input $\begin{bmatrix} u \\ y \end{bmatrix}$ given as

$$\Delta \; : \; \begin{aligned} \dot{\xi} &= \alpha(\xi, u, y) \\ w &= \beta(\xi, u, y) \end{aligned} \tag{9.28}$$

having *finite L_2-gain*. Setting $e_p = 0$ in (9.27) yields the *perturbed system*

$$\Sigma_\Delta \; : \; \begin{aligned} \dot{x} &= f(x) + g(x)u - k(x)w \\ y &= h(x) - w \end{aligned} \tag{9.29}$$

with w being the output of (9.28). Note that the *size* of the perturbation Δ is measured via its L_2-gain.

Alternatively, recall that given a stable *right* factorization $G(s) = \tilde{N}(s)\tilde{D}^{-1}(s)$ the perturbed transfer matrix $G_{\tilde{\Delta}}(s)$ can be defined as $G_{\tilde{\Delta}}(s) := [\tilde{N}(s) + \Delta_{\tilde{N}}(s)][\tilde{D}(s) + \Delta_{\tilde{D}}(s)]^{-1}$, with $\Delta_{\tilde{N}}, \Delta_{\tilde{D}}$ stable transfer matrices. This corresponds to a *perturbed stable image representation*

$$\begin{bmatrix} y \\ u \end{bmatrix} = \begin{bmatrix} \tilde{N}(s) + \Delta_{\tilde{N}}(s) \\ \tilde{D}(s) + \Delta_{\tilde{D}}(s) \end{bmatrix} l = \begin{bmatrix} \tilde{N}(s) \\ \tilde{D}(s) \end{bmatrix} l + \begin{bmatrix} \Delta_{\tilde{N}}(s) \\ \Delta_{\tilde{D}}(s) \end{bmatrix} l \tag{9.30}$$

Analogously, based on the stable image representation I_Σ given in (9.17) we consider the *perturbed stable image representation*

$$I_{\Sigma,\Delta} \; : \; \begin{aligned} \dot{x} &= f(x) - g(x)g^T(x)V_x^T(x) + g(x)s \\ y &= h(x) + w_1 \\ u &= s - g^T(x)V_x^T(x) + w_2, \end{aligned} \tag{9.31}$$

where $w = (w_1, w_2)$ is the output of a nonlinear system $\tilde{\Delta}$ with input s given by

$$\tilde{\Delta} \; : \; \begin{aligned} \dot{\xi} &= \tilde{\alpha}(\xi, s) \\ w_1 &= \tilde{\beta}_1(\xi, s) \\ w_2 &= \tilde{\beta}_2(\xi, s) \end{aligned} \tag{9.32}$$

and having *finite L_2-gain* (from s to w). Elimination of the auxiliary variables s in (9.31) leads to the *perturbed system*

$$\Sigma_{\tilde{\Delta}} \; : \; \begin{aligned} \dot{x} &= f(x) + g(x)u - g(x)w_2 \\ y &= h(x) + w_1 \end{aligned} \tag{9.33}$$

with $w = (w_1, w_2)$ the output of the system $\tilde{\Delta}$, with its input s replaced by

$$s = u + g^T(x)V_x^T(x) - w_2 \tag{9.34}$$

Note that again the *size* of the perturbation $\tilde{\Delta}$ is measured by its L_2-gain (from s to w).

Example 9.2.1 Consider a port-Hamiltonian system (6.1) without dissipation $(R(x) = 0)$, given as

$$\Sigma \ : \quad \begin{matrix} \dot{x} = J(x)\frac{\partial H}{\partial x}(x) + g(x)u \\ \\ y = g^T(x)\frac{\partial H}{\partial x}(x) \end{matrix} \quad , \quad J(x) = -J^T(x), \quad (9.35)$$

with internal energy $H \geq 0$. Nonnegative solutions of the Hamilton–Jacobi–Bellman equations (9.8) and (9.9) are both given by $V = H$ and $W = H$. Note furthermore that the solution $k(x)$ to (9.10) is simply $k(x) = g(x)$. Thus based on the stable kernel representation we obtain the perturbed system

$$\Sigma_\Delta \ : \quad \begin{matrix} \dot{x} = J(x)\frac{\partial H}{\partial x}(x) + g(x)u - g(x)w \\ \\ y = g^T(x)\frac{\partial H}{\partial x}(x) - w \end{matrix} \quad (9.36)$$

with w the output of Δ given by (9.28). On the other hand, the perturbed system based on the stable image representation is given as

$$\Sigma_{\tilde{\Delta}} \ : \quad \begin{matrix} \dot{x} = J(x)\frac{\partial H}{\partial x}(x) + g(x)u - g(x)w_2 \\ \\ y = g^T(x)\frac{\partial H}{\partial x}(x) - w_1 \end{matrix} \quad (9.37)$$

with (w_1, w_2) the output of $\tilde{\Delta}$ in (9.32) driven by the signal $s = u + g^T(x)\frac{\partial H}{\partial x}(x) - w_2$.

9.3 Stable Kernel Representations and Parametrization of Stabilizing Controllers

Given a nonlinear *plant* system together with a *stabilizing controller* system, we shall derive in this section a parametrization of *all* stabilizing controllers. This parametrization generalizes the well-known Youla–Kucera parametrization in the linear case. Key element is the stable kernel representation of the plant and controller system. On the other hand, we need a strengthened notion of closed-loop stability, which in the nonlinear case may be different from the classical one given in Chap. 1 and also different from internal state space (Lyapunov) stability.

Consider a pair of nonlinear state space systems

$$\Sigma_i \ : \quad \begin{matrix} \dot{x}_i = f_i(x_i) + g_i(x)u_i \ , & x_i \in \mathcal{X}_i, \ u_i \in \mathbb{R}^{m_i} \\ y_i = h_i(x_i) \ , & y_i \in \mathbb{R}^{p_i} \ , \quad i = 1, 2 \end{matrix} \quad (9.38)$$

Assume both systems admit stable kernel representations and suppose for simplicity of exposition that these stable kernel representations are of the affine form as in (9.11), i.e.,

Fig. 9.2 Closed-loop system
$\{\Sigma_1, \Sigma_2\}$

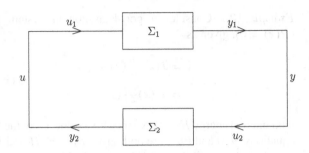

$$K_{\Sigma_i} : \quad \begin{array}{l} \dot{x}_i = [f_i(x_i) - k_i(x_i)h_i(x_i)] + g_i(x_i)u_i + k_i(x_i)y_i \\ z_i = y_i - h_i(x_i) \, , \, z_i \in \mathbb{R}^{p_i} \, , \, i = 1, 2 \end{array} \qquad (9.39)$$

However, k_i need not necessarily be constructed as in (9.10), and thus K_{Σ_i} need not necessarily have L_2-gain equal to one, but are only required to be L_2-stable. In fact, the approach works equally well for general stable kernel representations as defined in Definition 9.1.1.

Let $m := m_1 = p_2$ and $p := p_1 = m_2$, and consider the *feedback interconnection* of Σ_1 and Σ_2. Different from the previous chapters we throughout consider (for simplicity of notation and without loss of generality), the *positive feedback* interconnection. Furthermore, we will *not* consider external inputs e_1, e_2. Hence we consider throughout the interconnection equations

$$u := u_1 = y_2, \quad y := y_1 = u_2 \qquad (9.40)$$

The resulting (autonomous) closed system as given in Fig. 9.2 will be denoted by $\{\Sigma_1, \Sigma_2\}$. We will sometimes compare $\{\Sigma_1, \Sigma_2\}$ with the closed-loop system $\Sigma_1 \|_f \Sigma_2$ with external inputs e_1, e_2 as defined before (with the difference that we consider the *positive* feedback interconnection $u_1 = y_2, u_2 = y_1$).

A stable kernel representation of the closed-loop system $\{\Sigma_1, \Sigma_2\}$ (with outputs y and u, and without inputs) is obtained by substituting $u := u_1 = y_2, y := y_1 = u_2$ into (9.39) :

$$K_{\{\Sigma_1, \Sigma_2\}} : \quad \begin{array}{l} \dot{x}_1 = [f_1(x_1) - k_1(x_1)h_1(x_1)] + g_1(x_1)u + k_1(x_1)y \\[1.5ex] \dot{x}_2 = [f_2(x_2) - k_2(x_2)h_2(x_2)] + g_2(x_2)y + k_2(x_2)u \\[1.5ex] z_1 = y - h_1(x_1) \\[1.5ex] z_2 = u - h_2(x_2) \end{array} \qquad (9.41)$$

Indeed, by setting $z_1 = 0, z_2 = 0$ one recovers the closed-loop system $\{\Sigma_1, \Sigma_2\}$ of Fig. 9.2. On the other hand we may invert in (9.41) the map from y, u to z_1, z_2 by solving y and u as $y = z_1 + h_1(x_1), u = z_2 + h_2(x_2)$ to obtain the *inverse system* of (9.41), denoted as $\{K_{\Sigma_1}, K_{\Sigma_2}\}$, given by

$$\dot{x}_1 = f_1(x_1) + g_1(x_1)h_2(x_2) + g_1(x_1)z_2 + k_1(x_1)z_1$$

$$\{K_{\Sigma_1}, K_{\Sigma_2}\} \ : \quad \begin{aligned} \dot{x}_2 &= f_2(x_2) + g_2(x_2)h_1(x_1) + g_2(x_2)z_1 + k_2(x_2)z_2 \\ y &= z_1 + h_1(x_1) \\ u &= z_2 + h_2(x_2) \end{aligned} \tag{9.42}$$

Note that this inverse system can also be regarded as some kind of *closed-loop system* of Σ_1 and Σ_2, with *external* signals z_1 and z_2. Indeed, if $k_1 = 0$ and $k_2 = 0$ then (9.42) is exactly the closed-loop system $\Sigma_1 \|_f \Sigma_2$ of Fig. 4.1 (for the *positive feedback interconnection*) with $e_1 = z_2, e_2 = z_1$. However if $k_1 \neq 0$ or $k_2 \neq 0$ then $\{K_{\Sigma_1}, K_{\Sigma_2}\}$ will be in general different from $\Sigma_1 \|_f \Sigma_2$.

Just as the definition of $\Sigma_1 \|_f \Sigma_2$ leads to a notion of closed-loop stability (cf. Definition 1.2.11), the definition of $\{K_{\Sigma_1}, K_{\Sigma_2}\}$ leads to another notion of closed-loop stability:

Definition 9.3.1 The closed-loop system $\{K_{\Sigma_1}, K_{\Sigma_2}\}$ of Σ_1, Σ_2 with stable kernel representations K_{Σ_1} and K_{Σ_2} given by (9.42), is called *strongly (L_2-) stable* if for every pair of initial conditions $x_i(0) = x_{i0} \in \mathcal{X}_i, i = 1, 2$, and every pair of functions $z_1(\cdot) \in L_2(\mathbb{R}^p), z_2(\cdot) \in L_2(\mathbb{R}^m)$, the solutions $u(\cdot)$ and $y(\cdot)$ to (9.42) are in $L_2(\mathbb{R}^m)$, respectively, L_2^p. (That is, the system (9.42) with inputs z_1, z_2 and outputs y, u is L_2-stable.) If the above property holds for initial conditions x_{i0} in a *subset* $\overline{\mathcal{X}}_i$ of $\mathcal{X}_i, i = 1, 2$, then $\{K_{\Sigma_1}, K_{\Sigma_2}\}$ is called *strongly stable over* $\overline{\mathcal{X}}_1 \times \overline{\mathcal{X}}_2$.

Remark 9.3.2 It should be stressed that if $k_1 = 0, k_2 = 0$ in (9.42), then $\{K_{\Sigma_1}, K_{\Sigma_2}\} = \Sigma_1 \|_f \Sigma_2$ and strong L_2-stability equals L_2-stability of $\Sigma_1 \|_f \Sigma_2$. For instance, see (6.16), if Σ_1 and Σ_2 are themselves L_2-stable, then k_1 and k_2 in (9.42) may be taken equal to zero.

A special case arises if one of the systems is the *zero-system* given by the zero input–output map

$$O \ : \ u \mapsto y = 0 \tag{9.43}$$

with empty state space $\mathcal{X} = \emptyset$. A stable kernel representation K_O of O is simply

$$K_O \ : \ (y, u) \mapsto z = y \tag{9.44}$$

Proposition 9.3.3 *Consider a state space system Σ with stable kernel representation*

$$K_\Sigma \ : \quad \begin{aligned} \dot{x} &= [f(x) - k(x)h(x)] + g(x)u + k(x)y \\ z &= y - h(x) \end{aligned} \tag{9.45}$$

Then $\{K_\Sigma, K_O\}$ is strongly L_2-stable if and only if the system

$$\begin{aligned} \dot{x} &= f(x) + g(x)z_2 + k(x)z_1 \\ w &= h(x) \end{aligned} \tag{9.46}$$

is L_2-stable from (z_1, z_2) to w.

Proof Consider (9.42) with $\Sigma_1 = \Sigma$ and $\Sigma_2 = O$. Then (9.42) reduces to

$$
\begin{aligned}
\dot{x} &= f(x) + g(x)z_2 + k(x)z_1 \\
y &= z_1 + h(x) \\
u &= z_2
\end{aligned}
\tag{9.47}
$$

which is strongly L_2-stable if and only if (9.46) is L_2-stable, as follows immediately. $\qquad\square$

Corollary 9.3.4 *If $k = 0$ in K_Σ, then $\{K_\Sigma, K_O\}$ is strongly L_2-stable if and only if Σ is L_2-stable.*

Now we interpret Σ_1 as the *plant* and Σ_2 as the *controller*. We will say that Σ_2 is *strongly stabilizing* for Σ_1 if $\{K_{\Sigma_1}, K_{\Sigma_2}\}$ is strongly stable.

Based on a strongly stabilizing Σ_2 we wish to parametrize *all* the strongly stabilizing controllers. The key idea is to consider the external signals z_1 and z_2 in (9.39) as input and output signals for *another* state space system

$$
Q : \begin{array}{l} \dot{\xi} = \varphi(\xi) + \psi(\xi)z_1 \\ z_2 = \theta(\xi) \end{array} \quad , \xi \in \mathcal{X}_Q,
\tag{9.48}
$$

which we assume to be L_2-stable, and thus having a stable kernel representation

$$
K_Q : \begin{array}{l} \dot{\xi} = \varphi(\xi) + \psi(\xi)z_1 \\ \omega = z_2 - \theta(\xi) \end{array}
\tag{9.49}
$$

with external signal ω. Substituting for z_1 and z_2 the expressions from the stable kernel representations K_{Σ_i} in (9.39) (with $u = u_1 = y_2$ and $y = y_1 = u_2$) one obtains

$$
K_{\Sigma_2^Q} : \begin{array}{l} \dot{\xi} = \varphi(\xi) - \psi(\xi)h_1(x_1) + \psi(\xi)y \\ \dot{x}_1 = f_1(x_1) - k_1(x_1)h_1(x_1) + g_1(x_1)u + k_1(x_1)y \\ \dot{x}_2 = f_2(x_2) - k_2(x_2)h_2(x_2) + g_2(x_2)y + k_2(x_2)u \\ \omega = u - h_2(x_2) - \theta(\xi) \end{array}
\tag{9.50}
$$

By setting $\omega = 0$, and solving for u, it follows that (9.50) is a stable kernel representation of the following input-state-output system (with input y and output u):

$$
\Sigma_2^Q : \begin{array}{l} \dot{\xi} = \varphi(\xi) - \psi(\xi)h_1(x_1) + \psi(\xi)y \\ \dot{x}_1 = f_1(x_1) - k_1(x_1)h_1(x_1) \\ \qquad + g_1(x_1)h_2(x_2) + g_1(x_1)\theta(\xi) + k_1(x_1)y \\ \dot{x}_2 = f_2(x_2) - k_2(x_2)h_2(x_2) \\ \qquad + k_2(x_2)h_2(x_2) + k_2(x_2)\theta(\xi) + g_2(x_2)y \\ u = \theta(\xi) + h_2(x_2) \end{array}
\tag{9.51}
$$

Note that the stable kernel representation $K_{\Sigma_2^Q}$ corresponds to Fig. 9.3.

Fig. 9.3 Stable kernel representation of Σ_2^Q

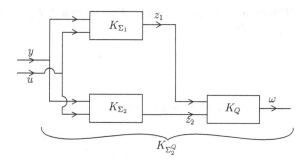

Fig. 9.4 Closed-loop system for Σ_2^Q

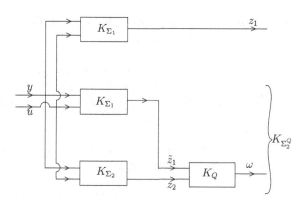

We interpret Σ_2^Q as a *perturbation* of the controller system Σ_2. Note that the state space of the perturbed controller system Σ_2^Q, defined by Σ_1 and Q, equals $\mathcal{X}_1 \times \mathcal{X}_2 \times \mathcal{X}_Q$.

Theorem 9.3.5 *Suppose* $\{K_{\Sigma_1}, K_{\Sigma_2}\}$ *is strongly L_2-stable. Then for every L_2-stable system Q the closed-loop system* $\{K_{\Sigma_1}, K_{\Sigma_2^Q}\}$, *with state space* $\mathcal{X}_1 \times (\mathcal{X}_1 \times \mathcal{X}_2 \times \mathcal{X}_Q)$, *is strongly L_2-stable over* $\mathrm{diag}(\mathcal{X}_1 \times \mathcal{X}_1) \times \mathcal{X}_2 \times \mathcal{X}_Q$.

Proof Consider K_{Σ_1} (with $u_1 = u$, $y_1 = y$) and $K_{\Sigma_2^Q}$. By Definition 9.3.1 strong L_2-stability follows if for all initial conditions in $\mathrm{diag}(\mathcal{X}_1 \times \mathcal{X}_1) \times \mathcal{X}_2 \times \mathcal{X}_Q$, and all $z_1, \omega \in L_2$ the signals y and u are in L_2. Thus let $z_1, \omega \in L_2$ and consider initial conditions in $\mathrm{diag}(\mathcal{X}_1 \times \mathcal{X}_1) \times \mathcal{X}_2 \times \mathcal{X}_Q$. Since Q is L_2-stable we obtain by Corollary 9.3.4 that z_1, z_2 are in L_2. Since $\{K_{\Sigma_1}, K_{\Sigma_2}\}$ is assumed to be strongly L_2-stable, this implies that y and u are in L_2. (Note that the state of the closed-loop system remains in $\mathrm{diag}(\mathcal{X}_1 \times \mathcal{X}_1) \times \mathcal{X}_2 \times \mathcal{X}_Q$, and that $\tilde{z}_1 = z_1$ in Fig. 9.4.)   □

Loosely speaking, we may conclude that if Σ_2 is a strongly stabilizing controller of Σ_1, then also Σ_2^Q will be a strongly stabilizing controller of Σ_1 for *every* stable system Q. Moreover, we obtain in this way *all* the strongly stabilizing controllers, in the following sense.

Theorem 9.3.6 *Suppose* $\{K_{\Sigma_1}, K_{\Sigma_2}\}$ *is strongly* L_2-*stable. Consider a controller* Σ^* *different from* Σ_2

$$\Sigma^* : \begin{array}{l} \dot{x}^* = f^*(x^*) + g^*(x^*)y \\ u = h^*(x^*) \end{array} \qquad x^* \in \mathcal{X}^* \qquad (9.52)$$

with stable kernel representation

$$K_{\Sigma^*} : \begin{array}{l} \dot{x}^* = f^*(x^*) + g^*(x^*)y + k^*(x^*)u \\ z^* = u - h^*(x^*) \end{array} \qquad (9.53)$$

Suppose also $\{K_{\Sigma_1}, K_{\Sigma^*}\}$ *is strongly* L_2-*stable. Then define* K_{Q^*} *by composing* K_{Σ^*} *with* $\{K_{\Sigma_1}, K_{\Sigma_2}\}$ *given by (9.42), i.e.,*

$$K_{Q^*} : \begin{array}{l} \dot{x} = f^*(x^*) + g^*(x^*)(z_1 + h_1(x_1)) + k^*(x^*)(z_2 + h_2(x_2)) \\ \dot{x}_1 = f_1(x_1) + g_1(x_1)h_2(x_2) + g_1(x_1)z_2 + k_1(x_1)z_1 \\ \dot{x}_2 = f_2(x_2) + g_2(x_2)h_1(x_1) + g_2(x_2)z_1 + k_2(x_2)z_2 \\ z^* = z_2 + h_2(x_2) - h^*(x^*) \end{array} \qquad (9.54)$$

This is a stable kernel representation (set $z^* = 0$ *and solve for* z_2*) of the following system with input* z_1 *and output* z_2:

$$Q^* : \begin{array}{l} \dot{x}^* = f^*(x^*) + g^*(x^*)(z_1 + h_1(x_1)) + k^*(x^*)h^*(x^*)) \\ \dot{x}_1 = f_1(x_1) + g_1(x_1)h^*(x^*) + k_1(x_1)z_1 \\ \dot{x}_2 = f_2(x_2) + g_2(x_2)h_1(x_1) + g_2(x_2)z_1 \\ \qquad + k_2(x_2)(h^*(x^*) - h_2(x_2)) \\ z_2 = h^*(x^*) - h_2(x_2) \end{array} \qquad (9.55)$$

Note that the state space of Q^* *is* $\mathcal{X}_{Q^*} = \mathcal{X}_1 \times \mathcal{X}_2 \times \mathcal{X}^*$. *Consider as in (9.51) the system* $\Sigma_2^{Q^*}$ *with state space* $\mathcal{X}_1 \times \mathcal{X}_2 \times \mathcal{X}_{Q^*} = \mathcal{X}_1 \times \mathcal{X}_2 \times \mathcal{X}_1 \times \mathcal{X}_2 \times \mathcal{X}^*$. *Then the input–output map of* $\Sigma_2^{Q^*}$ *for initial condition* $(x_{10}, x_{20}, x_{10}, x_{20}, x_0^*)$ *equals the input–output map of* Σ^* *for initial condition* x_0^*, *and this holds for all* $(x_{10}, x_{20}) \in \mathcal{X}_1 \times \mathcal{X}_2$ *and all* $x_0^* \in \mathcal{X}^*$.

Proof The input–output map of the kernel representation $K_{\Sigma_2^{Q^*}}$ for initial condition $(x_{10}, x_{20}, x_{10}, x_{20}, x_0^*)$ is given by

$$K_{\Sigma^*}^{x_0^*} \circ \left[K_{\{\Sigma_1, \Sigma_2\}}^{(x_{10}, x_{20})} \right]^{-1} \circ K_{\{\Sigma_1, \Sigma_2\}}^{(x_{10}, x_{20})} = K_{\Sigma^*}^{x_0^*}, \qquad (9.56)$$

with the superscripts denoting the initial conditions for the respective input–output maps.   □

Remark 9.3.7 In the linear case one recovers the Youla–Kucera parametrization of all stabilizing controllers as follows. Take all initial conditions to be equal to zero. Let Σ_1 be given by the transfer matrix $P(s) = D^{-1}(s)N(s)$, and let Σ_2 be a stabilizing controller given by $C(s) = X^{-1}(s)Y(s)$, where $D(s), N(s), X(s)$, and $Y(s)$ are stable rational matrices. Equivalently, the systems Σ_1 and Σ_2 are associated with the kernels of $[D(s) \,\vdots\, -N(s)]$, respectively, $[Y(s) \,\vdots\, -X(s)]$. Let $Q(s)$ be a stable rational matrix, corresponding to a stable input–output map $z_2 = Q(s)z_1$, or, equivalently, to the kernel of $[I \,\vdots\, -Q(s)]$. It follows that the set of all linear stabilizing controllers is given by the kernels of

$$(I \,\vdots\, -Q(s)) \begin{pmatrix} D(s) & -N(s) \\ -Y(s) & X(s) \end{pmatrix} =$$
$$(D(s) + Q(s)Y(s) \,\vdots\, -N(s) - Q(s)X(s)) \tag{9.57}$$

or, equivalently, by the transfer matrices $(D(s) + Q(s)Y(s))^{-1}(N(s) + Q(s)X(s))$.

Finally, let us consider as a special case of Theorem 9.3.5 the situation that Σ_1 is *itself* already L_2-*stable*. Then, as noted before in (6.16), a stable kernel representation K_{Σ_1} of Σ_1 is given as

$$K_{\Sigma_1} : \begin{array}{l} \dot{x}_1 = f_1(x_1) + g_1(x_1)u_1 \\ z_1 = y_1 - h_1(x_1) \end{array}$$

Furthermore, in this case the zero-controller $\Sigma_2 = 0$ with stable kernel representation K_O given by (9.44) yields a closed-loop system $\{K_\Sigma, K_O\}$ which is by Corollary 9.3.4 strongly L_2-stable.

Now, consider any L_2-stable system Q, given by (9.48) with stable kernel representation K_Q as in (9.49). It follows that the stabilizing controller Σ_2^Q is given as (cf. (9.51))

$$\Sigma_2^Q : \begin{array}{l} \dot{\xi} = \varphi(\xi) + \psi(\xi)(y_1 - h_1(x_1)) \\ u = \theta(\xi) \\ \dot{x}_1 = f_1(x_1) + g_1(x_1)u \end{array} \tag{9.58}$$

Hence every stabilizing controller Σ_2^Q contains a *model* (or *copy*) of the plant, which can be regarded as a generalization of the concept of *Internal Model Control* (see, e.g., [222]) to the nonlinear setting.

9.4 All-Pass Factorizations

Throughout this section, we consider state space systems Σ satisfying

$$\Sigma : \begin{array}{l} \dot{x} = f(x, u), \; f(0, 0) = 0, \; u \in \mathbb{R}^m, \; x \in \mathcal{X} \\ y = h(x, u), \; h(0, 0) = 0, \; y \in \mathbb{R}^p \end{array} \tag{9.59}$$

Recall, cf. Definitions 3.1.6 and 8.1.1, that a system Σ is *inner* if it is conservative with respect to the L_2-gain supply rate $s(u, y) = \frac{1}{2}||u||^2 - \frac{1}{2}||y||^2$, that is, there

Fig. 9.5 All-pass
factorization $\Sigma = \Theta \circ \bar{\Sigma}$

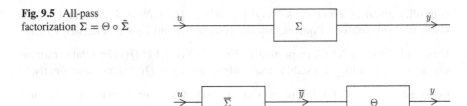

exists $S : \mathcal{X} \to \mathbb{R}^+$ such that along every trajectory of (9.59)

$$S(x(t_1)) - S(x(t_0)) = \frac{1}{2} \int_{t_0}^{t_1} (||u(t)||^2 - ||y(t)||^2)dt \qquad (9.60)$$

The topic of this section is to *factorize* the nonlinear system Σ as a *series interconnection* $\Theta \circ \bar{\Sigma}$ of an inner system Θ preceded by another nonlinear system $\bar{\Sigma}$, in the
sense that for every initial condition of Σ there should exist initial conditions of Θ
and $\bar{\Sigma}$ such that the corresponding input–output map of Σ equals the composition
of the respective input–output maps of $\bar{\Sigma}$ and Θ (see Fig. 9.5).

We will call this factorization an *all-pass factorization*.[1] An important motivation
for this type of factorization is that in view of (9.60) the asymptotic properties of Σ
and $\bar{\Sigma}$ are similar, while $\bar{\Sigma}$ may be simpler to control than Σ. In such a case, the
control of Σ may be based on the control of $\bar{\Sigma}$, as we will see at the end of this
section.

The all-pass factorizations of Σ are based on the following *reversed dissipation
inequality* for Σ

$$V(x(t_1)) - V(x(t_0)) + \frac{1}{2} \int_{t_0}^{t_1} ||h(x(t), u(t))||^2 dt \geq 0 \qquad (9.61)$$

in the unknown $V \geq 0$, or its differential version (V assumed to be C^1)

$$V_x(x)f(x, u) + \frac{1}{2}||h(x, u)||^2 \geq 0 , \qquad \text{for all } x, u \qquad (9.62)$$

Suppose there exists[2] $V \geq 0$ satisfying (9.62). Then consider the positive function

$$K_V(x, u) := V_x(x)f(x, u) + \frac{1}{2}||h(x, u)||^2 \qquad (9.63)$$

---

[1]Called "all-pass" since a single-input single-output linear system is inner if and only if its transfer
function has amplitude 1 for all frequencies.

[2]Many of the subsequent developments continue to hold for any C^1 function V satisfying (9.62)
(not necessarily ≥ 0). However, in this case the factor system Θ will not be inner anymore; but only
cyclo-conservative with respect to the supply rate $\frac{1}{2}||\bar{y}||^2 - \frac{1}{2}||y||^2$.

Assumption 9.4.1 There exists a C^1 map $\overline{h} : \mathcal{X} \times \mathbb{R}^m \to \mathbb{R}^{\overline{p}}$ such that

$$K_V(x, u) = \frac{1}{2}||\overline{h}(x, u)||^2 \tag{9.64}$$

Sufficient conditions for the *local* existence of a smooth $\overline{h}$ satisfying (9.64) are provided by the next lemma which follows from an application of Morse's lemma; see the Notes at the end of this chapter for a sketch of the proof.

Lemma 9.4.2 *Suppose that K_V is C^2, and that its Hessian matrix, i.e.,*

$$\begin{bmatrix} \frac{\partial^2 K_V}{\partial x^2}(x, u) & \frac{\partial^2 K_V}{\partial x \partial u}(x, u) \\ \frac{\partial^2 K_V}{\partial u \partial x}(x, u) & \frac{\partial^2 K_V}{\partial u^2}(x, u) \end{bmatrix} \tag{9.65}$$

has constant rank $\overline{p}$ on a neighborhood of $(x, u) = (0, 0)$. Then locally near $(0, 0)$ there exists a C^1 mapping $\overline{h} : \mathcal{X} \times \mathcal{R}^m \to \mathcal{R}^{\overline{p}}$ such that (9.64) holds.

Then define the *factor system*

$$\overline{\Sigma} : \begin{array}{l} \dot{\overline{x}} = f(\overline{x}, u), \ \overline{x} \in \mathcal{X}, \ u \in \mathbb{R}^m \\ \overline{y} = \overline{h}(\overline{x}, u), \ \overline{y} \in \mathbb{R}^{\overline{p}} \end{array} \tag{9.66}$$

Obviously $\overline{h}(0, 0) = 0$, and thus $\overline{\Sigma}$ is also within the class (8.1).

Furthermore, define the system Θ with inputs $\overline{y}$ and outputs y given in image representation (cf. Definition 9.1.1) as

$$I_\Theta : \begin{array}{l} \dot{\xi} = f(\xi, u) \\ \overline{y} = \overline{h}(\xi, u) \\ y = h(\xi, u) \end{array} \tag{9.67}$$

It immediately follows from (9.63) and (9.64) that Θ has the property

$$V(\xi(t_1)) - V(\xi(t_0)) = \frac{1}{2} \int_{t_0}^{t_1} (||\overline{y}(t)|| - ||y(t)||^2) dt \tag{9.68}$$

along every solution $\xi(\cdot)$. Furthermore, if in (9.59) (9.66) (9.67), $x(t_0) = \overline{x}(t_0) = \xi(t_0)$, then it immediately follows that $x(t) = \overline{x}(t) = \xi(t), t \geq t_0$, so that indeed $\Sigma = \Theta \circ \overline{\Sigma}$.

Now, if the equation $\overline{y} = \overline{h}(\xi, u)$ can be solved for u as $u = \alpha(\xi, \overline{y})$, then the image representation (9.67) can be reduced to a standard input-state-output system

$$\Theta : \begin{array}{l} \dot{\xi} = f(\xi, \alpha(\xi, \overline{y})) \\ y = h(\xi, \alpha(\xi, \overline{y})) \end{array} \tag{9.69}$$

defining an *inner* system.

We obtain the following asymptotic relation between Σ and $\overline{\Sigma}$.

Definition 9.4.3 Define for every (x_e, u_e) such that $f(x_e, u_e) = 0$, the *steady-state output amplitude*[3] as $||h(x_e, u_e)||$.

Proposition 9.4.4 *Consider Σ and a factor system $\overline{\Sigma}$. The set of steady-state output amplitudes of Σ and $\overline{\Sigma}$ are equal.*

Proof Consider (9.63) and (9.64), and substitute $f(x_e, u_e) = 0$.   □

All-pass factorizations of Σ have an immediate interpretation in terms of the Hamiltonian input–output system Σ^H associated to Σ (Definition 8.1.2) given by

$$\Sigma^H : \quad \begin{aligned} \dot{x} &= \frac{\partial H}{\partial p}(x, p, u) \\ \dot{p} &= -\frac{\partial H}{\partial x}(x, p, u) \\ y_a &= \frac{\partial H}{\partial u}(x, p, u), \end{aligned} \tag{9.70}$$

where

$$H(x, p, u) = p^T f(x, u) + \frac{1}{2} h^T(x, u) h(x, u) \tag{9.71}$$

Recall from Proposition 8.1.4 that the input–output map $G^H_{(0,0)}$ of Σ^H for initial state $(x, p) = (0, 0)$ is equal to $G^H_{(0,0)}(u) = (DG_0(u))^* \circ G_0(u)$, where $(DG_0(u))^*$ is the adjoint of the Fréchet derivative of the input–output map G_0 of Σ.

   Now consider an all-pass factorization $\Sigma = \Theta \circ \overline{\Sigma}$ as above, with Θ inner. Denote the input–output map for zero initial condition of Σ by G_0, of $\overline{\Sigma}$ by $\overline{G}_0$, and the input–output map of Θ for zero initial condition by A_0. It follows that

$$\begin{aligned} (DG_0(u))^* \circ G_0(u) &= \left(D(A_0 \circ \overline{G}_0)(u)\right)^* \circ (A_0 \circ \overline{G}_0)(u) \\ &= \left(D\overline{G}_0(u)\right)^* \left(DA_0(\overline{G}_0(u))\right)^* \circ A_0 \circ \overline{G}_0(u) \\ &= \left(D\overline{G}_0(u)\right)^* \circ \overline{G}_0(u) \end{aligned} \tag{9.72}$$

The last equality holds since A_0 is inner, and thus, in view of (1.24) in Proposition 1.2.9,

$$\left(DA_0(\overline{G}_0(u))\right)^* \circ A_0(\overline{G}_0(u)) = \overline{G}_0(u) \tag{9.73}$$

Hence the input–output map $G^H_{(0,0)}$ of the Hamiltonian input–output system Σ^H associated to Σ is *equal* to the input–output map $\overline{G}^H_{(0,0)}$ of the Hamiltonian input–output system $\overline{\Sigma}^H$ associated to $\overline{\Sigma}$.

   This motivates the following state space level interpretation. Consider any C^1 solution V to the dissipation inequality (9.62), with corresponding factor system $\overline{\Sigma}$. Define the following change of coordinates for Σ^H:

---

[3]If (x_e, u_e) is an asymptotically stable steady state then $||h(x_e, u_e)||$ is the steady-state value of the norm of the output y for a step input with magnitude u_e.

$$\bar{x} = x, \quad \bar{p} = p - V_x^T(x) \tag{9.74}$$

In the new coordinates for $T^*\mathcal{X}$ the Hamiltonian H transforms into

$$\begin{aligned}\overline{H}(\bar{x}, \bar{p}, u) &= (\bar{p} + V_x^T(x))^T f(x, u) + \tfrac{1}{2} h^T(x, u) h(x, u) \\ &= \bar{p}^T f(x, u) + \tfrac{1}{2}\overline{h}^T(x, u)\overline{h}(x, u)\end{aligned} \tag{9.75}$$

Therefore, the Hamiltonian input–output system Σ^H given by (9.70) transforms into

$$\begin{aligned}\dot{\bar{x}} &= \tfrac{\partial \overline{H}}{\partial \bar{p}}(\bar{x}, \bar{p}, u) \\ \dot{\bar{p}} &= -\tfrac{\partial \overline{H}}{\partial \bar{x}}(\bar{x}, \bar{p}, u) \\ y_a &= \tfrac{\partial \overline{H}}{\partial u}(\bar{x}, \bar{p}, u)\end{aligned} \tag{9.76}$$

which is exactly the associated Hamiltonian input–output system $\overline{\Sigma}^H$. Thus the Hamiltonian input–output systems associated to Σ and $\overline{\Sigma}$ are *equal*, in accordance with the equality of their input–output maps discussed above. This is summarized in the following proposition.

Proposition 9.4.5 *Consider a system Σ as in (9.59) and a solution V to the reversed dissipation inequality (9.62), together with a resulting all-pass factorization $\Sigma = \Theta \circ \overline{\Sigma}$ as in Fig. 9.5. Then $\Sigma^H = \overline{\Sigma}^H$.*

Remark 9.4.6 In case of a linear system $\Sigma : \dot{x} = Ax + Bu, y = Cx + Du$ the transformation of Σ to $\overline{\Sigma}$ corresponds to a *spectral factorization* of the "spectral density" matrix $\left(D^T + B^T(-Is - A^T)^{-1}C^T\right)\left(C(Is - A)^{-1}B + D\right)$. In the same spirit, the transformation from a *nonlinear* system Σ to $\overline{\Sigma}$ can be interpreted as a "*nonlinear spectral factorization*" of the associated input–output Hamiltonian system Σ^H.

The reversed dissipation inequality (9.62) is intimately related to the *optimal control problem* of minimizing for every initial condition x_0 of Σ the cost-functional

$$J(x_0, u) = \int_0^\infty \|y(t)\|^2 dt \tag{9.77}$$

along the trajectories of Σ starting at $t = 0$ at x_0. Define for every $x_0 \in \mathcal{X}$ the *value function* (see Sect. 11.2 for further information)

$$V^*(x_0) = \inf_u \{J(x_0, u) \mid u \text{ admissible}, \lim_{t \to \infty} x(t) = 0\}, \tag{9.78}$$

where $u : \mathbb{R}^+ \to \mathbb{R}^m$ is called *admissible* if the right-hand side of (9.77) is well-defined.

Assumption 9.4.7 $V^*(x_0)$ exists for every $x_0 \in \mathcal{X}$ (i.e., the system is stabilizable), and V^* is a C^1 function on $\mathcal{X}$.

Proposition 9.4.8 *(i) V^* satisfies the reversed dissipation inequality (9.62), and $V^*(0) = 0$.*

(ii) Let V satisfy (9.62) and $V(0) = 0$, then $V(x) \leq V^*(x)$ for every $x \in \mathcal{X}$.

Proof (i) Follows directly from the definition of V^*.

(ii) Consider (9.61) with $t_0 = 0$ and $t_1 = T$, and let u be such that $\lim_{T \to \infty} x(T) = 0$. Then it follows from (9.61) for $T \to \infty$ that

$$\frac{1}{2} \int_0^\infty ||y(t)||^2 dt \geq V(x(0)),$$

and thus by definition of V^* we obtain $V^*(x(0)) \geq V(x(0))$.   □

Let us denote the factor system $\overline{\Sigma}$ obtained by considering V^* as $\overline{\Sigma}^*$, that is

$$\Sigma^* \; : \; \begin{aligned} \dot{x} &= f(x, u), & f(0, 0) &= 0 \\ y^* &= h^*(x, u), & h^*(0, 0) &= 0 \end{aligned} \tag{9.79}$$

with h^* satisfying

$$V_x^*(x) f(x, u) + \frac{1}{2} ||h(x, u)||^2 = \frac{1}{2} ||h^*(x, u)||^2 \tag{9.80}$$

The factor system $\overline{\Sigma}^*$ has special properties among the set of all factor systems $\overline{\Sigma}$. First, it is characterized by the following appealing property. Denote the inner system corresponding to $\overline{\Sigma}^*$ by Θ^*, given by the image representation (see (9.67))

$$I_{\Theta^*} \; : \; \begin{aligned} \dot{\xi} &= f(\xi, u) \\ y^* &= h^*(\xi, u) \\ y &= h(\xi, u) \end{aligned} \tag{9.81}$$

Consider any other all-pass factorization $\Sigma = \Theta \circ \overline{\Sigma}$, with $\overline{\Sigma}$ and I_Θ as in (9.66), respectively (9.67). Corresponding to Θ there exists a function $V \geq 0$ satisfying (9.68), and corresponding to Θ^* there is the function V^* satisfying

$$V^*(\xi(t_1)) - V^*(\xi(t_0)) = \frac{1}{2} \int_{t_0}^{t_1} (||y^*(t)||^2 - ||y(t)||^2) dt \tag{9.82}$$

Assume as before that V and V^* are C^1, and without loss of generality let $V(0) = V^*(0) = 0$. Then it follows that V and V^* are both solutions to the reversed dissipation inequality (9.62), with V^* being the maximal solution (Proposition 9.4.8). Thus, subtracting (9.68) from (9.82), one obtains

$$(V^* - V)(\xi(t_1)) - (V^* - V)(\xi(t_0)) = \frac{1}{2} \int_{t_0}^{t_1} (||y^*(t)||^2 - ||\overline{y}(t)||^2) dt \tag{9.83}$$

Therefore, if we take $t_0 = 0, t_1 = T, \xi(t_0) = 0$, then

$$\int_0^T ||y^*(t)||^2 dt \geq \int_0^T ||\bar{y}(t)||^2 dt, \text{ for all } T \geq 0 \qquad (9.84)$$

In this sense, the all-pass factorization $\Sigma = \Theta^* \circ \overline{\Sigma}^*$ is the "*minimal delay factorization*," since for zero initial conditions the truncated L_2-norm of the output of the factor $\overline{\Sigma}^*$ is *maximal* among the outputs of all other factor systems $\overline{\Sigma}$; and this holds for *every* input function u.

Remark 9.4.9 In the linear case the above interpretation of $\overline{\Sigma}^*$ amounts to the classical definition of $\overline{\Sigma}^*$ being *minimum phase*.

A very much related characterization of $\overline{\Sigma}^*$ is by the *stability of its zero-output constrained dynamics*. Consider first the case that the optimal control problem of minimizing $J(x_0, u)$ given in (9.77) is *regular*, in the sense that there exists a unique solution $u = u^*(x, p)$ to

$$\frac{\partial}{\partial u}\left[p^T f(x, u) + \frac{1}{2}||h(x, u)||^2\right] = 0, \qquad (9.85)$$

which minimizes the expression $p^T f(x, u) + \frac{1}{2}||h(x, u)||^2$ as a function of u for every x, p. Then (by dynamic programming) it follows that V^* is the unique solution of the *Hamilton–Jacobi–Bellman* equation

$$V_x(x) f(x, u^*(x, V_x^T(x))) + \frac{1}{2}||h(x, u^*(x, V_x^T(x)))||^2 = 0, \qquad (9.86)$$

such that $V(0) = 0$ and

$$\dot{x} = f(x, u^*(x, V_x^T(x))) \qquad (9.87)$$

is globally asymptotic stable with regard to $x = 0$.

Proposition 9.4.10 *The zero-output constrained dynamics of $\overline{\Sigma}^*$ (see Sect. 5.1) is globally asymptotically stable with regard to the equilibrium $x = 0$.*

Proof Consider (9.80) and (9.86) for $V = V^*$. It follows from the regularity of the optimal control problem that $u = u^*(x, V_x^T(x))$ is the unique input such that $h^*(x, u) = 0$, as follows from (9.86) and (9.80). Thus the zero-output constrained dynamics of $\overline{\Sigma}^*$ is given by the asymptotically stable dynamics (9.87). □

Remark 9.4.11 In the case of an *affine* nonlinear system

$$\begin{aligned}
\dot{x} &= f(x) + g(x)u, \quad f(0) = 0, \quad u \in \mathbb{R}^m, \quad x \in \mathcal{X} \\
y &= h(x) + d(x)u, \quad h(0) = 0, \quad y \in \mathbb{R}^p
\end{aligned} \qquad (9.88)$$

the optimal control problem is regular if and only if the $m \times m$ matrix $E(x) :=$ $d^T(x)d(x)$ is invertible for all x. Furthermore, in this case the Hamilton–Jacobi–Bellman equation (9.86) reduces to

$$V_x(x)[f(x) - g(x)E^{-1}(x)d^T(x)h(x)] - \tfrac{1}{2}V_x(x)g(x)E^{-1}(x)g^T(x)V_x^T(x) \\ + \tfrac{1}{2}h^T(x)[I_p - d(x)E^{-1}(x)d^T(x)]h(x) = 0 \tag{9.89}$$

Furthermore, the map $h^*(x, u)$ is explicitly given as

$$h^*(x, u) = \overline{d}(x)E^{-1}(x)[d^T(x)h(x) + g^T(x)V_x^{*T}(x)] + \overline{d}(x)u, \tag{9.90}$$

where V^* is the unique solution to (9.89) such that

$$\dot{x} = f(x) - g(x)g^T(x)V_x^{*T}(x) \tag{9.91}$$

is asymptotically stable with respect to $x = 0$, and where $\overline{d}(x)$ is any $m \times m$ matrix such that

$$d^T(x)d(x) = \overline{d}^T(x)\overline{d}(x) \tag{9.92}$$

(for example, $\overline{d}(x) = d(x)$).

In case the optimal control problem is *not* regular we proceed as follows. The *zero-output constrained dynamics* (or output-nulling dynamics) for a general system Σ given by (9.59) is defined to be the set of all state trajectories $x(\cdot)$ generated by some input trajectory $u(\cdot)$ such that $y(t) = h(x(t), u(t))$ is identically zero. Under some regularity conditions (see [135, 233]) the zero-output constrained dynamics can be computed as follows. First, we compute the maximal *controlled invariant* output-nulling submanifold $\mathcal{N}^* \subset \mathcal{X}$ (if it exists) as the maximal submanifold $\mathcal{N} \subset \mathcal{X}$, for which there exists a smooth feedback $u = \alpha(x)$, $\alpha(0) = 0$, such that

$$f(x, \alpha(x)) \in T_x\mathcal{N}, \quad h(x, \alpha(x)) = 0, \quad \text{for all } x \in \mathcal{N} \tag{9.93}$$

In general such a smooth feedback $\alpha(x)$ is not unique (even not restricted to $\mathcal{N}$), and (again under some regularity conditions; see [233, Chap. 11]) the whole family of feedbacks $u = \alpha(x)$ satisfying (9.93) for $\mathcal{N} = \mathcal{N}^*$ can be parametrized as $\alpha(x, v)$, with $v \in \mathbb{R}^{\tilde{m}}$, $\tilde{m} \leq m$. Then the zero-output constrained dynamics is generated by the lower dimensional dynamics

$$\tilde{\Sigma}: \quad \dot{\tilde{x}} = \tilde{f}(\tilde{x}, v), \quad \tilde{x} \in \mathcal{N}^*, \quad \tilde{f}(0, 0) = 0, \tag{9.94}$$

where $\tilde{f}(\tilde{x}, v) := f(\tilde{x}, \alpha(\tilde{x}, v))$, and $v \in \mathbb{R}^{\tilde{m}}$ is a remaining input vector.

Definition 9.4.12 Σ in (9.59) is *weakly minimum phase* if its zero-output constrained dynamics $\tilde{\Sigma}$ on $\mathcal{N}^*$ given (9.94) can be rendered *Lyapunov stable* with regard to $x = 0$ by a smooth feedback $v = \tilde{\alpha}(\tilde{x})$. If the zero-output constrained dynamics $\tilde{\Sigma}$

can be rendered locally *asymptotically stable* with respect to $x = 0$, then Σ is called *minimum phase*.

Remark 9.4.13 For linear systems Σ, minimum phase is equivalent to the requirement that the *transmission zeros* of Σ are all in the open left half plane.

Now return to the singular optimal control problem (9.78). Recall Assumption 9.4.7, and consider the factorization (9.80), with the resulting system $\overline{\Sigma}^*$ given by (9.79). For this system, we can again consider the singular optimal control problem (9.77), but now with y replaced by the new output y^*. Equivalently, we can consider the reversed dissipation inequality (9.62) with h replaced by h^*.

Lemma 9.4.14 *The maximal solution to the reversed dissipation inequality (9.62) with $h(x, u)$ replaced by $h^*(x, u)$ is the zero function.*

Proof Clearly $\overline{V} = 0$ satisfies (9.62) with $h(x, u)$ replaced by $h^*(x, u)$. Let now $\overline{V}$ be an arbitrary solution to (9.62) with $h(x, u)$ replaced by $h^*(x, u)$. Then it is immediately verified that $\overline{V} + \overline{V}^*$ is a solution to the original dissipation inequality (9.62) for $h(x, u)$. Since $\overline{V}^*$ is maximal this implies that $\overline{V} \leq 0$. $\square$

An equivalent formulation of Lemma 9.4.14 is the statement that

$$P(x_0) := \inf_{u(\cdot)} \left\{ \frac{1}{2} \int_0^\infty \|y^*(t)\|^2 dt \mid u \text{ admissible s.t. } \lim_{t \to \infty} x(t) = 0 \right\} \qquad (9.95)$$

is zero for all $x_0 \in \mathcal{X}$.

Note that if we would be allowed to replace in (9.95) "inf" by "min" for all $x_0 \in \mathcal{N}^*$ then we could directly conclude that $\overline{\Sigma}^*$ is minimum phase, since in this case there exists an admissible control u such that $\|y^*(t)\| = 0, t \geq 0$, and $\lim_{t \to \infty} x(t) = 0$. Thus the remaining question is how in general $P(x_0) = 0$ for all x_0 implies that $\overline{\Sigma}^*$ is minimum phase.

To answer this question, we take recourse to the *linearization* of $\overline{\Sigma}^*$ and the linear theory for the singular LQ optimal control problem developed in [104, 311]. First, we consider for $\overline{\Sigma}^*$ the regularized cost criterion

$$\frac{1}{2} \int_0^\infty \left(\|y(t)\|^2 + \epsilon^2 \|u(t)\|^2 \right) dt \qquad (9.96)$$

for ϵ small. This is a *regular* optimal control problem, and thus the maximal solution P^ϵ of the regularized reversed dissipation inequality

$$P_x^\epsilon(x) f(x, u) + \frac{1}{2} \|h^*(x, u)\|^2 + \frac{1}{2} \epsilon^2 \|u\|^2 \geq 0 \qquad (9.97)$$

is given as the stabilizing solution of the Hamilton–Jacobi–Bellman equation

$$H_{\text{opt}}^\epsilon(x, (P_x^\epsilon(x))^T) = 0, \qquad (9.98)$$

Here $H_{\text{opt}}^\epsilon(x, p) := H^\epsilon(x, p, u^*(x, p))$, where $u^*(x, p)$ is the solution to $\frac{\partial H^\epsilon}{\partial u}(x, p, u) = 0$, and

$$H^\epsilon(x, p, u) = p^T f(x, u) + \frac{1}{2}\|h^*(x, u)\|^2 + \frac{1}{2}\epsilon^2\|u\|^2 \qquad (9.99)$$

Since all data $f(x, u), h^*(x, u)$ are *smooth* near 0, also the solution P^ϵ of the Hamilton–Jacobi–Bellmann equation (9.98) is a smooth function near $x = 0$ (cf. Chap. 11). It follows from the interpretation as an optimal control problem that $P^\epsilon \geq 0$, and that P^ϵ is nondecreasing as a function of ϵ. Hence the pointwise limit $P^{\lim}(x) := \lim_{\epsilon \downarrow 0} P^\epsilon(x) \geq 0$ exists. We obtain the following theorem.

Theorem 9.4.15 *Assume that the zero-output constrained dynamics of $\overline{\Sigma}^*$ does not have invariant eigenvalues on the imaginary axis (that is, it is hyperbolic). Furthermore assume that $P^{\lim}$ is a C^1 function. Then $\overline{\Sigma}$ is minimum phase.*

Proof Suppose $\overline{\Sigma}$ is not minimum phase. By the assumption of hyperbolicity this means that the zero-output constrained dynamics is *exponentially* unstable. Thus for every smooth feedback $u = \alpha(x)$ such that the output $y^* = h^*(\tilde{x}, \alpha(\tilde{x}))$ is identically zero along the closed-loop dynamics $\dot{\tilde{x}} = f(\tilde{x}, \alpha(\tilde{x}))$, $\tilde{x} \in \mathcal{N}^*$, this closed-loop dynamics is exponentially unstable. Linearizing at $\tilde{x} = 0$, this yields that for any such α, $\exists \lambda \in \mathbb{C}$, $\text{Re}\,\lambda > 0$, and $\exists v \in \mathbb{C}^n$, $v \neq 0$, such that

$$\begin{aligned}
\left[\tfrac{\partial f}{\partial x}(0, 0) + \tfrac{\partial f}{\partial u}(0, 0)\tfrac{\partial \alpha}{\partial x}(0)\right] v &= \lambda v \\
\left[\tfrac{\partial h^*}{\partial x}(0, 0) + \tfrac{\partial h^*}{\partial u}(0, 0)\tfrac{\partial \alpha}{\partial x}(0)\right] v &= 0
\end{aligned} \qquad (9.100)$$

However, this implies that the *linear* system

$$\begin{aligned}
\dot{x} &= \tfrac{\partial f}{\partial x}(0, 0)x + \tfrac{\partial f}{\partial u}(0, 0)u =: Ax + Bu \\
\bar{y} &= \tfrac{\partial h^*}{\partial x}(0, 0)x + \tfrac{\partial h^*}{\partial u}(0, 0)u =: Cx + Du
\end{aligned} \qquad (9.101)$$

is not weakly minimum phase (transmission zeros in the open right half plane). However by *linear* theory [104, 311] this means that there exists a solution $X = X^T \geq 0$, with $X \neq 0$, to the quadratic inequality corresponding to (9.97), i.e.,

$$x^T X(Ax + Bu) + \frac{1}{2}(Cx + Du)^T(Cx + Du) \geq 0 \qquad (9.102)$$

By continuity the regularized dissipation inequality (9.97) converges for $\epsilon \downarrow 0$ to

$$P_x^{\lim}(x) f(x, u) + \frac{1}{2}\|h^*(x, u)\|^2 \geq 0 \qquad (9.103)$$

and hence $P^{\lim} = P = 0$. On the other hand, by linearization at $x = 0$

Fig. 9.6 Controller $C_{\bar{\Sigma}}$ based on factor system $\bar{\Sigma}$

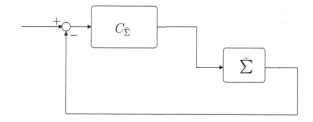

$$\frac{\partial^2 P^\epsilon}{\partial x^2}(0) \geq X \neq 0 \qquad (9.104)$$

for all $\epsilon > 0$, which yields a contradiction. □

Finally, we will discuss the use of the all-pass factorization $\Sigma = \Theta^* \circ \bar{\Sigma}^*$ for *control*, as already alluded to before. Throughout let us assume that the nonlinear system Σ given by (9.59) is *asymptotically stable*[4] with respect to $x = 0$.

As noted before in Proposition 9.4.4, a useful property of $\bar{\Sigma}^*$ as related to the original system Σ is that $\bar{\Sigma}^*$ and Σ have the same *steady-state output amplitudes*. This has the important consequence that the norm of the steady-state step responses of Σ and $\bar{\Sigma}$ are equal for every constant u. This suggests that output set-point control of Σ may be based on the factor system $\bar{\Sigma}$. Furthermore, since $\bar{\Sigma}^*$ is minimum phase, *inversion* techniques for its control can be used (e.g., by means of nonlinear input–output decoupling [135, 233]). A main open question is then how, once a suitable controller for the factor system $\bar{\Sigma}^*$ has been obtained, this can be translated to a suitable controller for the *original* system Σ. This problem will be solved by a nonlinear version of the structure of the classical *Smith predictor*.[5] In fact, this allows to convert a controller $C_{\bar{\Sigma}}$ for *any* factor system $\bar{\Sigma}$ (not necessarily $\bar{\Sigma}^*$) to a controller C for the original system Σ.

Thus consider the plant system Σ and an arbitrary factor system $\bar{\Sigma}$. Suppose that we have constructed for $\bar{\Sigma}$ a controller system $C_{\bar{\Sigma}}$, resulting in the closed-loop system as depicted in Fig. 9.6. How do we derive from this configuration a controller for the original system Σ? We use the following argument stemming from the derivation of the classical Smith predictor for linear systems. We add to the configuration of Fig. 9.6 two additional signal flows which exactly compensate each other, leading to Fig. 9.7. Subsequently, we shift the signal flow of y to the left-hand side of the block diagram, in order to obtain Fig. 9.8. The system within the dotted lines is now seen to be a controller for the original system Σ.

Note that the above construction based on the nonlinear Smith predictor may lead to problems in case Σ is *not* a stable system. Indeed, in this case the transition from

---

[4]In this case $\bar{\Sigma}^*$ is called an *outer system*: asymptotically stable with asymptotic stable zero-output constrained dynamics.

[5]I thank Gjerrit Meinsma for an illuminating discussion on the derivation of the Smith predictor.

Fig. 9.7 Addition of compensating signal flows

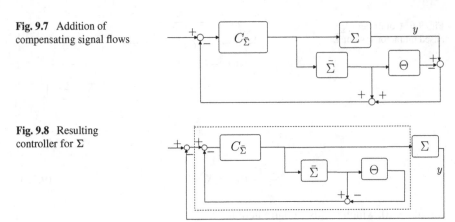

Fig. 9.8 Resulting controller for Σ

Figs. 9.6 to 9.7 may lead to diverging signal behavior, in the sense that although in the absence of disturbances the two added signal flows exactly compensate each other, any disturbance or mismatch may cause instabilities.

9.5 Notes for Chapter 9

1. Sections 9.1 and 9.2 are largely based on Scherpen & van der Schaft [310], van der Schaft [279], as well as Paice & van der Schaft [248, 249]. For the parametrization of stabilizing controllers in observer–controller configuration we refer to Viswanadham & Vidyasagar [345], and Banos [27].

2. Section 9.3 is based on Paice & van der Schaft [247, 249]. In these papers the more general and symmetric result is obtained of parametrizing all the *pairs* of state space systems Σ_1 and Σ_2 which are in stable closed-loop configuration.

3. With regard to the Youla–Kucera parametrization of stabilizing linear controllers of a linear system we refer, e.g., to Francis [103], and Green & Limebeer [117], and the references quoted therein. Apparently, the interpretation given in Remark 9.3.7, due to Paice & van der Schaft [247, 249], is new for linear systems.

4. Section 9.4 is based on van der Schaft & Ball [289]; see also van der Schaft & Ball [288], Ball & van der Schaft [24], which in its turn was motivated by work of Ball and Helton on inner–outer factorization of nonlinear operators, see, e.g., [21], and the theory of linear inner–outer factorization, see, e.g., Francis [103]. The approach taken in van der Schaft & Ball [288] and Ball & van der Schaft [24] emphasizes the "nonlinear spectral factorization" point of view of Proposition 9.4.5 and Remark 9.4.6.

5. The interpretation of the minimum phase factor as a "minimal delay factor," see Remark 9.4.9, can be found in Robinson [260]; see also Anderson [6].

6. The Morse Lemma (see, e.g., Abraham & Marsden [1]) can be stated as follows. Consider a C^2 function $H : \mathcal{X} \to \mathbb{R}$, with $H_x(0) = 0$ and $n \times n$ Hessian matrix $H_{xx}(0)$ which is supposed to be *nonsingular*. Then locally near 0 there exist coordinates $z_1, \ldots, z_n$ such that

$$H(z) = H(0) + \frac{1}{2}z_1^2 + \cdots + \frac{1}{2}z_k^2 - \frac{1}{2}z_{k+1}^2 - \cdots - \frac{1}{2}z_n^2, \qquad (9.105)$$

with k the number of positive eigenvalues of $H_{xx}(0)$. The following slight generalization can be easily proven (see, e.g., Maas [188]): Consider a C^2 function $H : \mathbb{R}^n \to \mathbb{R}$, with $H_x(0) = 0$, and $H_{xx}(x)$ of *constant* rank $\bar{n}$ in a neighborhood of $x = 0$. Then locally near 0 there exist coordinates $z_1, \ldots, z_k, z_{k+1}, \ldots, z_n$ such that

$$H(z) = H(0) + \frac{1}{2}z_1^2 + \cdots + \frac{1}{2}z_k^2 - \frac{1}{2}z_{k+1}^2 - \cdots - \frac{1}{2}z_{\bar{n}}^2 \qquad (9.106)$$

with, as above, k the number of positive eigenvalues of $H_{xx}(0)$. The application of this result to the factorization in (9.64) is done by considering the function K_V, and assuming that its partial derivatives with regard to x and u are zero at $(x, u) = (0, 0)$, and that its Hessian matrix with regard to x and u has constant rank. Since $K_V \geq 0$ and $K_V(0) = 0$ it then follows that there exists new coordinates z for $\mathcal{X} \times \mathbb{R}^m$ with $z_i = \bar{h}_i(x, u), i = 1, \ldots, \bar{p}$, where $\bar{p}$ the rank of the Hessian, such that locally (9.64) holds.

7. The problem of parametrizing the class of stabilizing controllers for a nonlinear plant is also addressed, using related approaches, in Lu [184], Imura & Yoshikawa [133], Fujimoto & Sugie [109, 110].

8. For studies of kernel representations in the context of closed-loop identification of nonlinear systems we refer to Linard, Anderson & De Bruyne [181] and Fujimoto, Anderson & De Bruyne [107].

9. The approach to all-pass factorization in Sect. 9.4 is largely based on Ball, Petersen & van der Schaft [25]; see also Ball & van der Schaft [24] for the regular case.

10. The idea of basing the control of a stable but nonminimum phase system on its minimum phase factor is well known in linear process control, see, e.g., Morari & Zafiriou [222], and Green & Limebeer [117]. The extension of this idea to nonlinear systems has been discussed in Doyle, Allgöwer & Morari [87], and Wright & Kravaris [357]. For a description of the Smith predictor in the linear case see, e.g., Morari & Zafiriou [222] and Meinsma & Zwart [216]. In this last reference also the treatment of the modified Smith predictor in case the linear plant is *not* asymptotically stable can be found.

11. For a study of nonlinear controllability and observability Gramians, their relation
    to the associated Hamiltonian system Σ^H, and their use for nonlinear Hankel
    singular values and balancing of nonlinear systems, we refer to, e.g., Scherpen
    [308], Scherpen & van der Schaft [310], Fujimoto, Scherpen & Gray [108].

Chapter 10
Nonlinear $\mathcal{H}_\infty$ Control

Consider the following standard control configuration, see Fig. 10.1. Let Σ be a nonlinear system on an n-dimensional state space manifold $\mathcal{X}$ of the general form

$$\Sigma \; : \; \begin{aligned} \dot{x} &= f(x, u, d) \\ y &= g(x, u, d) \\ z &= h(x, u, d) \end{aligned} \qquad (10.1)$$

with two sets of inputs u and d, two sets of outputs y and z, and state x.

Here u stands for the vector of *control* inputs, d are the *exogenous* inputs (disturbances-to-be-rejected or reference signals to-be-tracked), y are the *measured* outputs, and finally z denote the *to-be-controlled* outputs (tracking errors, cost variables). The *optimal $\mathcal{H}_\infty$ control problem*,[1] roughly speaking, is to find a controller C, processing the measurements y and producing the control inputs u, such that in the closed-loop configuration of Fig. 10.2 the L_2-gain from exogenous inputs d to to-be-controlled outputs z is minimized, while furthermore the closed-loop system is stable.

Instead of directly minimizing the L_2-gain the more feasible *suboptimal $\mathcal{H}_\infty$ control problem* is addressed. This consists in finding, if possible, for a given disturbance attenuation level γ a controller C such that the closed-loop system has L_2-gain $\leq \gamma$, and is stable. The solution to the optimal $\mathcal{H}_\infty$ control problem can then be approximated by an iteration of the suboptimal $\mathcal{H}_\infty$ control problem (successively decreasing γ toward the optimal disturbance attenuation level).

---

[1] The terminology $\mathcal{H}_\infty$ stems from the fact that in the linear case the L_2-gain of a stable system is equal to the $\mathcal{H}_\infty$ norm of its transfer matrix.

© Springer International Publishing AG 2017

A. van der Schaft, L_2-*Gain and Passivity Techniques in Nonlinear Control*,
Communications and Control Engineering, DOI 10.1007/978-3-319-49992-5_10

Fig. 10.1 Standard control configuration

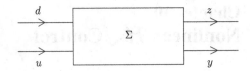

Fig. 10.2 Standard closed-loop configuration

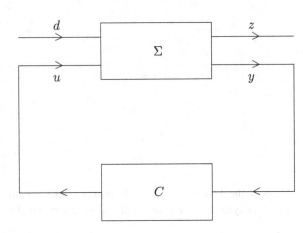

10.1 State Feedback $\mathcal{H}_\infty$ Control

In the *state feedback suboptimal* $\mathcal{H}_\infty$ *control problem* we assume that the whole state is available for feedback; or equivalently $y = x$ in (10.1).

We first study the state feedback problem for nonlinear systems of the special affine form

$$\Sigma : \begin{aligned} \dot{x} &= a(x) + b(x)u + g(x)d\,, \quad u \in \mathbb{R}^m,\, d \in \mathbb{R}^r \\ z &= \begin{bmatrix} h(x) \\ u \end{bmatrix} \qquad\qquad\qquad\quad x \in \mathcal{X},\, z \in \mathbb{R}^s \end{aligned} \qquad (10.2)$$

All data will be assumed to be C^k, with $k \geq 2$. Recall the notation $P_x(x)$ for the row vector $\left(\frac{\partial P}{\partial x_1}(x), \ldots, \frac{\partial P}{\partial x_n}(x) \right)$ of partial derivatives of a function $P : \mathcal{X} \to \mathbb{R}$. We start by stating the main result.

Theorem 10.1.1 *Let $\gamma > 0$. Suppose there exists a $C^r\,(k \geq r > 1)$ solution $P \geq 0$ to the Hamilton–Jacobi inequality (HJ1a) ("a" standing for affine):*

$$\textbf{(HJ1a)} \qquad \begin{aligned} P_x(x)a(x) + \tfrac{1}{2}P_x(x)&\left[\tfrac{1}{\gamma^2}g(x)g^T(x) - b(x)b^T(x) \right] P_x^T(x) \\ &+ \tfrac{1}{2}h^T(x)h(x) \leq 0 \end{aligned} \qquad (10.3)$$

Then with the C^{r-1} state feedback

$$u = -b^T(x)P_x^T(x) \tag{10.4}$$

the closed-loop system (10.2), (10.4), that is,

$$\dot{x} = a(x) - b(x)b^T(x)P_x^T(x) + g(x)d$$
$$z = \begin{bmatrix} h(x) \\ -b^T(x)P_x^T(x) \end{bmatrix} \tag{10.5}$$

has L_2-gain $\leq \gamma$ (from d to z).

Conversely, suppose there exists a C^{r-1} feedback

$$u = \ell(x) \tag{10.6}$$

such that there exists a C^1 storage function $P \geq 0$ for the closed-loop system (10.2), (10.6) with supply rate $\frac{1}{2}\gamma^2||d||^2 - \frac{1}{2}||z||^2$. Then $P \geq 0$ is also a solution of (HJ1a).

Proof Suppose $P \geq 0$ is solution to (HJ1a). Rewrite (HJ1a) as

$$P_x(x)\left[a(x) - b(x)b^T(x)P_x^T(x)\right] + \frac{1}{2}\frac{1}{\gamma^2}P_x(x)g(x)g^T(x)P_x^T(x)$$
$$+\frac{1}{2}P_x(x)b(x)b^T(x)P_x^T(x) + \frac{1}{2}h^T(x)h(x) \leq 0 \tag{10.7}$$

In view of (8.7) this means that P is a storage function for (10.5) with L_2-gain supply rate $\frac{1}{2}\gamma^2||d||^2 - \frac{1}{2}||z||^2$. Conversely, let $P \geq 0$ be a solution to

$$P_x(x)[a(x) + b(x)\ell(x)] + \frac{1}{2}\frac{1}{\gamma^2}P_x(x)g(x)g^T(x)P_x^T(x)$$
$$+\frac{1}{2}\ell^T(x)\ell(x) + \frac{1}{2}h^T(x)h(x) \leq 0 \tag{10.8}$$

Then by "completion of the squares"

$$P_x(x)\left[a(x) - b(x)b^T(x)P_x^T(x)\right]$$
$$= P_x(x)[a(x) + b(x)l(x)] - P_x(x)b(x)\left[b^T(x)P_x^T(x) + l(x)\right]$$
$$\leq -\frac{1}{2}\frac{1}{\gamma^2}P_x(x)g(x)g^T(x)P_x^T(x) - \frac{1}{2}h^T(x)h(x)$$
$$-\frac{1}{2}||l(x) + b^T(x)P_x^T(x)||^2 - \frac{1}{2}P_x(x)b(x)b^T(x)P_x^T(x) \tag{10.9}$$
$$\leq -\frac{1}{2}\frac{1}{\gamma^2}P_x(x)g(x)g^T(x)P_x^T(x) - \frac{1}{2}P_x(x)b(x)b^T(x)P_x^T(x)$$
$$-\frac{1}{2}h^T(x)h(x)$$

showing that P is a solution to (10.7), and thus to (10.3). $\qquad\square$

Note that we did not say anything sofar about the stability of the closed-loop system (10.5) resulting from the state feedback $u = -b^T(x)P_x^T(x)$. However, since $P \geq 0$ is a storage function for (10.5) we can simply invoke the results of Chap. 3, in particular Propositions 3.2.12 and 3.2.16.

Proposition 10.1.2 *Let $P \geq 0$ be a solution to (HJ1a).*

(i) Suppose the system

$$\dot{x} = a(x)$$
$$z = \begin{bmatrix} h(x) \\ -b^T(x)P_x^T(x) \end{bmatrix} \tag{10.10}$$

is zero-state observable, then $P(x) > 0$ for $x \neq 0$.

(ii) Suppose $P(x) > 0$ for all $x \neq 0$ and $P(0) = 0$, and suppose (10.10) is zero-state detectable. Then $x = 0$ is a locally asymptotically stable equilibrium of

$$\dot{x} = a(x) - b(x)b^T(x)P_x^T(x) \tag{10.11}$$

If additionally P is proper, *then $x = 0$ is a globally asymptotically stable equilibrium of (10.11).*

Proof Apply Propositions 3.2.12 and 3.2.16 to (10.5), using (HJ1a) rewritten as (10.7), and note that $a(x) - b(x)b^T(x)P_x^T(x) = a(x)$ whenever $z = 0$.   □

Remark 10.1.3 We leave the generalization of Proposition 10.1.2 to positive *semi*-definite $P \geq 0$, based on Theorem 3.2.17 and Proposition 3.2.19, to the reader.

At this point one may wonder where the Hamilton–Jacobi inequality (HJ1a) comes from, and how similar expressions may be obtained for systems of a more general form than (10.2). An answer to both questions is obtained from the theory of *differential games*. Indeed, one may look at the $\mathcal{H}_\infty$ suboptimal problem as a two-player zero-sum differential game with cost criterion

$$-\frac{1}{2}\gamma^2 ||d||^2 + \frac{1}{2}||z||^2 \tag{10.12}$$

(one player corresponds to the control input u, the other player to the exogenous input d).

The pre-Hamiltonian corresponding to this differential game is given as

$$K_\gamma(x, p, d, u) := p^T[a(x) + b(x)u + g(x)d] - \frac{1}{2}\gamma^2 ||d||^2 + \frac{1}{2}||z||^2 \tag{10.13}$$

with p being the co-state. From the equations $\frac{\partial K_\gamma}{\partial d} = 0$ and $\frac{\partial K_\gamma}{\partial u} = 0$ we determine, respectively,

$$d^*(x, p) = \frac{1}{\gamma^2}g^T(x)p$$
$$u^*(x, p) = -b^T(x)p, \tag{10.14}$$

which have the *saddle point* property

$$K_\gamma(x, p, d, u^*(x, p)) \leq K_\gamma(x, p, d^*(x, p), u^*(x, p))$$
$$\leq K_\gamma(x, p, d^*(x, p), u) \tag{10.15}$$

for every d, u and every (x, p). The input u^* may be called the *optimal control*, and d^* the *worst-case* exogenous input (disturbance).

Substitution of (10.14) in $K_\gamma(x, p, d, u)$ leads to the (optimal) Hamiltonian

$$H_\gamma(x, p) = p^T a(x) + \tfrac{1}{2} p^T \left[\tfrac{1}{\gamma^2} g(x) g^T(x) - b(x) b^T(x) \right] p$$
$$+ \tfrac{1}{2} h^T(x) h(x) \tag{10.16}$$

The Hamilton–Jacobi inequality (HJ1a) (10.3) is now also expressed as

$$H_\gamma(x, P_x^T(x)) \leq 0, \tag{10.17}$$

which is known in the theory of differential games as the *Hamilton–Jacobi–Isaacs equation*.

The same procedure may be followed for more general equations than (10.2), e.g.,

$$\dot{x} = f(x, u, d)$$
$$z = h(x, u, d) \tag{10.18}$$

by considering the pre-Hamiltonian

$$K_\gamma(x, p, d, u) := p^T f(x, u, d) - \frac{1}{2} \gamma^2 \|d\|^2 + \frac{1}{2} \|z\|^2 \tag{10.19}$$

Suppose, as above, that K_γ has a saddle point $u^*(x, p), d^*(x, p)$, i.e.,

$$K_\gamma(x, p, d, u^*(x, p)) \leq K_\gamma(x, p, d^*(x, p), u^*(x, p))$$
$$\leq K_\gamma(x, p, d^*(x, p), u) \tag{10.20}$$

for every d, u, and every (x, p). Then we consider the Hamilton–Jacobi inequality

$$(\mathbf{HJ1}) \, K_\gamma(x, P_x^T(x), d^*(x, P_x^T(x)), u^*(x, P_x^T(x))) \leq 0 \tag{10.21}$$

Analogously to Theorem 10.1.1 we obtain

Proposition 10.1.4 *Consider the system (10.18). Let $\gamma > 0$. Assume there exist $u^*(x, p), d^*(x, p)$ satisfying (10.20). Suppose there exists a C^r ($k \geq r > 1$) solution $P \geq 0$ to the Hamilton–Jacobi inequality (HJ1) given by (10.21). Then the C^{r-1} state feedback*

$$u = u^*(x, P_x^T(x)) \tag{10.22}$$

is such that the closed-loop system (10.18), (10.22), i.e.,

$$
\begin{aligned}
\dot{x} &= f(x, u^*(x, P_x^T(x)), d)\\
z &= h(x, u^*(x, P_x^T(x)), d)
\end{aligned}
\tag{10.23}
$$

has L_2-gain $\leq \gamma$.

  Conversely, suppose there exists a C^{r-1} feedback

$$
u = l(x)
\tag{10.24}
$$

such that there exists a C^1 storage function $P \geq 0$ for the closed-loop system (10.18), (10.24) with supply rate $\frac{1}{2}\gamma^2\|d\|^2 - \frac{1}{2}\|z\|^2$. Then $P \geq 0$ is also a solution of (HJ1).

Proof Let $P \geq 0$ satisfy (10.21). Substitute $p = P_x^T(x)$ in the first inequality in (10.20) to obtain

$$
\begin{aligned}
&P_x(x)f(x, u^*(x, P_x^T(x)), d) - \frac{1}{2}\gamma^2\|d\|^2 + \frac{1}{2}\|h(x, u^*(x, P_x^T(x)), d)\|^2\\
&\qquad \leq K_\gamma(x, P_x^T(x), d^*(x, P_x^T(x)), u^*(x, P_x^T(x))),
\end{aligned}
\tag{10.25}
$$

and thus by (10.21) for all d

$$
\begin{aligned}
&P_x(x)f(x, u^*(x, P_x^T(x)), d)\\
&\qquad \leq \frac{1}{2}\gamma^2\|d\|^2 - \frac{1}{2}\|h(x, u^*(x, P_x^T(x)), d)\|^2,
\end{aligned}
\tag{10.26}
$$

showing that (10.23) has L_2-gain $\leq \gamma$ with storage function P. Conversely, let $P \geq 0$ be a solution to

$$
P_x(x)f(x, l(x), d) \leq \frac{1}{2}\gamma^2\|d\|^2 - \frac{1}{2}\|h(x, l(x), d)\|^2
\tag{10.27}
$$

Then by substituting $p = P_x^T(x)$ and $u = l(x)$ in the *second* inequality in (10.20) we obtain

$$
\begin{aligned}
&K_\gamma(x, P_x^T(x), d^*(x, P_x^T(x)), u^*(x, P_x^T(x)))\\
&\qquad \leq K_\gamma(x, P_x^T(x), d^*(x, P_x^T(x)), l(x))
\end{aligned}
\tag{10.28}
$$

and thus by (10.27) with $d = d^*(x, P_x^T(x))$ we obtain (10.21). $\square$

Remark 10.1.5 The above proof reveals the essence of suboptimal state feedback $\mathcal{H}_\infty$ control. Solvability of the Hamilton–Jacobi inequality (HJ1) together with the first inequality in (10.20) yields the feedback (10.22) for which the closed-loop system (10.23) has L_2-gain $\leq \gamma$. Conversely, if the second inequality in (10.20) holds then the existence of $u = l(x)$ for which the closed-loop system has L_2-gain $\leq \gamma$ implies the solvability of (HJ1).

Remark 10.1.6 Note that in case of the affine system equations (10.2) the point $u^*(x, p), d^*(x, p)$ given by (10.14) not only satisfies the saddle point property (10.20) but in fact the stronger property

$$\begin{aligned} K_\gamma(x, p, d, u) &\leq K_\gamma(x, p, d^*(x, p), u) \\ K_\gamma(x, p, d, u^*(x, p)) &\leq K_\gamma(x, p, d, u), \end{aligned} \qquad \forall u, d \qquad (10.29)$$

We leave it to the reader to formulate the analog of Proposition 10.1.2 for the general case treated in Proposition 10.1.4.

Example 10.1.7 Consider the system

$$\dot{x} = (1 + x^2)u + d, \qquad z = \begin{bmatrix} x \\ u \end{bmatrix} \qquad (10.30)$$

The Hamilton–Jacobi inequality (HJ1a) reads

$$\left(\frac{dP}{dx}(x)\right)^2 \left((1 + x^2)^2 - \frac{1}{\gamma^2}\right) \geq x^2, \qquad (10.31)$$

which has a nonnegative solution for $\gamma > 1$, e.g.,

$$P(x) = \frac{1}{2}\ln\left(1 + x^2 + \sqrt{(1 + x^2)^2 - \frac{1}{\gamma^2}}\right), \qquad (10.32)$$

leading to the feedback

$$u = -(1 + x^2)\left((1 + x^2)^2 - \frac{1}{\gamma^2}\right)^{-\frac{1}{2}} x, \qquad (10.33)$$

stabilizing the system about $x = 0$.

Example 10.1.8 Consider the system

$$\dot{x} = u + (\arctan x)d, \qquad z = \begin{bmatrix} x \\ u \end{bmatrix} \qquad (10.34)$$

The Hamilton–Jacobi inequality (HJ1a) reads as

$$\left(\frac{dP}{dx}(x)\right)^2 \left(1 - \frac{\arctan^2 x}{\gamma^2}\right) \geq x^2, \qquad (10.35)$$

having solutions $P \geq 0$ for all γ such that

$$|\arctan x| < \gamma, \qquad \forall x \in \mathbb{R}, \qquad (10.36)$$

that is, for all $\gamma > \frac{\pi}{2}$. The feedback is given as

$$u = -x \left(1 - \frac{\arctan^2 x}{\gamma^2} \right)^{-\frac{1}{2}} \tag{10.37}$$

Example 10.1.9 Consider the system

$$\dot{x} = x(u + d), \qquad z = \begin{bmatrix} x \\ u \end{bmatrix} \tag{10.38}$$

The Hamilton–Jacobi inequality (HJ1a) is

$$\left(\frac{dP}{dx}(x) \right)^2 \left(\frac{1}{\gamma^2} - 1 \right) x^2 + x^2 \leq 0 \tag{10.39}$$

which has for $\gamma > 1$ the so-called *viscosity solution* (see the Notes at the end of this chapter)

$$\begin{aligned} P(x) &= \frac{\gamma}{\sqrt{\gamma^2 - 1}} |x| \\ u &= -\frac{\gamma}{\sqrt{\gamma^2 - 1}} |x| \end{aligned} \tag{10.40}$$

Note that P is not differentiable at $x = 0$, however the closed-loop system

$$\dot{x} = -\frac{\gamma}{\sqrt{\gamma^2 - 1}} |x| x + xd \tag{10.41}$$

is asymptotically stable about $x = 0$.

Now let us come back to the L_2-gain perturbation models formulated in Sect. 9.2, based on stable kernel or stable image representations. Consider an affine nonlinear input-state-output system

$$\Sigma : \begin{aligned} \dot{x} &= f(x) + g(x)u, \ x \in \mathcal{X}, \quad u \in \mathbb{R}^m \\ y &= h(x), \qquad\qquad\qquad\quad y \in \mathbb{R}^p \end{aligned} \tag{10.42}$$

First, we consider the perturbation Σ_Δ based on the *stable kernel representation* K_Σ, given in (9.29):

$$\Sigma_\Delta : \begin{aligned} \dot{x} &= f(x) + g(x)u - k(x)w \\ y &= h(x) - w \end{aligned} \qquad w \in \mathbb{R}^p, \tag{10.43}$$

with w being the output of some nonlinear system Δ with input $\begin{bmatrix} u \\ y \end{bmatrix}$

Fig. 10.3 Perturbed
nonlinear system

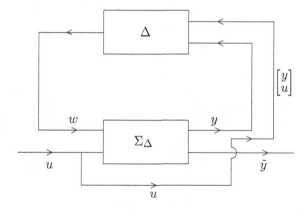

$$\Delta \; : \quad \begin{aligned} \dot{\xi} &= \alpha(\xi, u, y) \\ w &= \beta(\xi, u, y) \end{aligned} \qquad (10.44)$$

having finite L_2-gain γ_Δ (see Fig. 10.3). Here the matrix $k(x)$ is determined by Eq. (9.10), that is, $W_x(x)k(x) = h^T(x)$, with W being a nonnegative solution to the "filter" Hamilton–Jacobi equation (9.9).

We now formulate the *robust stabilization problem* as the problem of constructing a controller C processing the measurements $\tilde{y}$ and producing the control u such that the L_2-gain (from w to z) of Σ_Δ in closed-loop with C is minimized, say equal to γ^*. In the present section, we consider the robust stabilization problem by *state feedback* where we take $\tilde{y} = x$, and in the next section we will consider the *measurement feedback* case where $\tilde{y} = y$.

Once we have solved the robust stabilization problem, it follows from the small-gain theorem that the closed-loop system given in Fig. 10.4 will be closed-loop stable for all perturbations Δ with $\gamma_\Delta \cdot \gamma^* < 1$.

That is, in an input–output map context Theorem 2.1.1 implies that whenever the signals $e_1, e_2 \in L_2$ are such that the signals w and z are in L_{2e}, then actually w and z in L_2. Furthermore, in a state space context Theorem 8.2.1 will ensure internal stability of the controlled perturbed nonlinear system for $e_1 = e_2 = 0$.

Clearly, the problem of minimizing the L_2-gain (from w to z) of Σ_Δ by state feedback is a nonlinear $\mathcal{H}_\infty$ optimal control problem. The suboptimal version (for decreasing values of $\gamma > 0$) is addressed, as above, by considering the pre-Hamiltonian corresponding to (10.43):

$$\begin{aligned} K_\gamma(x, p, u, w) = {}& p^T[f(x) + g(x)u - k(x)w] - \tfrac{1}{2}\gamma^2\|w\|^2 \\ & + \tfrac{1}{2}\|u\|^2 + \tfrac{1}{2}\|h(x) - w\|^2 \end{aligned} \qquad (10.45)$$

As in (10.14) we obtain a saddle point $u^*(x, p)$, $w^*(x, p)$ given as

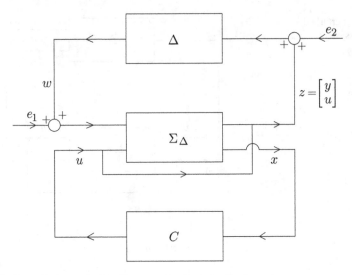

Fig. 10.4 Controlled perturbed nonlinear system (state feedback case)

$$u^* = -g^T(x)p, \quad w^* = \frac{-1}{\gamma^2 - 1}(k^T(x)p + h(x)) \qquad (10.46)$$

Necessarily $\gamma \geq 1$, since $y = h(x) + w$ and hence the L_2-gain from w to z is always ≥ 1. Substitution of (10.46) into (10.45) leads to the Hamilton–Jacobi inequality (HJ1), written out as

$$P_x(x)[f(x) + (\gamma^2 - 1)^{-1}k(x)h(x)] + \tfrac{1}{2}\gamma^2(\gamma^2 - 1)^{-1}h^T(x)h(x)$$
$$+ \tfrac{1}{2}P_x(x)[(\gamma^2 - 1)^{-1}k(x)k^T(x) - g(x)g^T(x)]P_x^T(x) \leq 0 \qquad (10.47)$$

Proposition 10.1.10 *Let $\gamma > 1$. Suppose there exists a solution $P \geq 0$ to (10.47). Then the state feedback*

$$u = -g^T(x)P_x^T(x) \qquad (10.48)$$

will be such that the controlled system

$$\dot{x} = [f(x) - g(x)g^T(x)P_x^T(x)] - k(x)w$$
$$z = \begin{bmatrix} h(x) - w \\ -g^T(x)P_x^T(x) \end{bmatrix} \qquad (10.49)$$

has L_2-gain $\leq \gamma$ (from w to z).

Furthermore, let Δ have L_2-gain $\gamma_\Delta < \frac{1}{\gamma}$ with storage function P_Δ. Suppose P and P_Δ have strict local minima at $x^* = 0$, respectively, $\xi^* = 0$, and Σ_Δ and Δ are zero-state detectable. Then (x^*, ξ^*) is a locally asymptotically stable equilibrium of

*the closed-loop system (10.49), (10.44), which is globally asymptotically stable if P
and P_Δ have global minima at $x^* = 0$ and $\xi^* = 0$ and are proper.*

Proof The first part immediately follows from Proposition 10.1.4. The second part
follows from the small-gain theorem in the state space setting (Theorem 8.2.1). □

Remark 10.1.11 By taking fixed initial conditions for the controlled system as well
as for the perturbation system Δ and considering the resulting input–output maps
one may also formulate closed-loop stability statements in the spirit of the small-gain
theorem in the input–output mapping setting (Theorem 2.1.1). We leave this to the
reader.

Alternatively, we may consider the perturbation $\Sigma_{\tilde\Delta}$ based on the *stable image representation* I_Σ of Σ, given in (9.33), i.e.,

$$\Sigma_{\tilde\Delta} : \quad \begin{aligned} \dot{x} &= f(x) + g(x)u - g(x)w_2 \\ y &= h(x) + w_1 \end{aligned} \qquad (10.50)$$

with (w_1, w_2) the output of some nonlinear system $\tilde\Delta$

$$\dot{\xi} = \tilde\alpha(\xi, s)$$

$$\tilde\Delta : \quad w_1 = \tilde\beta_1(\xi, s) \qquad (10.51)$$

$$w_2 = \tilde\beta_2(\xi, s)$$

with $s = u + g^T(x)V_x^T(x) - w_2$, having finite L_2-gain (from s to (w_1, w_2)). The
robust stabilization problem for this perturbation model is to minimize the L_2-gain
(from (w_1, w_2) to s) of

$$\begin{aligned} \dot{x} &= f(x) + g(x)u - g(x)w_2 \\ s &= u + g^T(x)V_x^T(x) - w_2 \end{aligned} \qquad (10.52)$$

In the state feedback case this problem admits an easy solution.

Proposition 10.1.12 *The state feedback*

$$u = -g^T(x)V_x^T(x) \qquad (10.53)$$

*renders the L_2-gain of (10.52) equal to 1. Furthermore, for every other controller C
the L_2-gain from (w_1, w_2) to s is ≥ 1.*

Proof Since $s = u + g^T(x)V_x^T(x) - w_2$ it is easily seen that the L_2-gain from w_2
to s is always ≥ 1. On the other hand, (10.53) clearly renders the L_2-gain equal
to 1.   □

The above result has a clear interpretation. Recall (see Remark 9.1.9 and Sect. 11.2) that (10.53) is the feedback that minimizes the cost criterion

$$\int_0^\infty (||u(t)||^2 + ||y(t)||^2)dt \tag{10.54}$$

for $\dot{x} = f(x) + g(x)u$, $y = h(x)$. Proposition 10.1.12 shows that (10.53) also optimizes the robustness margin for the perturbed system $\Sigma_{\tilde{\Delta}}$.

We leave the further translation of the statements of Proposition 10.1.10 to the perturbation model $\Sigma_{\tilde{\Delta}}$ to the reader.

Example 10.1.13 (Example 9.2.1 continued) Consider a port-Hamiltonian system with energy $H \geq 0$ and without energy dissipation

$$\begin{aligned}
\dot{x} &= J(x)H_x^T(x) + g(x)u, \quad J(x) = -J^T(x) \\
y &= g^T(x)H_x^T(x)
\end{aligned} \tag{10.55}$$

In Example 9.2.1, we have derived the perturbed systems Σ_Δ and $\Sigma_{\tilde{\Delta}}$, see (9.36), (9.37). With respect to the perturbed system Σ_Δ the Hamilton–Jacobi inequality (HJ1) takes the form

$$\begin{aligned}
&P_x(x)[J(x)H_x^T(x) + (\gamma^2 - 1)^{-1}g(x)g^T(x)H_x^T(x)] \\
&+ \tfrac{1}{2}[\gamma^2 - 1)^{-1} - 1]P_x(x)g(x)g^T(x)P_x^T(x) \\
&+ \tfrac{1}{2}\gamma^2(\gamma^2 - 1)^{-1}H_x(x)g(x)g^T(x)H_x^T(x) \leq 0,
\end{aligned} \tag{10.56}$$

having for $\gamma > \sqrt{2}$ the positive solution $P(x) = \frac{\gamma^2}{\gamma^2-2}H(x)$. Thus by Proposition 10.1.10 the state feedback

$$u = -\frac{\gamma^2}{\gamma^2 - 2}g^T(x)H_x^T(x), \quad \gamma > \sqrt{2}, \tag{10.57}$$

robustly stabilizes Σ_Δ for every perturbation Δ with L_2-gain $< \frac{1}{\gamma}$. Alternatively, cf. Proposition 10.1.12, the feedback

$$u = -g^T(x)H_x^T(x) \tag{10.58}$$

robustly stabilizes $\Sigma_{\tilde{\Delta}}$ for every perturbation $\tilde{\Delta}$ with L_2-gain < 1.

10.2   Output Feedback $\mathcal{H}_\infty$ Control

Consider the nonlinear system (10.1), that is,

$$\dot{x} = f(x, u, d), \ x \in \mathcal{X}, \ u \in \mathbb{R}^m, \ d \in \mathbb{R}^r$$

$$\Sigma: \quad y = g(x, u, d), \ y \in \mathbb{R}^p \tag{10.59}$$

$$z = h(x, u, d), \ z \in \mathbb{R}^s$$

In the *output feedback* suboptimal $\mathcal{H}_\infty$ control problem we want to construct, if possible, for a given attenuation level $\gamma \geq 0$ an output feedback controller

$$C: \quad \begin{aligned} \dot{\xi} &= \varphi(\xi, y) \\ u &= \alpha(\xi) \end{aligned} \tag{10.60}$$

such that the closed-loop system has L_2-gain $\leq \gamma$ (from d to z). Here $\xi = (\xi_1, \ldots, \xi_\nu)$ are local coordinates for the state space manifold $\mathcal{X}_c$ of the controller C, and φ and α are mappings whose degree of differentiability in general will depend on the degree of differentiability of the mappings f, g, and h. As before, we assume them to be at least C^2.

In this section, we will be primarily concerned with finding *necessary* conditions for the solvability of the output feedback suboptimal $\mathcal{H}_\infty$ problem, as well as with the analysis of the *structure* of controllers that solve the problem. In order to do so, let us assume that a controller C as in (10.60) solves the output feedback suboptimal $\mathcal{H}_\infty$ control problem for Σ given by (10.59), for a given disturbance attenuation level γ. Moreover, *assume* that there exists a *differentiable* storage function $S(x, \xi) \geq 0$ for the closed-loop system with respect to the supply rate $\frac{1}{2}\gamma^2\|d\|^2 - \frac{1}{2}\|z\|^2$, that is,

$$\begin{aligned} & S_x(x, \xi)f(x, \alpha(\xi), d) + S_\xi(x, \xi)\varphi(\xi, g(x, \alpha(\xi), d)) \\ & \quad -\tfrac{1}{2}\gamma^2\|d\|^2 + \tfrac{1}{2}\|h(x, \alpha(\xi), d)\|^2 \leq 0, \quad \text{for all } d \end{aligned} \tag{10.61}$$

Now consider the equation

$$S_\xi(x, \xi) = 0, \tag{10.62}$$

and *assume* this equation has a differentiable solution $\xi = F(x)$. (By the Implicit Function theorem this will locally be the case if the partial Hessian matrix $S_{\xi\xi}(x, \xi)$ is nonsingular for every (x, ξ) satisfying (10.62).) Define

$$P(x) := S(x, F(x)) \tag{10.63}$$

Substitution of $\xi = F(x)$ into (10.61) yields (noting that $P_x(x) = S_x(x, F(x))$ since $S_\xi(x, F(x)) = 0$)

$$P_x(x)f(x, \alpha(F(x)), d) - \frac{1}{2}\gamma^2\|d\|^2 + \frac{1}{2}\|h(x, \alpha(F(x)), d)\|^2 \leq 0 \tag{10.64}$$

for all d. Hence the *state feedback* $u = \alpha(F(x))$ solves the state feedback suboptimal $\mathcal{H}_\infty$ control problem for Σ, with storage function P. Therefore, P is solution of (HJ1).

Thus we have obtained a logical necessary condition for solvability of the output feedback suboptimal $\mathcal{H}_\infty$ control problem, namely solvability of the corresponding state feedback problem.

A second necessary condition is obtained by restricting to the (natural) class of controllers C which produce zero control u for measurements y being identically zero, at least for "zero initial condition". More specifically, we assume that the controller C satisfies

$$\varphi(0,0) = 0, \ \alpha(0) = 0 \tag{10.65}$$

Defining

$$R(x) := S(x,0), \tag{10.66}$$

substitution of $\xi = 0$ and $y = 0$ in (10.61) then yields

$$\textbf{(HJ2)} \ R_x(x)f(x,0,d) - \frac{1}{2}\gamma^2||d||^2 + \frac{1}{2}||h(x,0,d)||^2 \leq 0 \tag{10.67}$$

for all disturbances d *such that the measurements $y = g(x,0,d)$ remain zero.*

Thus a second necessary condition for solvability of the output feedback $\mathcal{H}_\infty$ suboptimal control problem is the existence of a solution $R \geq 0$ to (HJ2). This necessary condition also admits an obvious interpretation; it tells us that if we wish to render Σ dissipative by a controller C satisfying (10.65), then Σ constrained by $u = 0$ and $y = 0$ already has to be dissipative.

If we specialize (as in the previous Sect. 10.1) the equations for Σ and C to

$$\Sigma_a : \quad \begin{aligned} \dot{x} &= a(x) + b(x)u + g(x)d_1 \\ y &= c(x) + d_2 \\ z &= \begin{bmatrix} h(x) \\ u \end{bmatrix}, \end{aligned} \tag{10.68}$$

respectively,

$$C_a : \quad \begin{aligned} \dot{\xi} &= k(\xi) + \ell(\xi)y, \quad k(0) = 0 \\ u &= m(\xi), \qquad\qquad m(0) = 0, \end{aligned} \tag{10.69}$$

then (10.67) reduces to

$$R_x(x)[a(x) + g(x)d_1] - \frac{1}{2}\gamma^2(||d_1||^2 + ||d_2||^2) + \frac{1}{2}||h(x)||^2 \leq 0 \tag{10.70}$$

for all $d = \begin{bmatrix} d_1 \\ d_2 \end{bmatrix}$ such that $y = c(x) + d_2$ is zero; implying that $d_2 = -c(x)$. Computing the maximizing disturbance $d_1^* = \frac{1}{\gamma^2}g^T(x)R_x^T(x)$, it follows that (10.70) is equivalent to the Hamilton–Jacobi inequality

$$\textbf{(HJ2a)} \quad \begin{aligned} & R_x(x)a(x) + \tfrac{1}{2}\tfrac{1}{\gamma^2}R_x(x)g(x)g^T(x)R_x^T(x) \\ & + \tfrac{1}{2}h^T(x)h(x) - \tfrac{1}{2}\gamma^2 c^T(x)c(x) \le 0, \quad x \in \mathcal{X} \end{aligned} \qquad (10.71)$$

Remark 10.2.1 In cases different from (10.68) the constraint $y = 0$ may impose constraints on the state space $\mathcal{X}$, in which case we would obtain, contrary to (10.71), a Hamilton–Jacobi inequality defined on a *subset* of $\mathcal{X}$.

Thus we have derived, under mild technical assumptions, two necessary conditions for the solvability of the output feedback suboptimal $\mathcal{H}_\infty$ control problem; namely the existence of a solution $P \ge 0$ to the first Hamilton–Jacobi inequality (HJ1), and the existence of a solution $R \ge 0$ to the second Hamilton–Jacobi inequality (HJ2).

Furthermore, it is clear from the way we have derived the solutions $P \ge 0$ and $R \ge 0$ to these two Hamilton–Jacobi inequalities, that P and R are not unrelated. In fact, as we now will show, the solutions P and R have to satisfy a certain *coupling condition*. The easiest way of obtaining this coupling condition is to consider, as above, $P(x) = S(x, F(x))$ and $R(x) = S(x, 0)$ and to additionally *assume* that S has a minimum at $(0, 0)$, i.e.,

$$S(0, 0) = 0, \; S_x(0, 0) = 0, \; S_\xi(0, 0) = 0, \; S(x, \xi) \ge 0, \quad \forall x, \xi, \qquad (10.72)$$

and furthermore that the Hessian matrix of S at $(0, 0)$

$$\begin{bmatrix} S_{xx}(0, 0) & S_{x\xi}(0, 0) \\ S_{\xi x}(0, 0) & S_{\xi\xi}(0, 0) \end{bmatrix} =: \begin{bmatrix} S_{11} & S_{12} \\ S_{12}^T & S_{22} \end{bmatrix} \qquad (10.73)$$

satisfies $S_{22} = S_{\xi\xi}(0, 0) > 0$. By the Implicit Function theorem this will imply the existence of a unique $F(x)$ defined near $x = 0$ such that $S_\xi(x, F(x)) = 0$. It immediately follows that

$$P(0) = 0, \; P_x(0) = 0, \; R(0) = 0, \; R_x(0) = 0, \qquad (10.74)$$

and moreover from the definition of P and R it can be seen that

$$P_{xx}(0) = S_{11} - S_{12}S_{22}^{-1}S_{12}^T, \; R_{xx}(0) = S_{11} \qquad (10.75)$$

As a consequence one obtains the *"weak" coupling condition*

$$P_{xx}(0) \le R_{xx}(0), \qquad (10.76)$$

with strict inequality if $S_{12} \ne 0$, in which case $P(x) < R(x)$ for x near 0. In fact, in case of a *linear system* this amounts to the same coupling condition as obtained in linear $\mathcal{H}_\infty$ control theory [117]. See also Corollary 11.4.9 in the next Chap. 11.

A stronger, and more intrinsic, coupling condition can be derived as follows. Suppose there exists a controller C satisfying (10.65) which solves the output feedback suboptimal $\mathcal{H}_\infty$ control problem with a storage function S for the closed-loop system satisfying $S(0,0) = 0$ (or, equivalently, S has a *minimum* at $(0,0)$). Then the closed-loop system satisfies, whenever $x(-T_1) = 0$, $\xi(-T_1) = 0$

$$\tfrac{1}{2} \int_{-T_1}^0 (\gamma^2 \|d(t)\|^2 - \|z(t)\|^2)\, dt$$

$$+ \tfrac{1}{2} \int_0^{T_2} (\gamma^2 \|d(t)\|^2 - \|z(t)\|^2)\, dt \geq 0$$

for all $T_1 \geq 0$, $T_2 \geq 0$, and all disturbance functions $d(\cdot)$ on $[-T_1, T_2]$. In particular, we obtain for $x(-T_1) = 0$, $\xi(-T_1) = 0$

$$-\int_0^{T_2} \tfrac{1}{2}(\gamma^2 \|d(t)\|^2 - \|z(t)\|^2)\, dt$$

$$\leq \int_{-T_1}^0 \tfrac{1}{2}(\gamma^2 \|d(t)\|^2 - \|z(t)\|^2)\, dt \qquad (10.77)$$

for all $T_1 \geq 0$, $T_2 \geq 0$, and all disturbance functions $d(\cdot)$ on $[-T_1, T_2]$ *such that the measurements* $y(\cdot)$ *are zero on* $[-T_1, 0]$. Taking in (10.78) the supremum on the left-hand side, and taking the infimum on the right-hand side we obtain

$$\sup_{d(\cdot) \text{ on } [0,T_2],\, T_2 \geq 0} -\int_0^{T_2} \tfrac{1}{2}(\gamma^2 \|d(t)\|^2 - \|z(t)\|^2)\, dt$$

$$\leq \inf \int_{-T_1}^0 \tfrac{1}{2}(\gamma^2 \|d(t)\|^2 - \|z(t)\|^2)\, dt, \qquad (10.78)$$

where the infimum on the right-hand side is taken over all $T_1 \geq 0$ and all disturbance functions $d(\cdot)$ on $[-T_1, 0]$ such that $y(t) = 0$ for $t \in [-T_1, 0]$. We note that the right-hand side of (10.78) is precisely equal to $S_r(x(0))$; the *required supply* from the state $x^* = 0$ (cf. Theorem 3.1.16), *with regard to all* $d(\cdot)$ *on* $[-T_1, 0]$ *such that* $y(t) = 0$, $t \in [-T_1, 0]$. Furthermore, the left-hand side of (10.78) is considered for one *fixed* control strategy $u(\cdot)$ on $[0, T_2]$, namely the one delivered by the output feedback controller C. It follows that for all $T_2 \geq 0$

$$\inf_{u(\cdot)} \sup_{T_2 \geq 0,\, d(\cdot) \text{ on } [0,T_2]} \frac{1}{2} \int_0^{T_2} \|z(t)\|^2 - \gamma^2 \|d(t)\|^2\, dt \leq S_r(x(0)) \qquad (10.79)$$

for all $x(0)$ which can be reached from $x(-T_1) = 0$ by a disturbance function $d(\cdot)$ on $[-T_1, 0]$. Finally, from the theory of (zero-sum) differential games (see, e.g., [325], and the references quoted in the Notes for this chapter it follows that the function $P(x(0))$ determined by the left-hand side of (10.79) is, whenever it is differentiable, a solution of the first Hamilton–Jacobi inequality (in fact, *equality*) (HJ1). Thus we have reached the conclusion that there exists a solution $P \geq 0$ to the first Hamilton–Jacobi inequality (HJ1) such that

$$P(x(0)) \leq S_r(x(0)) \qquad (10.80)$$

for all $x(0)$ reachable from $x = 0$ by a disturbance $d(\cdot)$ leaving the measurements y equal to zero. From dissipative systems theory as treated in Chap. 3 it follows that the function $R(x) := S_r(x)$, whenever it is differentiable and defined for all x, is a solution of the second Hamilton–Jacobi inequality (HJ2). Thus we have derived as a *third* necessary condition for solvability of the output feedback suboptimal $\mathcal{H}_\infty$ control problem that there need to exist solutions $P \geq 0$ and $R \geq 0$ to (HJ1), respectively (HJ2), which satisfy the strong *coupling condition*

$$P(x) \leq R(x), \quad \text{for all } x \in \mathcal{X} \tag{10.81}$$

Furthermore, from Chap. 3 (Theorem 3.1.16) it follows that $R := S_r$ is in fact the *maximal* solution to (HJ2), while the theory of differential games indicates that the left-hand side of (10.79) is the *minimal* solution of the first Hamilton–Jacobi inequality (HJ1). (This has been explicitly proven for *linear* systems; see also Sect. 11.1.)

Example 10.2.2 Consider a system

$$\Sigma_a : \qquad \begin{aligned} \dot{x} &= f(x) + g(x)u, \ u, y \in \mathbb{R}^m, \ f(0) = 0 \\ y &= h(x), \qquad\qquad x \in \mathcal{X}, \qquad h(0) = 0 \end{aligned} \tag{10.82}$$

which is *lossless* (i.e., conservative with respect to the passivity supply rate $u^T y$). Hence there exists a nonnegative storage function (for clarity denoted by $H \geq 0$) such that

$$H_x(x)f(x) = 0, \ H_x(x)g(x) = h^T(x) \tag{10.83}$$

Furthermore assume that $H(0) = 0$, and that the system $\dot{x} = f(x), y = h(x)$ is *zero-state detectable* (Definition 3.2.15).

Consider the output feedback suboptimal $\mathcal{H}_\infty$ control problem associated with the equations

$$\begin{aligned} \dot{x} &= f(x) + g(x)u + g(x)d_1, \quad x \in \mathcal{X}, \ u \in \mathbb{R}^m, \ d_1 \in \mathbb{R}^m \\ y &= h(x) + d_2, \qquad\qquad\qquad y \in \mathbb{R}^m, \ d_2 \in \mathbb{R}^m \\ z &= \begin{bmatrix} h(x) + d_2 \\ u \end{bmatrix}, \end{aligned} \tag{10.84}$$

where the L_2-gain from $d = \begin{bmatrix} d_1 \\ d_2 \end{bmatrix}$ to z is sought to be minimized. Clearly, because of the direct feedthrough term in the z equations, the $\mathcal{H}_\infty$ control problem can never be solved for $\gamma < 1$. The Hamilton–Jacobi inequality (HJ1a) for $\gamma > 1$ takes the form

$$P_x(x)f(x) - \frac{1}{2}\frac{\gamma^2 - 1}{\gamma^2}P_x(x)g(x)g^T(x)P_x^T(x) + \frac{1}{2}\frac{\gamma^2}{\gamma^2 - 1}h^T(x)h(x) \leq 0 \tag{10.85}$$

The stabilizing solution $P \geq 0$ with $P(0) = 0$ to (10.85) is given as

$$P(x) = \frac{\gamma^2}{\gamma^2 - 1} H(x) \qquad (10.86)$$

Indeed, by zero-state detectability the vector fields

$$\dot{x} = f(x) - \alpha b(x) b^T(x) H_x^T(x) \qquad (10.87)$$

for $\alpha > 0$ are all asymptotically stable. In Sect. 11.1 it will be shown that the stabilizing solution to (10.85) is also the *minimal* solution.

The second Hamilton–Jacobi inequality (HJ2a) takes the following form (see (10.71)):

$$R_x(x) f(x) + \frac{1}{2} \frac{1}{\gamma^2} R_x(x) g(x) g^T(x) R_x^T(x) - \frac{1}{2} \gamma^2 h^T(x) h(x) \le 0 \qquad (10.88)$$

Using again zero-state detectability it is seen that

$$R(x) = \gamma^2 H(x) \qquad (10.89)$$

is the antistabilizing solution to (10.88), and thus (see Sect. 11.1) the *maximal* solution. Therefore, the coupling condition tells us that necessarily

$$\frac{\gamma^2}{\gamma^2 - 1} H(x) \le \gamma^2 H(x) \Leftrightarrow \gamma^2 \ge 2 \qquad (10.90)$$

On the other hand, the unity output feedback

$$u = -y \qquad (10.91)$$

yields a closed-loop system with L_2-gain $\le \sqrt{2}$. Indeed, by (10.91),

$$\begin{aligned} ||d_1||^2 + ||d_2||^2 &= ||d_1 + d_2 - y||^2 - (y - d_2)^T(y - d_1) + ||y||^2 \\ &= ||d_1 + d_2 - y||^2 + ||y||^2 + (y - d_2)^T(d_1 + u) \end{aligned} \qquad (10.92)$$

By integrating from 0 to T, and observing that, since the system is lossless, $\int_0^T (y - d_2)^T(d_1 + u) dt = H(x(T)) - H(x(0))$, it follows that

$$\int_0^T ||d(t)||^2 dt \ge \int_0^T ||y(t)||^2 dt + H(x(T)) - H(x(0)), \qquad (10.93)$$

and thus

$$\int_0^T ||z(t)||^2 dt \le 2 \int_0^T ||d(t)||^2 dt + 2H(x(0)) - 2H(x(T)) \qquad (10.94)$$

This implies that the closed-loop system has L_2-gain $\leq \sqrt{2}$ (with storage function equal to H). We conclude that the unity output feedback (10.91) in fact solves the *optimal* output feedback $\mathcal{H}_\infty$ problem, for the *optimal* disturbance attenuation level $\gamma^* = \sqrt{2}$.

Next we elaborate on the necessary *structure* of a controller C solving the output feedback suboptimal $\mathcal{H}_\infty$ control problem. Consider again the dissipation inequality (10.61) for the closed-loop system. Under quite general conditions the left-hand side of (10.61) as a function of the disturbance d is maximized by a certain $\bar{d}^*(x, \xi)$, in which case it reduces to the Hamilton–Jacobi inequality

$$S_x(x, \xi)f(x, \alpha(\xi), \bar{d}^*(x, \xi)) + S_\xi(x, \xi)\varphi(\xi, g(x, \alpha(\xi), \bar{d}^*(x, \xi)))$$
$$-\tfrac{1}{2}\gamma^2 \parallel \bar{d}^*(x, \xi) \parallel^2 + \tfrac{1}{2} \parallel h(x, \alpha(\xi), \bar{d}^*(x, \xi)) \parallel^2 \leq 0 \tag{10.95}$$

for $x \in \mathcal{X}, \xi \in \mathcal{X}_c$. Assume that (10.96) is satisfied with *equality* (this happens, for example, if S is the *available storage* or *required supply*, cf. Chap. 3 and/or Chap. 11). Assume furthermore, as before, that $S_\xi(x, \xi) = 0$ has a solution $\xi = F(x)$. Finally, assume that $P(x) = S(x, F(x))$ and that

$$\alpha(F(x)) = u^*(x, P_x^T(x)), \tag{10.96}$$

where $u^*(x, p)$ is the minimizing input of the *state feedback* problem, that is, satisfies (10.15). It is easily seen that $\bar{d}^*(x, F(x)) = d^*(x, P_x^T(x))$, with d^* satisfying (10.15). Under these three assumptions, differentiate (10.94) with equality with regard to ξ, and substitute afterward $\xi = F(x)$. Because of the maximizing, respectively minimizing, property of d^* and u^*, differentiating to ξ via d^* and α in $\xi = F(x)$ yields zero. Thus what remains is

$$S_{\xi x}(x, F(x))f(x, u^*(x, P_x^T(x)), d^*(x, P_x^T(x)))$$
$$+ S_{\xi\xi}(x, F(x))\varphi(F(x), g(x, u^*(x, P_x^T(x)), d^*(x, P_x^T(x)))) = 0 \tag{10.97}$$

Furthermore, since $S_\xi(x, F(x)) = 0$ for all x, differentiation of this expression with respect to x yields

$$S_{x\xi}(x, F(x)) + S_{\xi\xi}(x, F(x))F_x(x) = 0 \tag{10.98}$$

Combination of (10.97) and (10.98) leads to

$$S_{\xi\xi}(x, F(x))\varphi(F(x), g(x, u^*(x, P_x^T(x)), d^*(x, P_x^T(x)))) =$$
$$S_{\xi\xi}(x, F(x))F_x(x)f(x, u^*(x, P_x^T(x)), d^*(x, P_x^T(x))) \tag{10.99}$$

Thus, imposing the fourth assumption that the Hessian matrix $S_{\xi\xi}(x, F(x))$ is non-singular, it follows that

$$F_x(x)f(x, u^*(x, P_x^T(x)), d^*(x, P_x^T(x))) = \varphi(F(x), y) \tag{10.100}$$

with

$$y = g(x, u^*(x, P_x^T(x)), d^*(x, P_x^T(x))) \tag{10.101}$$

This constitutes an *invariance principle*. Indeed, whenever $\xi(t_0) = F(x(t_0))$, then during the occurrence of the *worst-case disturbance* $d^*(x(t), P_x^T(x(t)))$, it follows that $F_x(x(t))\dot{x}(t) = \dot{\xi}(t)$, and thus $\xi(t) = F(x(t))$ for $t \geq t_0$. Therefore, the controller C has a *certainty equivalence property*. This is especially clear if F is a diffeomorphism, in which case we may choose coordinates ξ for $\mathcal{X}_c$ such that F is the *identity* mapping. Then C is an *observer*, and $\xi(t)$ is an *estimate* for the actual state $x(t)$ (for the worst-case disturbance), while the control u is given as the state feedback $u^*(x, P_x^T(x))$ with the actual state x replaced by this estimate ξ.

Finally, in the last part of this section we will *transform* the output feedback suboptimal $\mathcal{H}_\infty$ control problem for Σ into an output feedback suboptimal $\mathcal{H}_\infty$ control problem for *another* system $\widetilde{\Sigma}$, based on the solvability of the *state feedback* suboptimal $\mathcal{H}_\infty$ control problem. By this we may focus on the inherent difficulties associated with the *output feedback* problem. Furthermore, the transformation will lead us to a *parametrization* of the controllers solving the output feedback problem. We will show that the transformation is very close to the all-pass factorization treated in Sect. 9.4.

We start by considering a general system Σ, as given by Eq. (10.59), and we assume the existence of a solution $P \geq 0$ to the state feedback Hamilton–Jacobi *equality* (HJ1), that is,

$$K_\gamma(x, P_x^T(x), d^*(x, P_x^T(x)), u^*(x, P_x^T(x))) = 0 \tag{10.102}$$

with d^*, u^* satisfying (10.15), and K_γ given as in (10.19). Now consider the function $K_\gamma(x, P_x^T(x), d, u)$. By (10.102) and (10.20), we have

$$K_\gamma(x, P_x^T(x), d, u^*(x, P_x^T(x))) \leq 0 \leq K_\gamma(x, P_x^T(x), d^*(x, P_x^T(x)), u) \tag{10.103}$$

Now assume that there exist mappings

$$\begin{aligned} r &= r(x, u, d) \\ v &= v(x, u, d) \end{aligned} \tag{10.104}$$

such that

$$K_\gamma(x, P_x^T(x), d, u) = -\frac{1}{2}\gamma^2||r||^2 + \frac{1}{2}||v||^2 \tag{10.105}$$

If the system equations are given by (10.68), then the factorization (10.105) is in fact easy to obtain by completing the squares, and is given as

$$r_1 = d_1 - \tfrac{1}{\gamma^2}g^T(x)P_x^T(x), \quad r_2 = d_2$$
$$v = u + b^T(x)P_x^T(x) \tag{10.106}$$

with $P \geq 0$ satisfying (HJ1a). In general, the factorization is of the same type as the one considered in Sect. 9.4 (the main difference being the *indefinite* sign). *Local* existence of the factorization can be again guaranteed by an application of the Morse Lemma if the Hessian matrix of $K_\gamma(x, P_x^T(x), d, u)$ with respect to d and u is nonsingular (see Lemma 9.4.2 in Chap. 9). Let us additionally assume that the mapping given by (10.104) is *invertible* in the sense that d can be expressed as function of u and r (and x), and u as function of v and d (and x), i.e.,

$$d = d(x, u, r)$$
$$u = u(x, v, d) \tag{10.107}$$

This assumption is trivially satisfied for (10.106), since in this case we may write

$$d_1 = r_1 + \tfrac{1}{\gamma^2}g^T(x)P_x^T(x), \quad d_2 = r_2$$
$$u = v - b^T(x)P_x^T(x) \tag{10.108}$$

The system Σ is now factorized as in Fig. 10.5, with $\widetilde{\Sigma}$ denoting the transformed system

$$\dot{\tilde{x}} = f(\tilde{x}, u, d(\tilde{x}, u, r)) =: \tilde{f}(\tilde{x}, u, r)$$

$$\widetilde{\Sigma} : \quad y = g(\tilde{x}, u, d(\tilde{x}, u, r)) =: \tilde{g}(\tilde{x}, u, r) \tag{10.109}$$

$$v = v(\tilde{x}, u, d(\tilde{x}, u, r)) =: \tilde{h}(\tilde{x}, u, r)$$

while the system Θ is given as

$$\dot{\theta} = f(\theta, u(\theta, v, d), d)$$

$$\Theta : \quad r = r(\theta, u(\theta, v, d), d) \tag{10.110}$$

$$z = h(\theta, u(\theta, v, d), d)$$

Fig. 10.5 Factorization of Σ

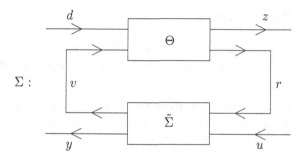

It can be readily seen that if $\tilde{x}(t_0) = \theta(t_0) = x(t_0)$, then also $\tilde{x}(t) = \theta(t) = x(t)$, $t \geq t_0$, and that $z(t)$ and $y(t)$ produced in Fig. 10.5 coincide with $z(t)$ and $y(t)$ produced by Σ for the same $u(t)$ and $d(t)$. Thus Fig. 10.5 indeed constitutes a valid factorization of Σ. Furthermore, the following proposition is immediately derived from (10.105).

Proposition 10.2.3 *Let $P \geq 0$ satisfy (10.102). Consider the factorization of Σ given in Fig. 10.5. Then for all $d(\cdot)$, $v(\cdot)$, and all $t_1 \geq t_0$ and $\theta(t_0)$*

$$P(\theta(t_1)) - P(\theta(t_0)) + \tfrac{1}{2}\int_{t_0}^{t_1}(||z(t)||^2 - \gamma^2||d(t)||^2)dt = \tfrac{1}{2}\int_{t_0}^{t_1}(||v(t)||^2 - \gamma^2||r(t)||^2)dt \tag{10.111}$$

or equivalently,

$$P(\theta(t_1)) - P(\theta(t_0)) = \tfrac{1}{2}\int_{t_0}^{t_1}(||v(t)||^2 + \gamma^2||d(t)||^2)\,dt - \tfrac{1}{2}\int_{t_0}^{t_1}(||z(t)||^2 + \gamma^2||r(t)||^2)\,dt \tag{10.112}$$

Thus Θ is inner (Definitions 3.1.6 and 8.1.1) from $\begin{bmatrix} \gamma d \\ v \end{bmatrix}$ to $\begin{bmatrix} z \\ \gamma r \end{bmatrix}$.

A first consequence of Proposition 10.2.3 is that the solution of the output feedback suboptimal $\mathcal{H}_\infty$ control problem for Σ can be reduced, in a certain sense, to the same problem for $\tilde{\Sigma}$:

Proposition 10.2.4 *Let $P \geq 0$ satisfy (10.102). Consider the factorization of Fig. 10.5.*

(i) A controller C which solves the output feedback suboptimal $\mathcal{H}_\infty$ control problem for $\tilde{\Sigma}$ also solves the same problem for Σ.

(ii) Suppose the controller C solves the output feedback suboptimal $\mathcal{H}_\infty$ control problem for Σ with a storage function $S(x, \xi)$ satisfying

$$S(x, \xi) - P(x) \geq 0, \quad x \in \mathcal{X}, \xi \in \mathcal{X}_c \tag{10.113}$$

Then the same controller also solves the output feedback suboptimal $\mathcal{H}_\infty$ control problem for $\tilde{\Sigma}$ with storage function $S(\tilde{x}, \xi) - P(\tilde{x})$.

Proof (i) Rewrite (10.111) as

$$\tfrac{1}{2}\int_0^T(||z(t)||^2 - \gamma^2||d(t)||^2)dt = \tfrac{1}{2}\int_0^T(||v(t)||^2 - \gamma^2||r(t)||^2)dt + P(\theta(0)) - P(\theta(T)) \tag{10.114}$$

If a controller C bounds the first term on the right-hand side by a constant depending on the initial conditions $\tilde{x}(0)$, $\xi(0)$, then the same holds for the left-hand side (since $P(\theta(T)) = P(x(T)) \geq 0$).

(ii) If C solves the output feedback suboptimal $\mathcal{H}_\infty$ control problem for Σ, then $\frac{1}{2}\int_0^T (\|z(t)\|^2 - \gamma^2\|d(t)\|^2)dt \le S(x(0), \xi(0))$ and thus by (10.114)

$$\frac{1}{2}\int_0^T (\|v(t)\|^2 - \gamma^2\|r(t)\|^2)dt \le S(\tilde{x}(0), \xi(0)) - P(\tilde{x}(0)) \qquad (10.115)$$

$\square$

Remark 10.2.5 Note that for part (i) P only needs to satisfy the state feedback Hamilton–Jacobi *inequality* (HJ1).

Since we have already used in the transformation from Σ to $\tilde{\Sigma}$ the knowledge of the existence of a solution $P \ge 0$ to the state feedback Hamilton–Jacobi equation (HJ1) we may *expect* that the output feedback suboptimal $\mathcal{H}_\infty$ control problem for $\tilde{\Sigma}$ will be "easier" than the same problem for Σ. At least the solution of the *state feedback* $\mathcal{H}_\infty$ control problem for $\tilde{\Sigma}$ has become trivial; $u = u^*(x, P_x^T(x))$ and $d^* = d^*(x, P_x^T(x))$ solve the equations

$$\begin{aligned} 0 &= v(x, u, d) \\ 0 &= r(x, u, d), \end{aligned} \qquad (10.116)$$

and thus yield a trivial solution to the state feedback $\mathcal{H}_\infty$ control problem for $\tilde{\Sigma}$. In particular, in the affine case, where r and v are given by (10.106), the state feedback $u = -b^T(x)P_x^T(x)$ renders v equal to zero and thus solves the *disturbance decoupling* problem for $\tilde{\Sigma}$.

A second consequence of the factorization in Fig. 10.5 and Proposition 10.2.3 concerns the parametrization of controllers solving the output feedback suboptimal $\mathcal{H}_\infty$ problem. Consider a controller C in closed-loop with the factorization of Fig. 10.6. Denote the system within dotted lines by K. Then it follows from Proposition 10.2.3 that if K has L_2-gain $\le \gamma$ (from r to v), then Θ in closed-loop with the "controller" K will also have L_2-gain $\le \gamma$ (from d to z). We may also *reverse* this relation. Consider an auxiliary system Q with inputs r and outputs v

$$Q : \begin{aligned} \dot{q} &= f_Q(q, r) \\ v &= h_Q(q, r) \end{aligned} \qquad (10.117)$$

and suppose Q has L_2-gain $\le \gamma$. Thus, there exists a storage function $S_Q(q) \ge 0$ such that along the trajectories of Q

$$S_Q(q(T)) - S_Q(q(0)) \le \frac{1}{2}\int_0^T (\gamma^2\|r(t)\|^2 - \|v(t)\|^2)dt \qquad (10.118)$$

Now consider this system Q in conjunction with the system $\tilde{\Sigma}$, i.e.,

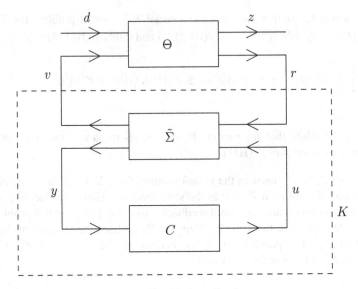

Fig. 10.6 Factorized system in closed-loop with controller C

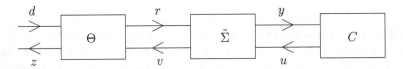

Fig. 10.7 Chain-scattering representation

$$\dot{q} = f_Q(q, r)$$
$$\dot{\tilde{x}} = \tilde{f}(\tilde{x}, u, r)$$
$$y = \tilde{g}(\tilde{x}, u, r) \tag{10.119}$$
$$h_Q(q, r) = \tilde{h}(\tilde{x}, u, r)$$

The idea is now to look at (10.119) as a *generalized form* of *image representation* (with driving variable r) of a *controller* C_Q (producing controls u on the basis of the measurements y). By construction this *implicitly defined* controller C_Q solves the output feedback suboptimal $\mathcal{H}_\infty$ control problem for Σ. Thus for every system Q with L_2-gain $\leq \gamma$ we obtain in this way a controller C_Q solving the $\mathcal{H}_\infty$ problem.

A more explicit way of describing these controllers C can be obtained by rewriting the configuration of Fig. 10.6 into the form of Fig. 10.7.

Assume that Θ and $\tilde{\Sigma}$ are *invertible* in the sense that Θ admits an input–output representation with inputs r and v and outputs d and z, and that $\tilde{\Sigma}$ admits an input–output representation with inputs y and u and outputs r and v. In this case one speaks about the *chain-scattering representation*. For more details, we refer to the literature cited in the Notes for this chapter.

10.3 Notes for Chapter 10

1. The computational complexity of obtaining (approximate) solutions to (HJ1) and/or (HJ2) is a major issue for the applicability of the theory. Power series solutions around an equilibrium are described in van der Schaft [275], Isidori and Kang [142], continuing on earlier similar approaches in nonlinear optimal control by Al'brekht [5] and Lukes [187]. See Krener, Aguilar & Hunt [172], Aguilar & Krener [4] for further developments, including the Toolbox by Krener [171]. Various other computational schemes have been proposed and analyzed in the literature; we mention Knobloch, Isidori & Flockerzi [166], McEneaney & Mease [210], Huang & Lin [130], Kraim, Kugelmann, Pesch & Breitner [169], Beard, Saridis & Wen [33], Beard, Saridis & Wen [34], Beard & McLain [32], Beeler, Tran & Banks [35], Sakamoto [263], Holzhüter [129], Osinga & Hauser [245], Navasca & Krener [229], Sakamoto & van der Schaft [265], Sakamoto & van der Schaft [266], Sassano & Astolfi [268], Sakamoto [264].

2. For general information concerning the $\mathcal{H}_\infty$ control problem for *linear* systems we refer to, e.g., Francis [103], Green & Limebeer [117], Kwakernaak [178], Scherer [305], Stoorvogel [327]. The state space solution to the linear suboptimal $\mathcal{H}_\infty$ control problem is due to Doyle, Glover, Khargonekar & Francis [88], see also, e.g., Khargonekar, Petersen & Rotea [162], Scherer [304], Tadmor [333]. For the differential game approach to linear $\mathcal{H}_\infty$ control theory see especially Basar & Bernhard [30].

3. The solution to the nonlinear suboptimal $\mathcal{H}_\infty$ control problem as described in Sect. 10.1 was given in van der Schaft [272, 273, 275], Isidori & Astolfi [140], Ball, Helton & Walker [23]. $\mathcal{H}_\infty$ control for general nonlinear systems (not necessarily affine in the inputs and disturbances) is treated in Isidori & Kang [142], Ball, Helton & Walker [23], see also van der Schaft [277]. For earlier work on nonlinear $\mathcal{H}_\infty$ control we refer to Ball & Helton [19, 20].

4. The treatment of the robust stabilization problem in Sect. 10.1 is based on van der Schaft [279]; see also Imura, Maeda, Sugie & Yoshikawa [131], and Astolfi & Guzzella [14], Pavel & Fairman [252] for other developments.

5. The existence of nonnegative solutions to (HJ1) and (HJ2), together with the (weak) coupling condition (10.76), as a necessary condition for the solvability of the nonlinear output feedback $\mathcal{H}_\infty$ control problem was shown in Ball, Helton & Walker [23], van der Schaft [276]. The key idea of deriving (HJ2) for linear systems via the dissipativity of the *constrained* system ($u = 0$, $y = 0$) is due to Khargonekar [161].

6. The invariance property of controllers solving the nonlinear suboptimal $\mathcal{H}_\infty$ problem, as described in Sect. 10.1, is due to Ball, Helton & Walker [23], with the present generalization given in van der Schaft [278].

7. As already shown in Example 10.1.9 it is often necessary, as indicated in Chaps. 3, 8 and 9, to consider *generalized* solutions of the Hamilton–Jacobi inequalities encountered in this chapter; see, e.g., Frankowska [105], James & Baras [145], Ball & Helton [22], Soravia [325], and Yuliar, James & Helton [361], and Day [80] for further information.

8. Much effort has been devoted to finding *sufficient* conditions for solvability of the output feedback suboptimal $\mathcal{H}_\infty$ control problem, but we have decided not to include these important contributions in this book. One line of research is devoted to finding sufficient conditions for the existence of output feedback controllers with dimension equal to the dimension of the plant and having an observer structure (compare with the invariance principle in Sect. 10.2), cf. Isidori & Astolfi [140], Ball, Helton & Walker [23], Isidori [136], Lu & Doyle [185, 186], Isidori & Kang [142]. Another approach is via the theory of differential games (see, e.g., Basar & Olsder [31]), interpreting the output feedback suboptimal $\mathcal{H}_\infty$ control problem as a two-player zero-sum differential game with partial information, see Basar & Bernhard [30], Bernhard [37], Soravia [325]. The resulting "central" controller, however, is in general infinite-dimensional, see Didinsky, Basar & Bernhard [84], van der Schaft [278], Bernhard [37], James & Baras [145]. Under the assumption of a "worst-case certainty equivalence principle" the equations for the central controller are derived in Didinsky, Basar & Bernhard [84], van der Schaft [278], Krener [170], Bernhard [37]. The methods and difficulties in this case are very similar to the ones encountered in nonlinear filtering theory, see, e.g., Mortensen [224], Hijab [121]. Another, more general, approach is to transform the differential game with partial information into an infinite-dimensional differential game with complete information, see, e.g., Helton & James [120]. This "information-state" approach has been detailed in the monograph by Helton & James [119]. For developments exploiting Max-Plus methods see McEneaney [209].

9. Example 10.2.2 is taken from van der Schaft [278].

10. If the system $\tilde{\Sigma}$ as obtained in Fig. 10.5 is stable and minimum phase, then the factorization obtained in Fig. 10.5 is also called an *J-inner-outer factorization* of Σ. A constructive approach to J-inner-outer factorization of Σ (under the assumption of invertibility from d to y) is given in Ball & van der Schaft [24], using the Hamiltonian extension Σ^H of Σ and "nonlinear spectral factorization"; cf. Sect. 9.4. Related work is Baramov & Kimura [29], Pavel & Fairman [253] and Baramov [28]. For further information on the chain-scattering representation, see Kimura [165], Ball & Helton [20], Ball & Verma [26], Ball & Helton [22]. The presentation of the parametrization of $\mathcal{H}_\infty$ controllers given in Sect. 10.2 is inspired by Ji & Gao [148]; see Doyle, Glover, Khargonekar & Francis [88], for

similar ideas in the linear case. For other related works on the parametrization of $\mathcal{H}_\infty$ controllers, see Lu & Doyle [186].

11. For work on nonlinear $\mathcal{H}_\infty$-filtering we refer to, e.g., Nguang & Fu [231], Berman & Shaked [36], Krener [170].

12. In case the control variables u do not enter the equations for z in an "injective way," or if the disturbance variables d do not enter the equations for y in a "surjective way," then we speak about the *singular* $\mathcal{H}_\infty$ control problem; see for the linear case Stoorvogel [327], and the references included therein. For a treatment of the nonlinear state feedback singular $\mathcal{H}_\infty$ control problem we refer to Maas & van der Schaft [189], Maas [188], Baramov [28]. The nonlinear $\mathcal{H}_\infty$ almost disturbance decoupling problem, which can be seen as a special case of the singular $\mathcal{H}_\infty$ control problem has been treated for a special class of systems in Marino, Respondek, van der Schaft & Tomei [196]. An interesting feature of the latter paper is that the solution to the dissipation inequality is *constructed* in an explicit recursive manner, thus avoiding the issue of solvability of multidimensional Hamilton–Jacobi inequalities. The results of [196] are vastly expanded in Isidori [137, 138]. The idea of solving Hamilton–Jacobi inequalities in a recursive manner for multiple cascaded systems is explored in Pan & Basar [250]; see also Dalsmo & Maas [77] for related work.

13. Some applications of nonlinear $\mathcal{H}_\infty$ control theory can be found in Kang, De & Isidori [157], Chen, Lee & Feng [65], Feng & Postlethwaite [96], Astolfi & Lanari [15], Kang [156], Dalsmo & Egeland [75, 76], Kugi & Schlacher [176].

14. The nonlinear suboptimal $\mathcal{H}_\infty$ control problem for $\gamma \geq 0$ is the problem of finding a controller such that the closed-loop system is dissipative with respect to the L_2-gain supply rate $\frac{1}{2}\gamma^2||d||^2 - \frac{1}{2}||z||^2$. This suggests to consider the *general dissipative control* problem of finding a controller which renders the closed-loop system dissipative with respect to a given supply rate $s(d, z)$; see Yuliar & James [360], Yuliar, James & Helton [361], Shishkin & Hill [314]. This includes the problem of rendering a system passive as treated in Chap. 5. The problem of considering "finite power" instead of finite L_2-gain is addressed in Dower & James [86].

15. Using the notion of robust L_2-gain (cf. Chap. 8, Corollary 8.2.6) one can also consider a "robust" nonlinear $\mathcal{H}_\infty$ control problem, see Shen & Tamura [313], Nguang [232]. A robustness result concerning the solution to the state feedback suboptimal $\mathcal{H}_\infty$ control problem with static perturbations on the inputs is derived in van der Schaft [278] (Proposition 4.7).

Chapter 11
Hamilton–Jacobi Inequalities

In the previous chapters, we have encountered at various places Hamilton–Jacobi equations, or, more generally, Hamilton–Jacobi inequalities. In this chapter, we take a closer look at conditions for *solvability* of Hamilton–Jacobi inequalities and the structure of their *solution set* using invariant manifold techniques for the corresponding Hamiltonian vector field (Sect. 11.1). In Sect. 11.2 we apply this to the nonlinear optimal control problem. An important theme will be the relation between Hamilton–Jacobi inequalities and the Riccati inequalities obtained by linearization, in particular for dissipativity (Sect. 11.3), and nonlinear $\mathcal{H}_\infty$ control (Sect. 11.4).

11.1 Solvability of Hamilton–Jacobi Inequalities

In Chaps. 3, 4, and 8 we considered differential dissipation inequalities

$$S_x(x) f(x, u) \leq s(u, h(x, u)), \quad \forall x, u \tag{11.1}$$

in an (unknown) storage function S. If the corresponding *pre-Hamiltonian*

$$K(x, p, u) := p^T f(x, u) - s(u, h(x, u)) \tag{11.2}$$

has a maximizing $u^*(x, p)$, that is

$$K(x, p, u) \leq K(x, p, u^*(x, p)), \quad \forall x, p, u, \tag{11.3}$$

then the dissipation inequality (11.1) is equivalent to

$$K(x, S_x^T(x), u^*(x, S_x^T(x))) \leq 0, \quad \forall x \tag{11.4}$$

© Springer International Publishing AG 2017
A. van der Schaft, *L2-Gain and Passivity Techniques in Nonlinear Control*,
Communications and Control Engineering, DOI 10.1007/978-3-319-49992-5_11

Defining in this case the *Hamiltonian*

$$H(x, p) := K(x, p, u^*(x, p)) \tag{11.5}$$

we are thus led to *Hamiltonian–Jacobi inequalities*

$$H(x, S_x^T(x)) \le 0, \quad \forall x \tag{11.6}$$

in the unknown S. Also in Chap. 9, we encountered Hamilton–Jacobi inequalities (in fact *equations*) in the factorization of nonlinear systems, while in Chap. 10 Hamilton–Jacobi inequalities turned out to be key in the study of the (sub-)optimal $\mathcal{H}_\infty$ control problem.

In this section, we will be concerned with deriving conditions for the solvability of Hamilton–Jacobi inequalities, and with the structure of their solution set. Many of the results presented in this section will not be proven here; proofs can be found in the references cited in the Notes at the end of this chapter.

We start on a general level. Consider an n-dimensional manifold M with local coordinates $x = (x_1, \ldots, x_n)$. The *cotangent bundle* T^*M is a $2n$-dimensional manifold, with *natural* local coordinates $(x, p) = (x_1, \ldots, x_n, p_1, \ldots, p_n)$ defined in the following way. Take any set of local coordinates $x = (x_1, \ldots, x_n)$ for M. Let σ be any one-form on M (i.e., $\sigma(q) \in T_q^*M$ is a cotangent vector for every $q \in M$), in the coordinates $x = (x_1, \ldots, x_n)$ for M expressed as

$$\sigma = \sigma_1 dx_1 + \sigma_2 dx_2 + \cdots + \sigma_n dx_n \tag{11.7}$$

for certain smooth functions $\sigma_1(x), \ldots, \sigma_n(x)$. Then, the natural coordinates $(x_1, \ldots, x_n, p_1, \ldots, p_n)$ for T^*M are defined by attaching to $\sigma(q)$ the coordinate values

$$(x_1(q), \ldots, x_n(q), \sigma_1(x(q)), \ldots, \sigma_n(x(q))),$$

$$\text{with } x_i(\sigma(q)) = x_i(q), \ p_i(\sigma(q)) = \sigma_i(q), \ i = 1, \ldots, n \tag{11.8}$$

Given the natural coordinates $(x_1, \ldots, x_n, p_1, \ldots, p_n)$ for T^*M we may locally define the *canonical two-form* ω on T^*M as

$$\omega = \sum_{i=1}^{n} dp_i \wedge dx_i \tag{11.9}$$

The two-form ω is called the (canonical) symplectic form on the cotangent bundle T^*M.

Definition 11.1.1 An n-dimensional submanifold N of T^*M is *Lagrangian* if ω restricted to N is zero.

Now consider any C^2 function $S : M \to \mathbb{R}$, and the n-dimensional submanifold $N_S \subset T^*M$, in local coordinates given as

$$N_S = \left\{ (x, p) \in T^*M \mid p_i = \frac{\partial S}{\partial x_i}(x), i = 1, \ldots, n \right\} \tag{11.10}$$

It can be immediately checked that N_S is *Lagrangian*. (In fact, this amounts to the property $\frac{\partial^2 S}{\partial x_i \partial x_j}(x) = \frac{\partial^2 S}{\partial x_j \partial x_i}(x), i, j = 1, \ldots, n$.) Conversely, defining the canonical projection

$$\pi : T^*M \to M, \quad (x, p) \overset{\pi}{\mapsto} x, \tag{11.11}$$

we obtain by Poincaré's lemma (see, e.g., [1]):

Proposition 11.1.2 *Let N be a C^{k-1} Lagrangian submanifold of T^*M such that $\pi : N \to M$ is a C^{k-1} diffeomorphism. Then locally (or globally if M is simply connected) there exists a C^k function $S : M \to \mathbb{R}$ such that $N = N_S$.*

The property that $\pi : N \to M$ is a C^{k-1} diffeomorphism will be referred as "*parametrizability of N by the x-coordinates*". Now take any C^k function $H : T^*M \to \mathbb{R}$ (not necessarily of the special type as obtained in (11.5)), and consider the Hamilton–Jacobi *equality* (equation)

$$H(x, S_x^T(x)) = 0 \tag{11.12}$$

in the unknown $S : M \to \mathbb{R}$. The Hamiltonian vector field X_H on T^*M corresponding to the Hamiltonian H is defined in natural coordinates as

$$\begin{aligned} \dot{x}_i &= \frac{\partial H}{\partial p_i}(x, p) \\ \dot{p}_i &= -\frac{\partial H}{\partial x_i}(x, p) \end{aligned} \qquad i = 1, \ldots, n \tag{11.13}$$

There is a close connection between solutions of the Hamilton–Jacobi equation (11.12) and invariant submanifolds of the Hamiltonian vector field (11.13). Recall (see Chap. 3, Theorem 3.2.8) that a submanifold $N \subset T^*M$ is an *invariant manifold* for X_H if the solutions of (11.13) starting on N *remain in N.*

Proposition 11.1.3 *Let $S : M \to \mathbb{R}$, and consider the submanifold $N_S \subset T^*M$ locally given by (11.10). Then*

$$H(x, S_x^T(x)) = constant, for all x \in M, \tag{11.14}$$

if and only if N_S is an invariant submanifold for X_H.

Note that by subtracting a constant value from H (not changing the Hamiltonian vector field X_H), we may always reduce (11.14) to (11.12). Solutions of (11.12) may thus be obtained by looking for invariant Lagrangian submanifolds of X_H which are parametrizable by the x-coordinates and thus, by Proposition 11.1.2, of the form N_S for some S. Not every n-dimensional invariant submanifold of X_H is Lagrangian, but the following two *special* invariant submanifolds of X_H always *are*. Consider an equilibrium (x_0, p_0) of X_H, that is

$$\frac{\partial H}{\partial x_i}(x_0, p_0) = \frac{\partial H}{\partial p_i}(x_0, p_0) = 0, \quad i = 1, \ldots, n. \tag{11.15}$$

Define $N^- \subset T^*M$ as the set of all points in T^*M converging along the solutions of the vector field X_H for $t \to \infty$ to (x_0, p_0), and N^+ as the set of all points converging in *negative* time to (x_0, p_0). Clearly, N^- and N^+ are invariant sets for X_H.

Proposition 11.1.4 N^- and N^+ are submanifolds of T^*M, called the stable invariant manifold of X_H, respectively the unstable invariant manifold of X_H. Furthermore ω restricted to N^- and to N^+ is zero. Hence if $\dim N^- = n$ ($\dim N^+ = n$), then N^- (respectively N^+) is a Lagrangian submanifold of T^*M.

Under additional conditions on the *linearization* of X_H at an equilibrium (x_0, p_0) we can be more explicit. Consider the linearization of X_H at (x_0, p_0), that is the $2n \times 2n$ matrix

$$DX_H(x_0, p_0) = \begin{bmatrix} \frac{\partial^2 H}{\partial x \partial p}(x_0, p_0) & \frac{\partial^2 H}{\partial p^2}(x_0, p_0) \\ -\frac{\partial^2 H}{\partial x^2}(x_0, p_0) & -\frac{\partial^2 H}{\partial p \partial x}(x_0, p_0) \end{bmatrix}$$

$$\tag{11.16}$$

$$=: \begin{bmatrix} A & P \\ -Q & -A^T \end{bmatrix} =: \mathcal{H}$$

The matrix $\mathcal{H} = DX_H(x_0, p_0)$ is a *Hamiltonian matrix*, that is

$$\mathcal{H}^T J + J \mathcal{H} = 0, \tag{11.17}$$

with J the linear symplectic form

$$J = \begin{bmatrix} 0 & -I_n \\ I_n & 0 \end{bmatrix} \tag{11.18}$$

It follows from (11.17) that if λ is an eigenvalue of $\mathcal{H}$, then so is $-\lambda$, and therefore the set of eigenvalues of $\mathcal{H}$ is symmetric with regard to the imaginary axis.

The equilibrium (x_0, p_0) is called *hyperbolic* if $\mathcal{H} = DX_H(x_0, p_0)$ does not have purely imaginary eigenvalues. This results in

Proposition 11.1.5 Let (x_0, p_0) be a hyperbolic equilibrium of X_H. Then N^- and N^+ are Lagrangian submanifolds, and N^- (respectively N^+) is tangent at (x_0, p_0) to the stable (respectively unstable) generalized eigenspace of $DX_H(x_0, p_0)$.

The linearization at (x_0, p_0) of the Hamiltonian vector field X_H given by (11.13) is given by

$$\begin{bmatrix} \dot{x} \\ \dot{p} \end{bmatrix} = \mathcal{H} \begin{bmatrix} x \\ p \end{bmatrix}, \tag{11.19}$$

which is by itself a linear Hamiltonian vector field with respect to the linear symplectic form J (which can be understood as the evaluation of ω at (x_0, p_0)) and the

Hamiltonian

$$p^T A x + \frac{1}{2} p^T P p + \frac{1}{2} x^T Q x, \tag{11.20}$$

consisting of the *quadratic* part of H. Equivalently, the symmetric matrix

$$\begin{bmatrix} Q & A^T \\ A & P \end{bmatrix} \tag{11.21}$$

is the Hessian of H at (x_0, p_0).

The Hamilton–Jacobi equation (11.12) for this quadratic Hamiltonian reduces to the *Riccati equation*

$$A^T X + X A + X P X + Q = 0, \tag{11.22}$$

in the unknown symmetric $n \times n$ matrix X (not to be confused with the Hamiltonian vector field X_H). Conversely, the solutions $S(x)$ to (11.12) for a quadratic H given by (11.20) may be restricted to quadratic functions $S(x) = \frac{1}{2} x^T X x$, thus leading to (11.22).

Similarly to Definition 11.1.1, a Lagrangian *subspace* L of $\mathbb{R}^{2n}$ is an n-dimensional subspace such that J restricted to L is zero. If the Lagrangian subspace L is parametrizable by the x-coordinates (meaning that L and span $\begin{bmatrix} 0 \\ I_n \end{bmatrix}$ are complementary, i.e., $L \oplus \operatorname{span} \begin{bmatrix} 0 \\ I_n \end{bmatrix} = \mathbb{R}^{2n}$), then as in Proposition 11.1.2 there exists $X = X^T$ such that

$$L = \operatorname{span} \begin{bmatrix} I \\ X \end{bmatrix} \tag{11.23}$$

Furthermore

Proposition 11.1.6 *Let (x_0, p_0) be an equilibrium of X_H. Suppose $N \subset T^*M$ is an invariant Lagrangian submanifold of X_H through (x_0, p_0). Then, the tangent space L to N at (x_0, p_0) is a Lagrangian subspace of $\mathbb{R}^{2n}$ which is an invariant subspace of $DX_H(x_0, p_0)$. In particular, if S is a solution to (11.12) with $\frac{\partial S}{\partial x_i}(x_0) = 0, i = 1, \dots, n$, then the Hessian matrix $X := S_{xx}(x_0)$ is a solution to (11.22).*

Suppose now we have found an invariant Lagrangian submanifold N of X_H through an equilibrium (x_0, p_0), for example, N^- or N^+. Then in view of Proposition 11.1.6 the question of parametrizability by the x-coordinates, and thus by Proposition 11.1.2 the existence of a solution S to the Hamilton–Jacobi equality (11.12), can be locally checked by investigating the parametrizability of the Lagrangian subspace L tangent to N at (x_0, p_0). For this *linear* problem we may invoke the following proposition.

Proposition 11.1.7 *Consider the Hamiltonian matrix $\mathcal{H}$ given by (11.18). Let P be either ≥ 0 or ≤ 0.*

(i) *If (A, P) is controllable, then every Lagrangian subspace of $\mathbb{R}^{2n}$ which is invariant for $\mathcal{H}$ is complementary to span $\begin{bmatrix} 0 \\ I_n \end{bmatrix}$ (that is, parametrizable by the x-coordinates).*

(ii) *Assume that $\mathcal{H}$ does not have purely imaginary eigenvalues, implying that the stable eigenspace L^- of $\mathcal{H}$, as well as the unstable eigenspace L^+, are Lagrangian. If (A, P) is stabilizable, then L^- and span $\begin{bmatrix} 0 \\ I_n \end{bmatrix}$ are complementary. If $(-A, P)$ is stabilizable, then L^+ and span $\begin{bmatrix} 0 \\ I_n \end{bmatrix}$ are complementary.*

With regard to the Hamilton–Jacobi equation (11.12) we derive the following corollary.

Corollary 11.1.8 *Let $H : T^*M \to \mathbb{R}$ be a Hamiltonian function, with Hamilton–Jacobi equation (11.12), and Hamiltonian vector field X_H, satisfying $\frac{\partial H}{\partial x}(x_0, p_0) = \frac{\partial H}{\partial p}(x_0, p_0) = 0$, and with linearization $DX_H(x_0, p_0) = \mathcal{H}$ given by (11.18). Assume $P = \frac{\partial^2 H}{\partial p^2}(x_0, p_0)$ is either ≥ 0 or ≤ 0.*

(i) *If (A, P) is controllable, then every Lagrangian invariant submanifold of X_H is locally near (x_0, p_0) of the form N_S for a certain function $S(x)$ defined for x near x_0.*

(ii) *Let (x_0, p_0) be a hyperbolic equilibrium. If (A, P) is stabilizable then locally near (x_0, p_0) the stable invariant manifold N^- is given as N_{S^-} for a certain function $S^-(x)$ defined near x_0. If $(-A, P)$ is stabilizable then N^+ is given as N_{S^+} for a certain $S^+(x)$ defined near x_0.*

Let us now assume that the solutions S^- and S^+ to the Hamilton–Jacobi equation (11.12), corresponding to the stable, respectively unstable, invariant manifold, exist *globally*. It is to be expected that they have special properties among the set of *all* solutions to the Hamilton–Jacobi equation or even among the set of all solutions to the Hamilton–Jacobi inequality (11.6). In fact, for Hamiltonians H arising from dissipation inequalities we obtain the following results.

Proposition 11.1.9 *Consider a Hamiltonian function H given as (cf. (11.5)) $H(x, p) := K(x, p, u^*(x, p))$, where*

$$K(x, p, u) = p^T f(x, u) - s(u, h(x, u)) \qquad (11.24)$$

satisfies (11.3). Additionally, let $f(0, 0) = 0$, $h(0, 0) = 0$, $s(0, 0) = 0$, and suppose $(0, 0)$ is a hyperbolic equilibrium of X_H. Suppose the stable and unstable invariant manifolds N^- and N^+ of X_H through $(0, 0)$ are globally parametrizable by the x-coordinates, leading to solutions S^- and S^+ to (11.12) with $S^-(0) = S^+(0) = 0$. Then every other solution S with $S(0) = 0$ to the dissipation inequality (11.1), or equivalently, to the Hamilton–Jacobi inequality (11.6), satisfies

$$S^-(x) \leq S(x) \leq S^+(x), \quad \forall x \in \mathcal{X} \qquad (11.25)$$

Proof S^- satisfies (leaving out for convenience all the "transpose" signs)

$$S_x^-(x) f(x, u^*(x, S_x^-(x))) = s(u^*(x, S_x^-(x)), h(x, u^*(x, S_x^-(x)))), \quad (11.26)$$

and since it corresponds to the stable invariant manifold

$$f(x, u^*(x, S_x^-(x))) \text{ is glob. asymptotically stable w.r.t. } x = 0 \quad (11.27)$$

Let S be an arbitrary solution to the dissipation inequality (11.1). Then

$$S_x(x) f(x, u^*(x, S_x^-(x))) \le s(u^*(x, S_x^-(x)), h(x, u^*(x, S_x^-(x)))) \quad (11.28)$$

Subtracting (11.26) from (11.28) yields

$$\left[S_x(x) - S_x^-(x) \right] f(x, u^*(x, S_x^-(x))) \le 0 \quad (11.29)$$

and thus by integration along $\dot{x} = f(x, u^*(x, S_x^-(x)))$

$$\left[S(x(t_1)) - S^-(x(t_1)) \right] \le \left[S(x(t_0)) - S^-(x(t_0)) \right] \quad (11.30)$$

Letting $t_1 \to \infty$, and using (11.27), it follows that $S(x(t_0)) \ge S^-(x(t_0))$ for every initial condition, proving the left-hand side of the inequality (11.25).

For proving the right-hand side we replace S^- by S^+, noting that by definition of S^+

$$- f(x, u^*(x, S_x^+(x))) \text{ is glob. asymptotically stable w.r.t. } x = 0 \quad (11.31)$$

and therefore letting $t_0 \to -\infty$ in (11.30), with S^- replaced by S^+, we obtain $S(x(t_1)) \le S^+(x(t_1))$ for every $x(t_1)$. $\qquad \square$

Remark 11.1.10 It can be also shown that $S^-(x) < S^+(x)$ for all $x \in \mathcal{X}, x \ne 0$.

Up to now we did not address the issue of *nonnegativity* of solutions of the Hamilton–Jacobi inequality (11.6), which is especially important if (11.6) arises from the dissipation inequality (11.1). The following sufficient conditions are straightforward.

Proposition 11.1.11 *Consider the dissipation inequality (11.1) and the Hamilton–Jacobi inequality (11.6). Suppose $f(0, 0) = 0, h(0, 0) = 0,$ and $s(0, 0) = 0$.*

(i) *If $\dot{x} = f(x, 0)$ is globally asymptotically stable with respect to $x = 0$, then every solution S to (11.1) or (11.6), with $S(0) = 0$, will be nonnegative whenever $s(0, y) \le 0$ for all y.*

(ii) *Suppose S^- exists globally, and $s(u^*(x, S_x^-(x)), h(x, u^*(x, S_x^-(x)))) \le 0$ for all x. Then every solution S to (11.1) or (11.6) with $S(0) = 0$ will be nonnegative.*

Proof (i) Consider $S_x(x) f(x, 0) \le s(0, h(x, 0)) \le 0$, and integrate along $\dot{x} = f(x, 0)$ to obtain $S(x(t_1)) \le S(x(t_0))$. Letting $t_1 \to \infty$, and thus $S(x(t_1)) \to S(0) = 0$ yields the result.

(ii) Consider the inequality

$$S_x(x) f(x, u^*(x, S_x^T(x))) \le s(u^*(x, S_x^T(x)), h(x, u^*(x, S_x^T(x)))) \le 0,$$

and integrate along the globally asymptotically vector field $\dot{x} = f(x, u^*(x, S_x^T(x)))]$. □

Let us now apply the results obtained so far to the Hamilton–Jacobi inequalities encountered in the L_2-gain analysis of nonlinear systems (Chap. 8) and in nonlinear $\mathcal{H}_\infty$-control (Chap. 10). (We postpone the treatment of the Hamilton–Jacobi-Bellman equations appearing in Chap. 9 to Sect. 11.2.) For the L_2-gain case we restrict attention to affine nonlinear systems

$$\Sigma_a : \quad \begin{aligned} \dot{x} &= f(x) + g(x)u, \quad f(0) = 0 \\ y &= h(x), \qquad\qquad h(0) = 0 \end{aligned} \tag{11.32}$$

with regard to the L_2-gain supply rate

$$s(u, y) = \frac{1}{2}\gamma^2 \|u\|^2 - \frac{1}{2}\|y\|^2 \tag{11.33}$$

The Hamiltonian H now takes the form

$$H(x, p) = p^T f(x) + \frac{1}{2}\frac{1}{\gamma^2} p^T g(x) g^T(x) p + \frac{1}{2} h^T(x) h(x), \tag{11.34}$$

leading to the Hamilton–Jacobi inequality

$$S_x(x) f(x) + \frac{1}{2}\frac{1}{\gamma^2} S_x(x) g(x) g^T(x) S_x^T(x) + \frac{1}{2} h^T(x) h(x) \le 0, \tag{11.35}$$

and the Hamiltonian matrix

$$\mathcal{H} = \begin{bmatrix} F & \frac{1}{\gamma^2} GG^T \\ -H^T H & -F^T \end{bmatrix}, \quad F = \frac{\partial f}{\partial x}(0), \, G = g(0), \, H = \frac{\partial h}{\partial x}(0) \tag{11.36}$$

Corollary 11.1.8 and Propositions 11.1.9, 11.1.11 all apply in this case, and we obtain:

Corollary 11.1.12 *Assume $\mathcal{H}$ does not have purely imaginary eigenvalues. If (F, G), respectively $(-F, G)$ is stabilizable, then locally about $x = 0$ there exists a solution S^-, respectively S^+, to (11.35) with equality, such that $S^-(0) = 0$, $S^+(0) = 0$, and*

$$f(x) + \tfrac{1}{\gamma^2} g(x) g^T(x) S_x^-(x) \ \text{asymptotically stable}$$

$$-\left[f(x) + \tfrac{1}{\gamma^2} g(x) g^T(x) S_x^+(x) \right] \ \text{asymptotically stable}$$

(11.37)

If S^- and S^+ exist globally, then any other solution S of (11.35) with $S(0) = 0$ satisfies

$$S^-(x) \leq S(x) \leq S^+(x), \quad \forall x \in \mathcal{X}$$

(11.38)

In particular, if $S^- \geq 0$, which is the case (see Proposition 11.1.11) if $\dot{x} = f(x)$ is globally asymptotically stable, then $S^- = S_a$ (the available storage, cf. Theorem 3.1.11), and whenever S_r (the required supply from $x^* = 0$) exists, then $S^+ = S_r$ (see Theorem 3.1.16).

Next let us consider the Hamilton–Jacobi inequality (HJ2a) arising in the output feedback $\mathcal{H}_\infty$ control problem, that is (see (10.71))

$$R_x(x)a(x) + \frac{1}{2}\frac{1}{\gamma^2} R_x(x)g(x)g^T(x)R_x^T(x) + \frac{1}{2}h^T(x)h(x) - \frac{1}{2}\gamma^2 c^T(x)c(x) \leq 0$$

corresponding to the dissipation inequality (see (11.103))

$$R_x(x)[a(x) + g(x)d_1] \leq \frac{1}{2}\gamma^2 \|d_1\|^2 + \frac{1}{2}\gamma^2 \|c(x)\|^2 - \frac{1}{2}\|h(x)\|^2$$

(11.39)

The Hamiltonian H for this case is given as

$$H(x, p) = p^T a(x) + \frac{1}{2}\frac{1}{\gamma^2} p^T g(x) g^T(x) p + \frac{1}{2}h^T(x)h(x) - \frac{1}{2}\gamma^2 c^T(x)c(x)$$

(11.40)

and leads to the Hamiltonian matrix

$$\mathcal{H} = \begin{bmatrix} A & \frac{1}{\gamma^2} G G^T \\ -H^T H + C^T C & -A^T \end{bmatrix},$$

(11.41)

$$A = \tfrac{\partial a}{\partial x}(0), \ G = g(0), \ H = \tfrac{\partial h}{\partial x}(0), \ C = \tfrac{\partial c}{\partial x}(0)$$

In this case only Corollary 11.1.8 and Proposition 11.1.9 apply, while the conditions of Proposition 11.1.11 will not be satisfied. Thus we obtain:

Corollary 11.1.13 Assume $\mathcal{H}$ given by (11.41) does not have purely imaginary eigenvalues. If (F, G) (resp. $(-F, G)$) is stabilizable then locally about $x = 0$ there exists a solution S^- (resp. S^+) to (11.35) with equality, such that $S^-(0) = 0$, $S^+(0) = 0$, and

$$f(x) + \frac{1}{\gamma^2}g(x)g^T(x)S_x^-(x) \text{ asymptotically stable}$$

(11.42)

$$-[f(x) + \frac{1}{\gamma^2}g(x)g^T(x)S_x^+(x)] \text{ asymptotically stable}$$

If S^- and S^+ exist globally then any other solution S of (11.35) with $S(0) = 0$ satisfies

$$S^-(x) \le S(x) \le S^+(x), \quad \forall x \in \mathcal{X}$$

(11.43)

In particular, if there exists a solution $S \ge 0$ to (11.35) with $S(0) = 0$, and if S_r (the required supply for ground state $x^* = 0$) for the dissipation inequality (11.39) exists (i.e., is finite) and is C^1, then $S^+ = S_r \ge 0$.

Proof Only the last statement needs some clarification. If there exists $S \ge 0$ to (11.39), and if S_r exists and is C^1 then S_r is solution of (11.39) and (11.35), and by Theorem 3.1.16 equals the maximal solution.   □

Finally, let us consider the Hamilton–Jacobi inequality (HJ1a) of the state feedback $\mathcal{H}_\infty$ control problem, for affine nonlinear systems given as (see (10.3))

$$P_x(x)a(x) + \frac{1}{2}P_x(x)\left[\frac{1}{\gamma^2}g(x)g^T(x) - b(x)b^T(x)\right]P_x^T(x)$$

$$+\frac{1}{2}h^T(x)h(x) \le 0 \qquad (11.44)$$

This Hamilton–Jacobi inequality does *not* correspond to a dissipation inequality. The Hamiltonian

$$H(x, p) = p^T a(x) + \frac{1}{2}p^T\left[\frac{1}{\gamma^2}g(x)g^T(x) - b(x)b^T(x)\right]p + \frac{1}{2}h^T(x)h(x)$$

(11.45)

leads to the Hamiltonian matrix

$$\mathcal{H} = \begin{bmatrix} A & \frac{1}{\gamma^2}GG^T - BB^T \\ -H^T H & -A^T \end{bmatrix},$$

(11.46)

$$A = \frac{\partial a}{\partial x}(0), \; G = g(0), \; B = b(0), \; H = \frac{\partial h}{\partial x}(0),$$

and we conclude that Corollary 11.1.8 and Propositions 11.1.9, 11.1.11 do not apply. Of course, the general Proposition 11.1.5 may still be invoked, guaranteeing local existence of a solution P to (HJ1a) if $\mathcal{H}$ does not have purely imaginary eigenvalues and its stable (or unstable) eigenspace can be parametrized by the x-coordinates. We will come back to this in Sect. 11.4. The same remarks, of course, can be made about the general Hamilton–Jacobi inequality (HJ1) of the state feedback $\mathcal{H}_\infty$ control problem, that is (see (10.21))

$$K_\gamma(x, P_x^T(x), d^*(x, P_x^T(x)), u^*(x, P_x^T(x))) \leq 0 \qquad (11.47)$$

where (see (10.19))

$$K_\gamma(x, p, d, u) = p^T f(x, u, d) - \frac{1}{2}\gamma^2 ||d||^2 + \frac{1}{2}||h(x, u, d)||^2 \qquad (11.48)$$

On the other hand, we will now show that, under some additional conditions, the solution set of (11.47), or, a fortiori, of (11.44), contains a *minimal* element which corresponds to the *stabilizing* solution of (11.47) or (11.44). Let us throughout assume that

$$f(0, 0, 0) = 0, \quad h(0, 0, 0) = 0, \qquad (11.49)$$

and that the Hamiltonian $H(x, p)$ corresponding to (11.48), i.e.,

$$H(x, p) := K_\gamma(x, p, d^*(x, p), u^*(x, p)) \qquad (11.50)$$

has a Hamiltonian matrix $\mathcal{H} = DX_H(0, 0)$, see (11.16), which does not have purely imaginary eigenvalues. Now consider an arbitrary solution $P \geq 0$ to (11.47) with $P(0) = 0$. Define the feedback

$$\alpha_0(x) := u^*(x, P_x^T(x)) \qquad (11.51)$$

It immediately follows that the closed-loop system

$$\Sigma_0 : \quad \begin{aligned} \dot{x} &= f(x, \alpha_0(x), d) \\ z &= h(x, \alpha_0(x), d) \end{aligned} \qquad (11.52)$$

has L_2-gain $\leq \gamma$ (from d to z), since

$$K_\gamma(x, P_x^T(x), d^*(x, P_x^T(x)), \alpha_0(x)) \leq 0 \qquad (11.53)$$

Now define the Hamiltonian

$$H_0(x, p) := K_\gamma(x, p, d^*(x, p), \alpha_0(x)) \qquad (11.54)$$

This Hamiltonian is of the type as considered in Proposition 11.1.9. Assume that the stabilizing solution $P^1 := S^-$ with $P^1(0) = 0$ to $H_0(x, S_x^T(x)) = 0$ exists globally, and that

$$\dot{x} = f(x, \alpha_0(x), 0) \text{ is globally asymptotically stable} \qquad (11.55)$$

Then it follows from Propositions 11.1.9 and 11.1.11(i) that

$$0 \leq P^1 \leq P \qquad (11.56)$$

Subsequently, define the feedback

$$\alpha_1(x) := u^*(x, P_x^{1T}(x)) \qquad (11.57)$$

Then, by using the second inequality of (10.20)

$$
\begin{aligned}
0 = H_0(x, P_x^{1T}(x)) &= K_\gamma(x, P_x^{1T}(x), d^*(x, P_x^{1T}(x)), \alpha_0(x)) \\
&\geq K_\gamma(x, P_x^{1T}(x), d^*(x, P_x^{1T}(x)), u^*(x, P_x^{1T}(x))) \\
&= K_\gamma(x, P_x^{1T}(x), d^*(x, P_x^{1T}(x)), \alpha_1(x)))
\end{aligned}
\qquad (11.58)
$$

Hence also the closed-loop system

$$
\Sigma_1 : \quad
\begin{aligned}
\dot{x} &= f(x, \alpha_1(x), d) \\
z &= h(x, \alpha_1(x), d)
\end{aligned}
\qquad (11.59)
$$

has L_2-gain $\leq \gamma$, with storage function P^1. Define the corresponding Hamiltonian

$$H_1(x, p) := K_\gamma(x, p, d^*(x, p), \alpha_1(x)), \qquad (11.60)$$

and assume again that the stabilizing solution $P^2 := S^-$ with $P^2(0) = 0$ to $H_1(x, S_x^T(x)) = 0$ exists globally, while

$$\dot{x} = f(x, \alpha_1(x), 0) \text{ is globally asymptotically stable} \qquad (11.61)$$

Application of Propositions 11.1.9 and 11.1.11 then yields

$$0 \leq P^2 \leq P^1 \leq P, \qquad (11.62)$$

and we subsequently define the feedback

$$\alpha_2(x) := u^*(x, P_x^{2T}(x)) \qquad (11.63)$$

Continuing this process we arrive at the following conclusion

Proposition 11.1.14 *Consider the Hamilton–Jacobi inequality (11.47), and assume that (11.49) holds, while $DX_H(0, 0)$ corresponding to (11.50) does not have purely imaginary eigenvalues. Let $P \geq 0$ be any solution to (11.47). Define as above inductively $\alpha_0, \alpha_1, \ldots$ and $H_0, H_1, \ldots$. Assume that the vector fields $\dot{x} = f(x, \alpha_i(x), 0)$, $i = 0, 1, \ldots$ are all globally asymptotically stable, and that the stabilizing solutions $P^1, P^2, \ldots$ to $H_i(x, S_x^T(x)) = 0$ with $P^i(0) = 0$ exist globally. Then*

$$P \geq P^1 \geq P^2 \geq \cdots \geq 0 \qquad (11.64)$$

Assume that the pointwise limit $P^*(x) := \lim_{i \to \infty} P^i(x)$ *is* C^1. *Then* $P^* \leq P$ *is the unique solution to (11.47) with equality, with the property that*

$$\dot{x} = f(x, u^*(x, P_x^{*T}(x)), d^*(x, P_x^{*T}(x))) \tag{11.65}$$

is globally asymptotically stable

Proof P^i is the unique solution to

$$K_\gamma(x, P_x^T(x), d^*(x, P_x^T(x)), u^*(x, P_x^{(i-1)T}(x))) = 0, \tag{11.66}$$

with the property that

$$\dot{x} = f(x, u^*(x, P_x^{(i-1)T}(x)), d^*(x, P_x^T(x))) \tag{11.67}$$

is globally asymptotically stable

Therefore P^* is a solution to (11.47) with equality, such that the vector field in (11.65) is at least *stable*. However by assumption $DX_H(0, 0)$ does not have imaginary eigenvalues, and so necessarily (11.65) holds.   □

Note that if the conditions of Proposition 11.1.14 are satisfied for all solutions $P \geq 0$ to (11.47), then P^*, the stabilizing solution of (11.47), is also the *minimal* solution of (11.47).

Remark 11.1.15 With a similar procedure and under similar conditions, one can show that the *anti-stabilizing* solution of (11.47) is also the *maximal* one.

11.2 An Aside on Optimal Control

In this section, we apply some of the techniques developed in the preceding section to the standard optimal control problem as already encountered in Chaps. 3 and 9. It will be seen that this problem is indeed very much related to the dissipation inequalities and Hamilton–Jacobi inequalities treated above.

Let us consider the *infinite horizon optimal control problem*

$$\min_u \int_0^\infty L(x(t), u(t))dt, \tag{11.68}$$
$$\dot{x} = f(x, u),$$

where $L(x, u)$ is a cost function, which we assume to be non-negative, that is

$$L(x, u) \geq 0, \quad \forall x, u \tag{11.69}$$

We associate with (11.68) the *reversed dissipation inequality*

$$V_x(x)f(x,u) + L(x,u) \geq 0, \quad \forall x, u \tag{11.70}$$

in the unknown V. The basic reason for introducing this inequality is that by the *principle of optimality* the *value function*

$$\bar{V}(x_0) := \min_u \left\{ \int_0^\infty L(x(t), u(t))dt \mid \dot{x} = f(x,u), x(0) = x_0 \right\} \tag{11.71}$$

satisfies (whenever it exists) for all $t_1 \geq t_0$, and for all $u(\cdot)$

$$\bar{V}(x(t_0)) \leq \bar{V}(x(t_1)) + \int_{t_0}^{t_1} L(x(t), u(t))dt, \tag{11.72}$$

and thus satisfies, if it is differentiable, the inequality (11.70).

In order to apply the invariant manifold techniques of Sect. 11.1 we will throughout assume that

$$f(0,0) = 0, L(0,0) = 0, \frac{\partial L}{\partial x}(0,0) = 0, \frac{\partial L}{\partial u}(0,0) = 0 \tag{11.73}$$

Define the linearizations

$$F = \tfrac{\partial f}{\partial x}(0), G = \tfrac{\partial f}{\partial u}(0,0),$$

$$Q = \tfrac{\partial^2 L}{\partial x^2}(0,0), R = \tfrac{\partial^2 L}{\partial u^2}(0,0), N = \tfrac{\partial^2 L}{\partial x \partial u}(0,0)$$

From (11.69) it follows that

$$\begin{bmatrix} Q & N \\ N^T & R \end{bmatrix} \geq 0, \tag{11.74}$$

and we will moreover assume the *regularity* condition

$$R > 0 \tag{11.75}$$

Defining the pre-Hamiltonian corresponding to (11.70) as

$$K(x, p, u) := p^T f(x,u) + L(x,u), \tag{11.76}$$

then by the Implicit Function theorem the condition (11.75) implies that at least locally near $(0,0)$ there exists $u^*(x,p)$ such that

$$K(x, p, u) \geq K(x, p, u^*(x,p)) =: H(x,p), \quad \forall x, p, \tag{11.77}$$

but we will assume that $u^*(x,p)$ exists *globally*. Under this assumption the dissipation inequality (11.70) is equivalent to the (reversed) Hamilton–Jacobi inequality

$$H(x, V_x^T(x)) \geq 0, \quad \forall x \in \mathcal{X} \tag{11.78}$$

The Hamiltonian matrix $\mathcal{H} = DX_H(0,0)$ corresponding to H can be immediately computed as

$$\mathcal{H} = \begin{bmatrix} F - GR^{-1}N^T & -GR^{-1}G^T \\ -Q + NR^{-1}N^T & -(F - GR^{-1}N^T)^T \end{bmatrix} \tag{11.79}$$

Because of (11.74) and (11.75) we see that not only the right upper block of $\mathcal{H}$ is non-positive but also the left-lower block (being a Schur complement of a non-positive matrix). This allows us to state the following strengthened version of Proposition 11.1.7(ii) (see the references in the notes at the end of this chapter).

Proposition 11.2.1 *Consider $\mathcal{H}$ given by (11.79). Assume the pair $(F - GR^{-1}N^T, GR^{-1}G^T)$ is stabilizable, and assume that the purely imaginary eigenvalues of $F - GR^{-1}N^T$ (if existing) are $(Q - NR^{-1}N^T)$-detectable. Then $\mathcal{H}$ does not have purely imaginary eigenvalues, and the stable eigenspace L^- of $\mathcal{H}$ is Lagrangian and complementary to span $\begin{bmatrix} 0 \\ I_n \end{bmatrix}$. Hence $L^- = $ span $\begin{bmatrix} I \\ X^- \end{bmatrix}$ for some symmetric matrix X^-, satisfying $X^- \geq 0$. If the pair $(Q - NR^{-1}N^T, F - GR^{-1}N^T)$ is observable then $X^- > 0$.*

As in Corollary 11.1.8 it follows that under the assumptions of Proposition 11.2.1 the stable invariant manifold N^- of X_H through the equilibrium $(0,0)$ is given as N_{V^-} for a certain function $V^-(x)$ defined near 0. Let us assume that V^- exists *globally*. Note that $\frac{\partial V^-}{\partial x}(0) = 0$, $\frac{\partial^2 V^-}{\partial x^2}(0) = X^-$, and without loss of generality $V^-(0) = 0$.

Proposition 11.2.2 *Assume that the stable invariant manifold of X_H through $(0,0)$ is given as N_{V^-} with $V^-(0) = 0$. Then $V^- \geq 0$, and every other solution V of (11.70) with $V(0) = 0$ satisfies*

$$V(x) \leq V^-(x), \quad \forall x \in \mathcal{X} \tag{11.80}$$

Proof Similar to Proposition 11.1.9 we consider

$$V_x^-(x)f(x, u^*(x, V_x^-(x)) + L(x, u^*(x, V_x^-(x))) = 0 \tag{11.81}$$

where

$$f(x, u^*(x, V_x^-(x))) \text{ is glob. asymptotically stable w.r.t. } x = 0 \tag{11.82}$$

By integration of (11.81) from 0 to T, and letting $T \to \infty$, using (11.82) and non-negativity of $L(x, u)$ (see (11.69)) it immediately follows that $V^- \geq 0$. Since any solution V to (11.70) satisfies

$$V_x(x)f(x, u^*(x, V_x^-(x))) + L(x, u^*(x, V_x^-(x))) \geq 0 \tag{11.83}$$

we obtain

$$\left[V_x^-(x) - V_x(x)\right] f(x, u^*(x, V_x^-(x))) \le 0, \tag{11.84}$$

and the result follows from integration from 0 to T, and letting $T \to \infty$, using again (11.82). $\qquad\square$

Remark 11.2.3 Analogously we may prove that every solution V to (11.70) satisfies $V^+(x) \le V(x)$, with N_{V^+} the *unstable* invariant manifold of X_H.

Remark 11.2.4 Similarly we may look at Hamilton–Jacobi–Bellman equations corresponding to optimal control problems in *reversed* time, where we replace the integral $\int_0^\infty$ by $\int_{-\infty}^0$. In particular, in Sect. 9.1 we studied the Hamilton–Jacobi–Bellman equation

$$W_x(x)f(x) + \frac{1}{2}W_x(x)g(x)g^T(x)W_x^T(x) - \frac{1}{2}h^T(x)h(x) = 0 \tag{11.85}$$

corresponding to a reversed-time optimal control problem. The Hamiltonian matrix in this case is

$$\mathcal{H} = \begin{bmatrix} F & GG^T \\ H^T H & -F^T \end{bmatrix}, \tag{11.86}$$

with F, G as in (11.74), and $H = \frac{\partial h}{\partial x}(0)$. By considering $-\mathcal{H}$ and applying Proposition 11.2.1 it immediately follows that if $(-F, G)$ is stabilizable, and the purely imaginary eigenvalues of F are H-detectable, then $\mathcal{H}$ will not have purely imaginary eigenvalues, and the *unstable* eigenspace L^+ of $\mathcal{H}$ is of the form span $\begin{bmatrix} I \\ X^+ \end{bmatrix}$ with $X^+ \ge 0$ and $X^+ > 0$ if (H, F) is observable. It follows that at least locally there exists a solution $W \ge 0$ to (11.85), satisfying $W(0) = 0$, $W_x(0) = 0$, $W_{xx}(0) = X^+$.

For the solution of the optimal control problem the following observation is crucial. Let V be any solution to (11.78) *with equality*, that is, the Hamilton–Jacobi(–Bellman) equation

$$H(x, V_x^T(x)) = 0 \tag{11.87}$$

Then by (11.77)

$$R(x, u) := K(x, V_x^T(x), u) \ge 0 \tag{11.88}$$

for all x, u. Since $K(x, V_x^T(x), u) = V_x(x)f(x, u) + L(x, u)$ we obtain by integration

$$\int_0^T L(x(t), u(t))dt = \int_0^T R(x(t), u(t))dt + V(x(0)) - V(x(T)) \tag{11.89}$$

for every input function $u(\cdot)$, and for every T.

Proposition 11.2.5 *Consider any globally defined solution V to (11.87) with the property that $V(0) = 0$ and the feedback $u = -g^T(x)V_x^T(x)$ is asymptotically stabilizing. Then*

$$V(x_0) = \min_u \left\{ \int_0^\infty L(x(t), u(t))\, dt \mid \dot{x} = f(x, u), x(0) = x_0, \lim_{t\to\infty} x(t) = 0 \right\}$$
(11.90)

Additionally, since $L(x, u) \geq 0$ this implies that for any such V we have $V \geq 0$. In particular, assume that V^- exists globally, then the above holds for V^-, with optimal stabilizing control given in feedback form as

$$u = u^*(x, V_x^{-T}(x))$$
(11.91)

Proof Since $\lim_{T\to\infty} x(T) = 0$ the right-hand side of (11.89) is minimized by substituting (11.91). $\qquad\square$

On the other hand, it is not clear if (11.91) is also the optimal control corresponding to the original optimal control problem (11.68) *without* the terminal constraint $\lim_{t\to\infty} x(t) = 0$. Indeed, let V be any other *non-negative* solution to (11.87). Then by considering (11.89) for this V, and noting that $V(x(T)) \geq 0$, while $R(x, u)$ can be rendered zero by choosing $u = u^*(x, V_x^T(x))$, it immediately follows that

$$\min_u \left\{ \int_0^\infty L(x(t), u(t))dt \mid \dot{x} = f(x, u), x(0) = x_0 \right\} \leq V(x_0)$$
(11.92)

Thus if there exists a *non-negative* solution V with $V(0) = 0$ to (11.87) *different* from V^-, that is (see Proposition 11.2.2), $V(x) < V^-(x)$ for some $x \in \mathcal{X}$, then (11.91) will *not* be the optimal control and V^- will *not* be the value function $\bar{V}$. The existence of non-negative solutions V to (11.87) different from V^- can be excluded by imposing a (nonlinear) *detectability* condition.

Proposition 11.2.6 *Let V^- exist globally, and let the optimal control problem (11.68) be solvable. Suppose $f(x, u), L(x, u)$ satisfy the following detectability property*

$$\lim_{t\to\infty} L(x(t), u(t)) = 0, \text{ along solutions of } \dot{x}(t) = f(x(t), u(t)),$$
(11.93)
$$\text{implies } \lim_{t\to\infty} x(t) = 0$$

Then the only non-negative solution V to (11.87) with $V(0) = 0$ is V^-. Furthermore, the solution to the optimal control problem (11.68) is given by (11.90) with value function $\bar{V} = V^-$.

Proof If $\min_u \int_0^\infty L(x(t), u(t))dt$ exists, then necessarily along the optimal trajectory $\lim_{t\to\infty} L(x(t), u(t)) = 0$, and thus by (11.93) $\lim_{t\to\infty} x(t) = 0$. Thus the optimal control problem (11.68) is the same as the optimal control problem (11.90), which has by Proposition 11.2.5 the solution (11.91) with value function V^-. By the reasoning preceding Proposition 11.2.6, see (11.92), it thus follows that there cannot exist any non-negative solution V to (11.87) with $V(0) = 0$ *different* from V^-. $\square$

Remark 11.2.7 Note that *without* the detectability condition (11.93) the value function $\bar{V}$ (which in this case may be different from V^-) still satisfies the property $\lim_{t\to\infty} \bar{V}(x(t)) = 0$ along optimal trajectories.

Finally, let us compare (11.89), which is derived under the *regularity* assumption of existence of $u^*(x, p)$ satisfying (11.77), with the general dissipation inequality (11.70), written in integral form as

$$\int_0^T L(x(t), u(t))dt \geq V(x(0)) - V(x(T)) \tag{11.94}$$

for every input function u. Clearly, the Eq. (11.94) contains much less information than (11.89), but it already suffices to draw the following conclusion, which is also valid if $u^*(x, p)$, and thus $H(x, p)$ in (11.77), can*not* be defined. The following proposition was already stated in a more restricted form in Sect. 9.4 as Proposition 9.4.8.

Proposition 11.2.8 *For every x_0 define $V^*(x_0)$ as*

$$\min_u \left\{ \int_0^\infty L(x(t), u(t))dt \mid \dot{x} = f(x, u), x(0) = x_0, \lim_{t\to\infty} x(t) = 0 \right\} \tag{11.95}$$

and assume that V^* exists for every x_0. Then

(i) $V^* \geq 0$ satisfies the dissipation inequality (11.94), and $V^*(0) = 0$.
(ii) Let V satisfy (11.94) and $V(0) = 0$, then $V(x) \leq V^*(x)$ for every $x \in \mathcal{X}$.

Proof Consider (11.94) for $u(\cdot)$ such that $\lim_{T\to 0} x(T) = 0$. Then

$$\int_0^\infty L(x(t), u(t))dt \geq V(x(0)),$$

and by definition of V^* we obtain $V \leq V^*$. $\square$

11.3 Dissipativity of a Nonlinear System and Its Linearization

In this section we will relate the dissipativity of a nonlinear state space system

$$\Sigma : \quad \begin{aligned} \dot{x} &= f(x, u), \quad f(0, 0) = 0 \\ y &= h(x, u), \quad h(0, 0) = 0 \end{aligned} \tag{11.96}$$

to the dissipativity of its *linearization* at $(x, u) = (0, 0)$, and vice versa.

We will do so for supply rates $s(u, y)$ satisfying

$$s(0, 0) = 0, \quad \frac{\partial s}{\partial u}(0, 0) = 0, \quad \frac{\partial s}{\partial y}(0, 0) = 0, \tag{11.97}$$

including the (output or input strict) passivity and L_2-gain supply rates. Moreover, we throughout assume that there exists a storage function $S \geq 0$ for the dissipation inequality

$$S_x(x) f(x, u) \leq s(u, h(x, u)), \tag{11.98}$$

which is C^2 and has a minimum at $x = 0$ with $S(0) = 0$, implying that

$$S(0) = 0, \quad \frac{\partial S}{\partial x}(0) = 0, \quad X := \frac{\partial^2 S}{\partial x^2}(0) \geq 0 \tag{11.99}$$

Defining

$$A = \tfrac{\partial f}{\partial x}(0, 0), \, B = \tfrac{\partial f}{\partial u}(0, 0), \, C = \tfrac{\partial h}{\partial x}(0, 0), \, D = \tfrac{\partial h}{\partial u}(0, 0)$$

$$P = \tfrac{\partial^2 s}{\partial u^2}(0, 0), \, Q = \tfrac{\partial^2 s}{\partial y^2}(0, 0), \, R = \tfrac{\partial^2 s}{\partial u \partial y}(0, 0) \tag{11.100}$$

it follows by collecting the quadratic terms in (11.98) that

$$\frac{1}{2}\bar{x}^T X (A\bar{x} + B\bar{u}) \leq \frac{1}{2} \begin{bmatrix} \bar{u}^T & \bar{y}^T \end{bmatrix} \begin{bmatrix} P & R \\ R^T & Q \end{bmatrix} \begin{bmatrix} \bar{u} \\ \bar{y} \end{bmatrix}, \quad \forall \bar{x}, \bar{u} \tag{11.101}$$

where $\bar{y} := C\bar{x} + D\bar{u}$.

This implies that the *linearized system*

$$\Sigma : \quad \begin{aligned} \dot{\bar{x}} &= A\bar{x} + B\bar{u} \\ \bar{y} &= C\bar{x} + D\bar{u} \end{aligned} \tag{11.102}$$

is *dissipative* with respect to the quadratic supply rate

$$\bar{s}(\bar{u}, \bar{y}) := \frac{1}{2} [\bar{u}^T \ \bar{y}^T] \begin{bmatrix} P & R \\ R^T & Q \end{bmatrix} \begin{bmatrix} \bar{u} \\ \bar{y} \end{bmatrix}, \tag{11.103}$$

with storage function $\frac{1}{2}\bar{x}^T X \bar{x} = \frac{1}{2}\bar{x}^T \frac{\partial^2 S}{\partial x^2}(0)\bar{x}$. In this sense, dissipativity of the non-linear system Σ with respect to the supply rate s implies dissipativity of its linearization $\overline{\Sigma}$ with respect to the supply rate $\bar{s}$.

Conversely, let us investigate under which conditions dissipativity of the linearized system $\overline{\Sigma}$ implies dissipativity of the nonlinear system Σ.

Proposition 11.3.1 *Consider the nonlinear system Σ given by (11.96), with linearization $\overline{\Sigma}$ given by (11.102). Consider the supply rate s satisfying (11.97) with quadratic part $\bar{s}$ defined by (11.103) and (11.100). Suppose that $\overline{\Sigma}$ is dissipative with respect to the supply rate $\bar{s}$. Assume $D = 0$ and $P > 0$, and that the Hamiltonian matrix*

$$\mathcal{H} = \begin{bmatrix} A - BP^{-1}RC & BP^{-1}B^T \\ C^T QC & -(A - BP^{-1}RC)^T \end{bmatrix} \tag{11.104}$$

does not have purely imaginary eigenvalues. Also assume that the pair $(A - BP^{-1}RC, BP^{-1}B^T)$ is stabilizable. Then there exists a neighborhood V of $x = 0$ and $U \subset \mathbb{R}^m$ of $u = 0$, and a function $S : V \subset \mathcal{X} \to \mathbb{R}$ with $S(0) = 0$, $\frac{\partial S}{\partial x}(0) = 0$, such that

$$S_x(x) f(x, u) \leq s(u, h(x, u)), \ \text{for all } x \in V, \ \text{all } u \in U \tag{11.105}$$

Thus if $S \geq 0$ then Σ is locally dissipative with respect to the supply rate s.

Proof The pre-Hamiltonian corresponding to (11.105) is

$$K(x, p, u) = p^T f(x, u) - s(u, h(x, u)) \tag{11.106}$$

By the Implicit Function theorem and $P > 0$ there exists locally a function $u^*(x, p)$ satisfying (11.3). Furthermore, the resulting Hamiltonian $H(x, p) = K(x, p, u^*(x, p))$ has corresponding Hamiltonian matrix $\mathcal{H}$ given by (11.104). By Corollary 11.1.8 there exists locally near $x = 0$ a function S with $S(0) = 0$, $\frac{\partial S}{\partial x}(0) = 0$ satisfying $H(x, S_x^T(x)) = 0$ (in fact, S corresponds to the stable invariant manifold of X_H). It follows that S satisfies (11.105). □

Remark 11.3.2 For $D \neq 0$ a similar statement can be proved, by replacing the assumption $P > 0$ by $P + RD + D^T R^T + D^T QD > 0$, and by defining a more complicated Hamiltonian matrix $\mathcal{H}$.

The main drawback of Proposition 11.3.1 is that it does not provide conditions which guarantee that the obtained function S satisfying (11.105) is non-negative on a neighborhood of the equilibrium $x = 0$, and so is a valid storage function. One possible set of sufficient conditions for non-negativity of S is given in the following corollary.

Corollary 11.3.3 *Consider a nonlinear system* Σ *satisfying (11.97), with linearization* $\overline{\Sigma}$ *having* $D = 0$ *and a supply rate* s *satisfying (11.97). Suppose* $\overline{\Sigma}$ *is dissipative with respect to the supply rate* $\bar{s}$ *given by (11.103). Assume* $P > 0$, *and assume that the Hamiltonian matrix* $\mathcal{H}$ *in (11.104) does not have purely imaginary eigenvalues. Also assume that* A *is asymptotically stable, and that* $s(0, y) \leq 0$ *for all* y. *Then there exists a neighborhood* V *of* $x = 0$, *and* $S : V \subset \mathcal{X} \to \mathbb{R}$ *with* $S(0) = 0$, $\frac{\partial S}{\partial x}(0) = 0$, *satisfying (11.105) and such that* $S(x) \geq 0$, $x \in V$. *Thus* Σ *is locally dissipative on* V *with respect to the supply rate* s.

Proof The proof of Proposition 11.3.1 yields locally a function S with $S(0) = 0$, $\frac{\partial S}{\partial x}(0) = 0$, and satisfying $H(x, S_x^T(x)) = 0$. By (11.3) it thus follows that $S_x(x)f(x, 0) \leq s(0, h(x)) \leq 0$. Since $A = \frac{\partial f}{\partial x}(0, 0)$ is asymptotically stable, locally near $x = 0$ also $\dot{x} = f(x, 0)$ is asymptotically stable, and $S \geq 0$ follows by integration from 0 to T, and letting $T \to \infty$, using asymptotic stability. $\square$

For the L_2-gain supply rate $s(u, y) = \frac{1}{2}\gamma^2||u||^2 - \frac{1}{2}||y||^2$ we obtain the following particularly pleasing corollary.

Corollary 11.3.4 *Consider the nonlinear system* Σ *given by (11.96), with linearization* $\overline{\Sigma}$ *having* $D = 0$. *Let* $\gamma > 0$, *and suppose that* $\overline{\Sigma}$ *has* L_2-gain $< \gamma$. *Assume that* $A = \frac{\partial f}{\partial x}(0, 0)$ *is asymptotically stable. Then there exists a neighborhood* V *of* $x = 0$ *and* U *of* $u = 0$ *such that* Σ *has locally* L_2-gain $< \gamma$ *for* $x \in V$ *and* $u \in U$.

Proof Since $\overline{\Sigma}$ has L_2-gain $< \gamma$ the corresponding Hamiltonian matrix $\mathcal{H}$ as in (11.104) does not have purely imaginary eigenvalues (see e.g., [117]). Thus we may apply Corollary 11.3.3 yielding neighborhoods V of $x = 0$ and U of $u = 0$, and a function $S : V \to \mathbb{R}^+$ such that

$$S_x(x)f(x, u) \leq \frac{1}{2}\gamma^2||u||^2 - \frac{1}{2}||h(x)||^2 \tag{11.107}$$

for all $x \in V \subset \mathcal{X}$ and all $u \in U \subset \mathbb{R}^m$. This shows that Σ has locally L_2-gain $\leq \gamma$. Since the same story can be repeated for some $\tilde{\gamma} < \gamma$ arbitrarily close to γ, it follows that actually Σ has locally L_2-gain $< \gamma$. $\square$

Remark 11.3.5 For an affine system $\dot{x} = f(x) + g(x)u$, $y = h(x)$ we may always take $U = \mathbb{R}^m$.

Remark 11.3.6 Since $\overline{\Sigma}$ is assumed to have L_2-gain $< \gamma$ there exists $X = X^T \geq 0$ such that $A^T X + XA + \frac{1}{\gamma^2}XBB^T + C^TC \leq 0$, and thus $A^TX + XA \leq -C^TC$. If (C, A) is detectable, then this actually *implies* that A is asymptotically stable [356].

Fig. 11.1 Mathematical
pendulum

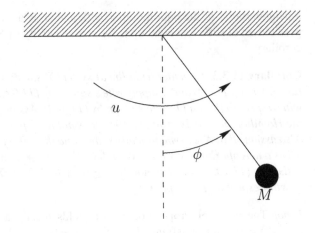

Remark 11.3.7 Since $\overline{\Sigma}$ is assumed to have L_2-gain $< \gamma$ it can be shown [117] that there exists $X = X^T \geq 0$ satisfying

$$A^T X + XA + C^T C + \frac{1}{\gamma^2} XBB^T X < 0 \qquad (11.108)$$

It readily follows that also $S(x) := \frac{1}{2} x^T Xx$ will satisfy (11.107), for *different* neighborhoods V and U, however. (In fact, we conjecture that the neighborhoods will be *smaller* than the ones obtained in Corollary 11.3.4; see also the similar discussion in Sect. 11.4.)

Contrary to the L_2-gain supply rate, the passivity supply rate $s(u, y) = u^T y$ does *not* satisfy the conditions of Proposition 11.3.1, since $P = 0$ in this case. The following physical example shows that, in fact, passivity of the linearized system does *not* imply (local) passivity of the nonlinear system.

Example 11.3.8 Consider a mathematical pendulum with input torque u as depicted in Fig. 11.1 (see also Examples 7.1.12 and 7.2.2). Taking as output the angular velocity $\dot{\varphi}$, the nonlinear system is passive (in fact, lossless) with storage function equal to the total energy

$$S(\varphi, \dot{\varphi}) = \frac{1}{2} m\ell^2 \dot{\varphi}^2 + \ell(1 - \cos \varphi) \qquad (11.109)$$

Indeed, $\frac{dS}{dt} = u\dot{\varphi}$. By the discussion preceding Proposition 11.3.1 the system linearized about $\varphi = 0$, $\dot{\varphi} = 0$, $u = 0$ is also passive. On the other hand, if we take as output the *horizontal* velocity of the endpoint, that is

$$y = \frac{d}{dt}(\ell \sin \varphi) = \ell \cos \varphi \cdot \dot{\varphi}, \qquad (11.110)$$

then the linearized output (about $\varphi = 0$) remains the same, and thus the linearized system is still passive. However, as we will now demonstrate, the nonlinear system is *not* passive anymore. Indeed, the equations of the nonlinear system with output $y = \ell \cos \varphi \cdot \dot{\varphi}$ are given as (take for simplicity $\ell = 1$, $m = 1$, and denote $q = \varphi$)

$$\dot{q} = p$$
$$\dot{p} = -\sin q + u \qquad (11.111)$$
$$y = p \cos q$$

Suppose $S(q, p)$ is a (locally defined) storage function with respect to the supply rate $s(u, y) = uy$. Then by (4.15) for all q, p close to zero

$$(i) \quad \frac{\partial S}{\partial q}p + \frac{\partial S}{\partial p} \cdot -\sin q \le 0, \quad (ii) \quad \frac{\partial S}{\partial p} = p\cos q \qquad (11.112)$$

From (ii) we infer $S(q, p) = \frac{1}{2}p^2 \cos q + F(q)$, for some function F. Substitution in (i) yields

$$(iii) \quad -\frac{1}{2}p^3 \sin q + p\,[\frac{dF}{dq} - \sin q \cos q] \le 0 \qquad (11.113)$$

For fixed q, the left hand side of (11.113) is a polynomial $ap^3 + bp$. For p small, the linear term dominates, and thus the inequality (11.113) implies that $b = \frac{\partial F}{\partial q} - \sin q \cos q = 0$. Hence, $-\frac{1}{2}p^3 \sin q \le 0$ for all p, q close to zero, which is a contradiction. We conclude that there are no storage functions, and thus the system is *not* passive.

11.4 $\mathcal{H}_\infty$ Control of a Nonlinear System and Its Linearization

In this section we investigate, in a spirit similar to the previous section, the relations between the (suboptimal) $\mathcal{H}_\infty$ control problem for a nonlinear system with given equilibrium on the one hand, and the same problem for its linearization on the other hand. In particular we show that if the state feedback strictly suboptimal $\mathcal{H}_\infty$ control problem for the linearization is solvable, then the same problem for the nonlinear system is solvable *on a neighborhood* of the equilibrium. Thus solvability of the "linearized problem" implies local solvability of the "nonlinear problem". Similar results will be shown to hold for the output feedback $\mathcal{H}_\infty$ problem. These results are quite useful since the solvability of the linear $\mathcal{H}_\infty$ problems is relatively easy to check (certainly compared with the nonlinear $\mathcal{H}_\infty$ problems). Also the optimal disturbance level for the linearized problem will provide a useful lower bound for the optimal nonlinear disturbance level.

Let us start with the *state feedback* $\mathcal{H}_\infty$ problem. For simplicity of exposition[1] we focus on affine non-linear systems as in (10.2), that is

$$\Sigma : \quad \begin{aligned} \dot{x} &= a(x) + b(x)u + g(x)d, \ u \in \mathbb{R}^m, \ d \in \mathbb{R}^r \\ z &= \begin{bmatrix} h(x) \\ u \end{bmatrix}, \qquad\qquad x \in \mathcal{X}, \ z \in \mathbb{R}^s \end{aligned} \qquad (11.114)$$

Since we wish to consider the linearization of Σ we assume the existence of an equilibrium $x = 0$, that is,

$$a(0) = 0, \quad h(0) = 0 \qquad (11.115)$$

We denote the linearization of Σ about $x = 0$ by

$$\overline{\Sigma} : \quad \begin{aligned} \dot{\bar{x}} &= A\bar{x} + B\bar{u} + G\bar{d}, \ \bar{u} \in \mathbb{R}^m, \ \bar{d} \in \mathbb{R}^r \\ \bar{z} &= \begin{bmatrix} H\bar{x} \\ \bar{u} \end{bmatrix}, \qquad\qquad \bar{x} \in \mathbb{R}^n, \ \bar{z} \in \mathbb{R}^s \end{aligned} \qquad (11.116)$$

with

$$A = \frac{\partial a}{\partial x}(0), \ B = b(0), \ G = g(0), \ H = \frac{\partial h}{\partial x}(0) \qquad (11.117)$$

Recall, see (10.3), that the state feedback Hamilton–Jacobi equation (HJ1a) of Σ is given as

$$P_x(x)a(x) + \frac{1}{2}P_x(x)\left[\frac{1}{\gamma^2}g(x)g^T(x) - b(x)b^T(x)\right]P_x^T(x) + \frac{1}{2}h^T(x)h(x) = 0 \qquad (11.118)$$

It immediately follows that if $P \geq 0$ is a solution to (11.118) satisfying

$$P(0) = 0 \qquad (11.119)$$

(implying also $P_x(0) = 0$), then the Hessian matrix $P_{xx}(0)$ of P at 0

$$X := P_{xx}(0) \geq 0 \qquad (11.120)$$

is a solution of the Riccati equation

$$A^T X + XA + X\left(\frac{1}{\gamma^2}GG^T - BB^T\right)X + H^T H = 0 \qquad (11.121)$$

Furthermore, if $P \geq 0$ satisfying (11.119) is a solution to the Hamilton–Jacobi *inequality* (HJ1a), then X is a solution to the Riccati *inequality* (11.121), with =

---

[1]However the results to be obtained are directly extendable to more general system classes.

replaced by $\leq$. This Riccati equation (or inequality) is governing the state feedback $\mathcal{H}_\infty$ control problem for the linear system (11.116). In fact we have the following specialization and refinement of Theorem 10.1.1 and Proposition 10.1.2 to the linear case.

Theorem 11.4.1 *Assume (H, A) is detectable. Let $\gamma > 0$. Then there exists a linear feedback*

$$\bar{u} = L\bar{x} \tag{11.122}$$

such that the closed-loop system (11.116, 11.122) has L_2-gain $\leq \gamma$ and is asymptotically stable (i.e., the linear state feedback suboptimal $\mathcal{H}_\infty$ control problem has been solved), if and only if there exists a solution $X \geq 0$ to (11.121). Furthermore, there exists a linear feedback $\bar{u} = L\bar{x}$ such that the closed-loop system has L_2-gain strictly less than γ and is asymptotically stable (i.e., the linear state feedback strict suboptimal $\mathcal{H}_\infty$ control problem has been solved), if and only if there exists a solution $X \geq 0$ to (11.121), satisfying additionally

$$\sigma\left(A - BB^T X + \frac{1}{\gamma^2} GG^T X\right) \subset \mathbb{C}^- \tag{11.123}$$

($\mathbb{C}^-$ being the open left half plane). Moreover, if $X \geq 0$ is a solution to (11.121) then the linear feedback $\bar{u} = L\bar{x}$ with

$$L = -B^T X \tag{11.124}$$

solves the linear state feedback suboptimal $\mathcal{H}_\infty$ control problem, respectively the strict suboptimal problem if X satisfies additionally (11.123).

We recall (see e.g., Sect. 10.1) that the Hamiltonian corresponding to (HJ1a) given by (11.118) is

$$H(x, p) = p^T a(x) + \frac{1}{2} p^T \left[\frac{1}{\gamma^2} g(x) g^T(x) - b(x) b^T(x)\right] p + \frac{1}{2} h^T(x) h(x), \tag{11.125}$$

with corresponding Hamiltonian matrix

$$\mathcal{H} = \begin{bmatrix} A & \frac{1}{\gamma^2} GG^T - BB^T \\ -H^T H & -A^T \end{bmatrix} \tag{11.126}$$

Following the approach of Sect. 11.1 we notice that X is a solution to (11.121, 11.123) if and only if the subspace span $\begin{bmatrix} I \\ X \end{bmatrix}$ is the stable generalized eigenspace of $\mathcal{H}$. Application of Propositions 11.1.5 and 11.1.6 yields the following result.

Theorem 11.4.2 *Assume* (H, A) *is detectable. Let* $X \geq 0$ *be a solution to (11.121, 11.123). Then there exists a neighborhood* W *of* $x = 0$ *and a smooth function* $P \geq 0$ *defined on* W *with* $P(0) = 0$, $P_x(0) = 0$, *such that* P *is a solution of (HJ1a) given by (11.118) on* W.

Proof The local existence of a solution P to (11.118) with $P(0) = 0$, $P_x(0) = 0$, $P_{xx}(0) = X$, immediately follows from Propositions 11.1.5, 11.1.6. Furthermore since by Theorem 11.4.1 $A - BB^T X$ is asymptotically stable the vector field $a - bb^T P_x^T$ is locally asymptotically stable. Rewriting (11.118) as

$$P_x[a - bb^T P_x^T] = -\frac{1}{2}\frac{1}{\gamma^2} P_x gg^T P_x^T - \frac{1}{2} P_x bb^T P_x^T - \frac{1}{2} h^T h,$$

this implies by integration from 0 to T along the asymptotically stable vector field $a - bb^T P_x^T$, letting $T \to \infty$, that locally about $x = 0$ $P \geq 0$. $\square$

The local existence of a solution $P \geq 0$ to (11.118) yields a local solution to the nonlinear $\mathcal{H}_\infty$ control problem, as formulated in the following corollary.

Corollary 11.4.3 *Let* $P \geq 0$ *defined on a neighborhood* W *of* $x = 0$ *be a solution to (11.118). Then with the locally defined feedback*

$$u = -b^T(x) P_x^T(x), \quad x \in W, \tag{11.127}$$

the closed-loop system has locally L_2*-gain* $\leq \gamma$, *in the sense that*

$$\int_0^T \|z(t)\|^2 dt \leq \gamma^2 \int_0^T \|d(t)\|^2 dt + 2P(x(0)) \tag{11.128}$$

for all $x(0) \in W$, *all* $T \geq 0$ *and all* L_2 *functions* $d(\cdot)$ *on* $[0, T]$ *such that the state space trajectories starting from* $x(0)$ *do not leave* W.

The locally defined feedback (11.127) corresponding to a local solution $P \geq 0$ to (11.118) is not the only feedback resulting in a closed-loop system having locally L_2-gain $\leq \gamma$. In fact for an arbitrary state feedback $u = \alpha(x)$ we may rewrite $H(x, p)$ given by (11.125) as

$$H(x, p) = p^T[a(x) + b(x)\alpha(x)] + \frac{1}{2}\frac{1}{\gamma^2} p^T g(x)g^T(x)p +$$
$$\frac{1}{2}\alpha^T(x)\alpha(x) + \frac{1}{2}h^T(x)h(x) - \frac{1}{2}\|b^T(x)p + \alpha(x)\|^2 \tag{11.129}$$

Thus if we take the *linear* feedback

$$\alpha(x) = -B^T X x \tag{11.130}$$

with $X = P_{xx}(0) \geq 0$ being the solution to (11.121), then the zero-th, first and second order terms of $\| b^T(x)p + \alpha(x) \|^2$ are all zero. Thus the Hamiltonian matrix corresponding to $\tilde{H}(x, p) := H(x, p) + \frac{1}{2} \| b^T(x)p + \alpha(x) \|^2$ given as

$$\tilde{H}(x, p) = p^T[a(x) + b(x)\alpha(x)] + \frac{1}{2}\frac{1}{\gamma^2}p^T g(x)g^T(x)p +$$
$$\frac{1}{2}\alpha^T(x)\alpha(x) + \frac{1}{2}h^T(x)h(x) \tag{11.131}$$

is equal to $\mathcal{H}$ in (11.126). Hence there exists locally also a solution $\tilde{P} \geq 0$ to the Hamilton–Jacobi equation

$$\tilde{P}_x(x)[a(x) - b(x)B^T Xx] + \frac{1}{2}\frac{1}{\gamma^2}\tilde{P}_x(x)g(x)g^T(x)\tilde{P}_x^T(x)$$
$$+ \frac{1}{2}x^T XBB^T Xx + \frac{1}{2}h^T(x)h(x) = 0 \tag{11.132}$$

This implies that the inequality (11.128) also holds for the closed-loop system resulting from the feedback $u = -B^T Xx$ if we replace P by $\tilde{P}$, and W by a neighborhood $\tilde{W}$ of $x = 0$ which in general will be *different* from W. Generalizing this observation a little further we obtain

Corollary 11.4.4 *Let $X \geq 0$ be a solution to (11.121, 11.123). Then any feedback $u = \alpha(x)$ with*

$$\alpha(0) = 0, \quad \frac{\partial \alpha}{\partial x}(0) = -B^T X, \tag{11.133}$$

yields a closed-loop system satisfying (11.128) for some neighborhood W of $x = 0$ and some solution $P \geq 0$ to the Hamilton–Jacobi inequality (HJ1a).

Example 11.4.5 Consider the system of Example 10.1.8

$$\dot{x} = u + (\arctan x)\, d, \quad z = \begin{bmatrix} x \\ u \end{bmatrix} \tag{11.134}$$

Clearly, its linearization at $x = 0$ is not affected by disturbances, and the Riccati equation (11.121) is given as $X^2 = 1$ (independent of γ), yielding the positive solution $X = 1$. The state-feedback Hamilton–Jacobi equation (HJ1a) takes the form

$$\left(\frac{dP}{dx}(x)\right)^2 \left[1 - \frac{1}{\gamma^2}\arctan^2 x\right] = x^2, \tag{11.135}$$

which has for every γ a solution $P_\gamma \geq 0$ on the neighborhood

$$W_\gamma = \{x \in \mathbb{R} \mid |\arctan x| < \gamma\} \tag{11.136}$$

yielding a feedback $u = \alpha_\gamma(x) = -x(1 - \frac{1}{\gamma^2}\arctan^2 x)^{-\frac{1}{2}}$. Note that the solution P_γ as well as the feedback α_γ become unbounded for x approaching the boundary of W_γ. Applying instead of $u = \alpha_\gamma(x)$ its linear part $u = -x$ one obtains the closed-loop system

$$\dot{x} = -x + (\arctan x)d, \quad z = \begin{bmatrix} x \\ -x \end{bmatrix} \tag{11.137}$$

with L_2-gain Hamilton–Jacobi equation given as

$$-\frac{d\tilde{P}}{dx}(x)x + \frac{1}{2}\frac{1}{\gamma^2}\left(\frac{d\tilde{P}}{dx}(x)\right)^2 \arctan^2 x + \frac{1}{2}x^2 + \frac{1}{2}x^2 = 0, \qquad (11.138)$$

having a solution $\tilde{P}_\gamma \geq 0$ on the neighborhood

$$\tilde{W}_\gamma = \{x \in \mathbb{R} \mid |\arctan x| < \frac{1}{2}\sqrt{2} \cdot \gamma\} \qquad (11.139)$$

It should be noted that the neighborhood $\tilde{W}_\gamma$ arising from applying the linearized feedback $u = -B^T X x$ in the above example is *smaller* than the neighborhood W_γ arising from applying the full nonlinear feedback $u = -b^T(x)P_x^T(x)$. This can be conjectured to be true in general; the nonlinear controller will solve the $\mathcal{H}_\infty$ control problem on a larger domain than its linearization. A possible starting point for proving such a conjecture is the observation that by (11.129) $\tilde{H}(x, p) \geq H(x, p)$ for all x, p, and thus every solution of $\tilde{H}(x, P_x^T(x)) \leq 0$ will be also a solution of $H(x, P_x^T(x)) \leq 0$, while the converse is not true. Since the main obstruction for extending a local solution $P \geq 0$ of $H(x, P_x^T(x)) \leq 0$ into a global one seems to be the fact that P can become infinite for finite x (see for instance Example 11.4.5), this suggests that P corresponding to the stable invariant manifold of $H(x, p)$ (which by Sect. 11.1 is the minimal solution to $H(x, P_x^T(x)) \leq 0$!) and the corresponding feedback $u = -b^T(x)P_x^T(x)$ have the largest domain of validity.

Example 11.4.5 also shows that in some cases we need to make a compromise between the achievable level of disturbance attenuation γ, and the domain of validity of the feedback. Indeed comparing with Example 10.1.8 we see that *global* disturbance attenuation is possible for every $\gamma > \frac{\pi}{2}$, while a disturbance attenuation level $0 \leq \gamma \leq \frac{\pi}{2}$ can be only met on the neighborhood W_γ (with W_γ shrinking to the origin for $\gamma \downarrow 0$).

The following example shows that the nonlinear feedback $u = -b^T(x)P_x(x)$ corresponding to a (local) solution $P \geq 0$ to the Hamilton–Jacobi inequality (HJ1a) may have other advantages when compared to its linear part $u = -B^T X x$.

Example 11.4.6 ([142]) Consider the system

$$\begin{aligned}\dot{x}_1 &= x_1 x_2 \\ \dot{x}_2 &= x_1^2 + u + d\end{aligned}, \quad z = \begin{bmatrix} x_2 - x_1^2 \\ u \end{bmatrix} \qquad (11.140)$$

The Hamilton–Jacobi inequality (HJ1a) takes the form

$$\frac{\partial P}{\partial x_1}x_1 x_2 + \frac{\partial P}{\partial x_2}x_1^2 + \frac{1}{2}\left(\frac{1}{\gamma^2} - 1\right)\left(\frac{\partial P}{\partial x_2}\right)^2 + \frac{1}{2}(x_2 - x_1^2)^2 \leq 0 \qquad (11.141)$$

This has for $\gamma > 1$ locally positive solutions

$$P(x_1, x_2) = ax_1^2 + bx_2^2 + cx_1^2 x_2, \tag{11.142}$$

provided $a > 0$, and b, c are large enough. This yields the feedback

$$u = -\frac{\partial P}{\partial x_2} = -2bx_2 - cx_1^2 \tag{11.143}$$

which, apart from rendering the L_2-gain of the closed-loop system locally $\leq \gamma$, also locally asymptotically stabilizes the system. On the other hand, the linearized feedback

$$u = -2bx_2 \tag{11.144}$$

does *not* locally asymptotically stabilize the system. In fact it may be proved that there does not exist any linear feedback which renders the system (11.140) locally asymptotically stable. (Note that the linearization of (11.140) does not exactly fit the assumptions of Theorem 11.4.2 and its Corollaries since it is not detectable.)

Next let us consider the *output feedback* case. Again, for simplicity of exposition we focus on affine systems (10.68), that is

$$\begin{aligned}
\dot{x} &= a(x) + b(x)u + g(x)d_1, & a(0) &= 0 \\
y &= c(x) + d_2, & c(0) &= 0 \\
z &= \begin{bmatrix} h(x) \\ u \end{bmatrix}, & h(0) &= 0
\end{aligned} \tag{11.145}$$

Apart from the state feedback Hamilton–Jacobi equation (HJ1a) given by (11.118) we also consider, see (10.71), the Hamilton–Jacobi equation (HJ2a) given as

$$R_x(x)a(x) + \frac{1}{2}\frac{1}{\gamma^2} R_x(x)g(x)g^T(x)R_x^T(x) + \frac{1}{2}h^T(x)h(x) - \frac{1}{2}\gamma^2 c^T(x)c(x) = 0 \tag{11.146}$$

By collecting second-order terms it follows that for any solution $R \geq 0$ of (11.146), with $R(0) = 0$ and thus $R_x(0) = 0$, the matrix

$$W := R_{xx}(0) \geq 0 \tag{11.147}$$

is a solution of the Riccati equation

$$A^T W + WA + \frac{1}{\gamma^2} WGG^T W + H^T H - \gamma^2 C^T C = 0, \tag{11.148}$$

where, as before $A = \frac{\partial a}{\partial x}(0)$, $G = g(0)$, $H = \frac{\partial h}{\partial x}(0)$, and additionally

$$C = \frac{\partial c}{\partial x}(0) \tag{11.149}$$

If W is invertible, then $Y := \gamma^2 W^{-1}$ is the solution to the "dual" Riccati equation

$$YA^T + AY + GG^T + \frac{1}{\gamma^2}YH^T HY - YC^T CY = 0 \qquad (11.150)$$

The Riccati equation (11.150), together with the Riccati equation (11.121), is governing the linear $\mathcal{H}_\infty$ output feedback problem for the system linearized at $x = 0$ given as

$$\dot{\bar{x}} = A\bar{x} + B\bar{u} + G\bar{d}_1$$

$$\bar{y} = C\bar{x} + \bar{d}_2 \qquad (11.151)$$

$$\bar{z} = \begin{bmatrix} H\bar{x} \\ \bar{u} \end{bmatrix}$$

Indeed we recall the following basic theorem from linear $\mathcal{H}_\infty$ control theory, see [88, 117].

Theorem 11.4.7 *Consider the linear system (11.151). Assume the triples (A, B, H) and (A, G, C) are stabilizable and detectable. Then there exists a linear dynamic controller such that the closed-loop system has L_2-gain $< \gamma$ and is asymptotically stable if and only if there exist solutions $X \geq 0$, $Y \geq 0$ to (11.121), respectively (11.150), satisfying additionally (11.123), respectively*

$$\sigma\left(A - YC^T C + \frac{1}{\gamma^2}YH^T H\right) \subset \mathbb{C}^-, \qquad (11.152)$$

together with the coupling condition

$$\sigma_{\max}(XY) < \gamma^2 \qquad (11.153)$$

where $\sigma_{\max}$ denotes the largest singular value. Furthermore one such controller (called the "central controller") is given as

$$\dot{z} = \left(A - BB^T X + \frac{1}{\gamma^2}GG^T X\right)z + \left(I - \frac{1}{\gamma^2}YX\right)^{-1} YC^T(y - Cz) \qquad (11.154)$$

$$u = -B^T Xz$$

We have seen before that the existence of a solution $X \geq 0$ to (11.121, 11.123) implies the local existence of a solution $P \geq 0$ to (HJ1a). Let us now additionally assume that the solution $Y \geq 0$ to (11.150, 11.152) is positive definite, that is, $Y > 0$. Then $W = \gamma^2 Y^{-1} > 0$ satisfies (11.148), while (11.152) can be rewritten as

$$\sigma\left(A + \frac{1}{\gamma^2}GG^T W\right) \subset \mathbb{C}^+ \qquad (11.155)$$

This means that span $\begin{bmatrix} I \\ W \end{bmatrix}$ is the *unstable* generalized eigenspace of the Hamiltonian matrix $\mathcal{H}$ corresponding to (HJ2a) and (11.150), given as

$$\mathcal{H} = \begin{bmatrix} A & \frac{1}{\gamma^2}GG^T \\ \gamma^2 C^T C - H^T H & -A^T \end{bmatrix} \tag{11.156}$$

Application of Propositions 11.1.5 and 11.1.6 (with respect to the unstable invariant manifold) yields

Proposition 11.4.8 *Let $Y > 0$ be a solution to (11.150, 11.152). Then there exists a neighborhood U of $x = 0$ and a smooth function $R > 0$ defined on U with $R(0) = 0$, $R_x(0) = 0$, such that R is a solution of (HJ2a) on U.*

Proof The local existence of a solution R to (HJ2a) with $R(0) = 0$, $R_x(0) = 0$ and $R_{xx}(0) = \gamma^2 Y^{-1}$ follows from Propositions 11.1.5, 11.1.6. Since $R_{xx}(0) = \gamma^2 Y^{-1} > 0$ it follows that $R > 0$ on a neighborhood of $x = 0$. □

Finally, note that the coupling condition $\sigma_{\max}(XY) < \gamma^2$ can be rewritten as

$$X < \gamma^2 Y^{-1} \tag{11.157}$$

Since $P_{xx}(0) = X$ and $R_{xx}(0) = \gamma^2 Y^{-1}$ it follows that for all x near 0

$$P(x) < R(x) \tag{11.158}$$

and so we recover, in a *strict* inequality but *local* form, the weak coupling derived in Sect. 10.2. This is summarized in the following form.

Corollary 11.4.9 *If there exist solutions $X \geq 0$, $Y > 0$ to (11.121), respectively (11.150), satisfying (11.123), respectively (11.152), then there exists a neighborhood U of $x = 0$ and smooth functions $P \geq 0$ and $R > 0$ on U with $P_{xx}(0) = X$ and $R_{xx}(0) = \gamma^2 Y^{-1}$, which are solutions of (HJ1a), respectively (HJ2a), and satisfy (11.158) on U.*

Therefore, loosely speaking, the solvability of the output feedback suboptimal $\mathcal{H}_\infty$ control problem for the linearization (11.151) implies that *locally* all the necessary conditions derived in Sect. 10.2 for solvability of the same problem for the nonlinear system are satisfied.

Furthermore (and more importantly), under the assumptions of Theorem 11.4.7 we can derive a controller which locally solves the *nonlinear* output feedback suboptimal $\mathcal{H}_\infty$ control problem.

Proposition 11.4.10 *Consider the nonlinear system (11.145), together with its linearization (11.151). Assume the triples (A, B, H) and (A, G, C) are stabilizable and detectable. Let $X \geq 0$, $Y > 0$ be solutions to (11.121, 11.123), respectively (11.150,*

*11.152), and satisfying (11.153), leading by Corollary 11.4.9 to local solutions $P \geq 0$
and $R > 0$ of (HJ1a), respectively (HJ2a). Then the nonlinear controller*

$$\dot{\xi} = a(\xi) - b(\xi)b^T(\xi)P_\xi^T(\xi) + \frac{1}{\gamma^2}g(\xi)g^T(\xi)P_\xi^T(\xi)$$

$$+ \gamma^2 \left[R_{\xi\xi}(\xi) - P_{\xi\xi}(\xi) \right]^{-1} \frac{\partial c^T}{\partial \xi}(\xi) \left[y(t) - c(\xi) \right] \qquad (11.159)$$

$$u = -b^T(\xi)P_\xi^T(\xi), \qquad \xi \in \mathcal{X}_c := \mathcal{X}$$

*locally solves the output feedback suboptimal $\mathcal{H}_\infty$ control problem for (11.145), in
the sense that there exists a neighborhood W of the equilibrium $(0,0) \in \mathcal{X} \times \mathcal{X}_c$
and a function $S \geq 0$ defined on it, such that the closed-loop system (11.145, 11.159)
satisfies*

$$\int_0^T \|z(t)\|^2 dt \leq \gamma^2 \int_0^T \|d(t)\|^2 dt + 2S(x(0), \xi(0)) \qquad (11.160)$$

*for all $(x(0), \xi(0)) \in W$, and all $T \geq 0$ and $d \in L_2(0, T)$ such that the $(x(t), \xi(t))$
trajectories do not leave W.*

Proof The linearization of (11.159) at $\xi = 0$ is precisely the central controller
(11.154). Hence the linearization of the closed-loop system (11.145, 11.159) equals
the linear closed-loop system (11.151, 11.154), which by Theorem 11.4.7 has L_2-
gain $< \gamma$ and is asymptotically stable. Hence by Corollary 11.3.4 also the nonlinear
closed-loop system has locally L_2-gain $< \gamma$ in the sense of (11.160) for some neighborhood W of $(x, \xi) = (0, 0)$. $\qquad \qquad \square$

Remark 11.4.11 Note that W may be smaller than the neighborhood $U \times U$ of
$(x, \xi) = (0, 0)$, with U as obtained in Corollary 11.4.9.

Similarly to Corollary 11.4.4 we derive from the proof of Proposition 11.4.10 the
following corollary.

Corollary 11.4.12 *Under the conditions of Proposition 11.4.10 every nonlinear
controller whose linearization equals the "central controller" (11.154) locally solves
the output feedback suboptimal $\mathcal{H}_\infty$ control problem.*

The important issue is thus to construct a controller which solves the nonlinear output
feedback suboptimal $\mathcal{H}_\infty$ control problem on a domain which is as large as possible.
Since the controller (11.159) incorporates in a "maximal way" all the nonlinear
characteristics of the nonlinear system under consideration (as compared with the
other controllers whose linearization equals the central controller), there is some
reason to believe that the controller (11.159) has a large domain of validity.

11.5   Notes for Chapter 11

1. Section 11.1 is an expanded and generalized version of results presented in van der Schaft [272, 275, 278].

2. The Hamilton–Jacobi–Isaacs equation is at the core of dynamic game theory; cf. Basar & Olsder [31].

3. Propositions 11.1.3, 11.1.5 and 11.1.6 can be found in van der Schaft [272].

4. Proposition 11.1.7(i) can be found e.g. in Kucera [175] (Lemma 3.2.1) and part (ii) in Molinari [221], Francis [103].

5. The main idea explored in Proposition 11.1.14 can be found in van der Schaft [275].

6. Section 11.2 is close in spirit to the treatment of the linear quadratic optimal control problem in Anderson & Moore [7], Willems [349].

7. Proposition 11.2.1 can be found in Kucera [174], Francis [103], Kucera [175].

8. For a result related to Proposition 11.2.6 we refer to Byrnes & Martin [57].

9. Using the invariant manifold techniques of Sect. 11.1 it can be also shown that solvability of the *linearized* optimal control problem (for the linearized system and the quadratic part of the cost criterion) implies *local* solvability of the nonlinear optimal control problem, see van der Schaft [274] (compare with Sect. 11.3).

10. Section 11.3 is a generalized version of some results presented in van der Schaft [272, 275] for the L_2-gain case, and Nijmeijer, Ortega, Ruiz & van der Schaft [234] for the passivity case.

11. The Hamiltonian matrix $\mathcal{H}$ as in (11.104) reduces for the L_2-gain supply rate $\frac{1}{2}\gamma^2\|u\|^2 - \frac{1}{2}\|y\|^2$ (see Corollary 11.3.4) to

$$\mathcal{H}_\gamma = \begin{bmatrix} A & \frac{1}{\gamma^2}BB^T \\ C^T C & -A^T \end{bmatrix} \tag{11.161}$$

with A assumed to be asymptotically stable.

It follows (see e.g., Green & Limebeer [117], Scherer [304]) that there exists $\gamma^* \geq 0$ such that if

(a) $\gamma > \gamma^*$, then $\mathcal{H}_\gamma$ does not have purely imaginary eigenvalues.
(b) $0 \leq \gamma \leq \gamma^*$, then $\mathcal{H}_\gamma$ *does* have purely imaginary eigenvalues.

In the dynamical systems literature this is known as a *Hamiltonian Hopf bifurcation* of the corresponding Hamiltonian vector field (van der Meer [214]): for γ smaller than γ^* and monotonously increasing the purely imaginary eigenvalues of $\mathcal{H}_\gamma$ become or remain of even multiplicity at $\gamma = \gamma^*$, and for $\gamma > \gamma^*$ split of into eigenvalues located in the open left-half plane and in the open right-half plane, symmetrically with respect to the imaginary axis.

12. Section 8.4 is largely based on van der Schaft [275, 278]. Example 11.4.6 is due to Isidori & Kang [142].

13. The relations between viscosity solutions of Hamilton–Jacobi equations and the properties of the corresponding Lagrangian submanifolds are explored e.g., in Day [80].

14. For computational schemes for solving Hamilton–Jacobi equations and inequalities we refer to the list of references given in Note 1 in Sect. 10.3 at the end of the previous Chap. 10.

References

1. R.A. Abraham & J.E. Marsden, *Foundations of Mechanics* (2nd edition), Reading, MA: Benjamin/Cummings, 1978.
2. D. Aeyels, 'On stabilization by means of the Energy-Casimir method', *Systems & Control Letters*, 18(5), pp. 325–328, 1992.
3. D. Aeyels, M. Szafranski, 'Comments on the stabilizability of the angular velocity of a rigid body', *Systems & Control Letters*, 10, pp. 35–40, 1988.
4. C. Aguilar, A.J. Krener, 'Numerical Solutions to the Bellman Equation of Optimal Control', *Journal of Optimization Theory and Applications*, 160(2), pp. 527–552, 2014.
5. E. G. Al'brekht, 'On the optimal stabilization of nonlinear systems', *Journal of Applied Mathematics and Mechanics*, 25(5), pp. 1254–1266, 1961.
6. B.D.O. Anderson, 'Algebraic properties of minimal degree spectral factors', *Automatica*, 9, pp. 491–500, 1973.
7. B.D.O. Anderson, J.B. Moore, *Optimal Control-Linear Quadratic Methods*, Prentice Hall Information and System Sciences Series, Prentice Hall, Englewood Cliffs, NJ, 1989.
8. D. Angeli, 'A Lyapunov approach to incremental stability properties', *IEEE Transactions on Automatic Control*, 47, pp. 410–422, 2002.
9. D. Angeli, E. D. Sontag, Y. Wang , 'A characterization of integral input-to-state stability', *IEEE Trans. on Automatic Control*, 45, pp. 1082–1097, 2000.
10. M. Arcak, 'Passivity as a design tool for group coordination', *IEEE Transactions on Automatic Control*, 52(8):1380 –1390, 2007.
11. M. Arcak, C. Meissen, A. Packard, *Networks of dissipative systems*, Springer Briefs in Electrical and Computer Engineering, Control, Automation and Control, 2016.
12. M. Arcak, E.D. Sontag, 'Diagonal stability of a class of cyclic systems and its connection with the secant criterion', *Automatica*, 42, pp. 1531–1537, 2006.
13. S. Arimoto, *Control Theory of Nonlinear Mechanical Systems: A Passivity-Based and Circuit-Theoretic Approach*, Oxford University Press, 1996.
14. A. Astolfi, L. Guzzella, 'Robust control of nonlinear systems: an $\mathcal{H}_\infty$ approach', in *Proc. 12th IFAC World Congress*, Sydney, pp. 281–284, 1993.
15. A. Astolfi, L. Lanari, 'Disturbance attenuation and setpoint regulation of rigid robots via $\mathcal{H}_\infty$ control', in *Proc. 33rd Conf. on Decision and Control*, Orlando, FL, pp. 2578–2583, 1994.
16. A. Astolfi, D. Karagiannis, R. Ortega, *Nonlinear and Adaptive Control with Applications*, Springer, 2008.
17. D. Auckly, L. Kapitanski, W. White, 'Control of nonlinear underactuated systems', *Comm. Pure Appl. Math.*, 53, pp. 354–369, 2000.

© Springer International Publishing AG 2017

A. van der Schaft, *L₂-Gain and Passivity Techniques in Nonlinear Control*,
Communications and Control Engineering, DOI 10.1007/978-3-319-49992-5

18. H. Bai, M. Arcak, J. Wen, *Cooperative control design: a systematic, passivity-based, approach*, Springer, 2011.
19. J.A. Ball, J.W. Helton, '$\mathcal{H}_\infty$ control for nonlinear plants: connections with differential games', in *Proc. 28th Conf. on Decision and Control*, Tampa, FL, pp. 956–962, 1989.
20. J.A. Ball, J.W. Helton, '$\mathcal{H}_\infty$ control for stable nonlinear plants', *Math. Contr. Sign. Syst.*, 5, pp. 233–262, 1992.
21. J.A. Ball, J.W. Helton, 'Inner-outer factorization of nonlinear operators', *J. Funct. Anal.*, 104, pp. 363–413, 1992.
22. J.A. Ball, J.W. Helton, 'Viscosity solutions of Hamilton-Jacobi equations arising in nonlinear $\mathcal{H}_\infty$ control', *J. Mathematical Systems, Estimation & Control*, 6, pp. 1–22, 1996.
23. J.A. Ball, J.W. Helton, M. Walker, '$\mathcal{H}_\infty$ control control for nonlinear systems via output feedback', *IEEE Trans. Aut. Contr.*, AC-38, pp. 546–559, 1993.
24. J.A. Ball, A.J. van der Schaft, 'J-inner-outer factorization, J-spectral factorization and robust control for nonlinear systems', *IEEE Trans. Aut. Contr.*, AC-41, pp. 379–392, 1996.
25. J.A. Ball, M.A. Petersen, A.J. van der Schaft, 'Inner-outer factorization for nonlinear noninvertible systems', *IEEE Transactions on Automatic Control*, AC-49, pp. 483–492, 2004.
26. J.A. Ball, M. Verma, 'Factorization and feedback stabilization for nonlinear systems', *Systems & Control Letters*, 23, pp. 187–196, 1994.
27. A. Banos, 'Parametrization of nonlinear stabilizing controllers: the observer-controller configuration', *IEEE Trans. Aut. Contr.*, 43, pp. 1268–1272, 1998.
28. L. Baramov, 'Solutions to a class of nonstandard nonlinear $\mathcal{H}_\infty$ control problems', *Int. J. Control*, 73, (4), pp. 276–291, 2000.
29. L. Baramov, H. Kimura, 'Nonlinear local J-lossness conjugation and factorization', *Int. J. Robust and Nonlinear Control*, pp. 869–894, 1996.
30. T. Basar, P. Bernhard, *$\mathcal{H}_\infty$ optimal control and related minimax design problems: A dynamic game approach*, Birkhauser, Boston, 1990, 2nd edition 1995.
31. T. Basar, G.J. Olsder, *Dynamic Noncooperative Game Theory*, Academic Press, New York, 1982.
32. R.W. Beard, T.W. McLain, 'Successive Galerkin approximation algorithms for nonlinear optimal and robust control', *Int. J. Control*, 71, pp. 717–744, 1998.
33. R. W. Beard, G. Saridis & J.T. Wen, 'Galerkin approximations of the generalized Hamilton-Jacobi-Bellman equation', *Automatica*, 33(12), pp. 2195–2177, 1997.
34. R. Beard, G. Saridis, J. Wen, 'Approximate solutions to the time-invariant Hamilton-Jacobi-Bellman equation', *J. Optimization Theory and Applications*, 96, pp. 589–626, 1998.
35. S. C. Beeler, H. T. Tran, H. T. Banks, 'Feedback control methodologies for nonlinear systems', *Journal of Optimization Theory and Applications*, 107(1), pp. 1–33, 2000.
36. N. Berman, U. Shaked, '$\mathcal{H}_\infty$ nonlinear filtering', *Int. J. Robust and Nonlinear Control*, 6, pp. 281–296, 1996.
37. P. Bernhard, 'Discrete and continuous time partial information minimax control', preprint 1994.
38. B. Besselink, N. van de Wouw, H. Nijmeijer, 'Model reduction for nonlinear systems with incremental gain or passivity properties', *Automatica*, 49(4), pp. 861–872, 20
39. G. Blankenstein, 'Geometric modeling of nonlinear RLC circuits', *IEEE Trans. Circ. Syst.*, Vol. 52(2), 2005.
40. G. Blankenstein, A.J. van der Schaft, 'Closedness of interconnected Dirac structures', *Proc. 4th IFAC NOLCOS*, Enschede, the Netherlands, pp. 381–386, 1998.
41. G. Blankenstein, A.J. van der Schaft, 'Symmetry and reduction in implicit generalized Hamiltonian systems', *Rep. Math. Phys.*, 47, pp. 57–100, 2001.
42. G. Blankenstein, R. Ortega, A.J. van der Schaft, 'The matching conditions of controlled Lagrangians and interconnection assignment passivity based control', *Int. Journal Robust and Nonlinear Control*, 75, pp. 645–665, 2002.
43. A.M. Bloch, *Nonholonomic Mechanics and Control*, 2nd edition, Interdisciplinary Applied Mathematics Series 24, Springer, New York, 2015.

44. A. Bloch, N. Leonard, J.E. Marsden, 'Matching and stabilization by the method of controlled Lagrangians', in *Proc. 37th IEEE Conf. on Decision and Control*, Tampa, FL, pp. 1446–1451, 1998.
45. A.M. Bloch, N.E. Leonard and J.E. Marsden, 'Controlled Lagrangians and the stabilization of mechanical systems I: The first matching theorem', *IEEE Trans. Automat. Control*, AC-45, pp. 2253–2270, 2000.
46. A.M. Bloch, D.E. Chang, N.E. Leonard and J.E. Marsden, 'Controlled Lagrangians and the stabilization of mechanical systems II: Potential shaping', *IEEE Trans. Automat. Control*, AC-46, pp. 1556–1571, 2001.
47. A.M. Bloch, P.E. Crouch, 'Representations of Dirac structures on vector spaces and nonlinear *LC* circuits', *Proc. Symposia in Pure Mathematics, Differential Geometry and Control Theory*, G. Ferreyra, R. Gardner, H. Hermes, H. Sussmann, eds., Vol. 64, pp. 103–117, AMS, 1999.
48. B. Bollobas, *Modern Graph Theory*, Graduate Texts in Mathematics 184, Springer, New York, 1998.
49. W.A. Boothby, *An Introduction to Differentiable Manifolds and Riemannian Geometry*, Academic Press, New York, 1975.
50. R.W. Brockett, 'Control theory and analytical mechanics', in *Geometric Control Theory*, (eds. C. Martin, R. Hermann), Vol. VII of Lie Groups: History, Frontiers and Applications, Math. Sci. Press, Brookline, pp. 1–46, 1977.
51. R.W. Brockett, 'Asymptotic stability and feedback stabilization', in *Differential Geometric Control Theory*, R.W. Brockett, R.S. Millmann, H.J. Sussmann, eds., Birkhauser, Boston, pp. 181–191, 1983.
52. B. Brogliato, R. Lozano, B. Maschke, O. Egeland, *Dissipative Systems Analysis and Control: Theory and Applications*, Springer, London, 2nd edition 2007.
53. F. Bullo. A.D. Lewis, *Geometric Control of Mechanical Systems*, Texts in Applied Mathematics, Springer, New York, 2004.
54. M. Bürger, C. De Persis, 'Dynamic coupling design for nonlinear output agreement and time-varying flow control', *Automatica*, 51, pp. 210–222, 2015.
55. M. Bürger, D. Zelazo, F. Allgöwer, 'Duality and network theory in passivity-based cooperative control', *Automatica*, 50(8), pp. 2051–2061, 2014.
56. C.I. Byrnes, A. Isidori, J.C. Willems, 'Passivity, feedback equivalence and the global stabilization of minimum phase nonlinear systems', *IEEE Trans. Aut. Contr.*, AC-36, pp. 1228–1240, 1991.
57. C.I. Byrnes, C.F. Martin, 'An integral-invariance principle for nonlinear systems', *IEEE Trans. Aut. Contr.*, AC-40, pp. 983–994, 1995.
58. M.K. Camlibel, M.N. Belur, J.C. Willems, 'On the dissipativity of uncontrollable systems', *Proc. 42th IEEE Conference on Decision and Control*, 2003, Hawaii, USA
59. M.K. Camlibel, L. Iannelli, F. Vasca. 'Passivity and complementarity', *Mathematical Programming*, 145(1-2), pp. 531–563, 2014.
60. F. Cantrijn, M. de Leon, M. de Diego, 'On almost-Poisson structures in nonholonomic mechanics', *Nonlinearity*, 12, pp. 721–737, 1999.
61. J. Carr, *Applications of centre manifold theory*, Springer-Verlag, New York, 1981.
62. F. Castaños, R. Ortega, A.J. van der Schaft, A. Astolfi, 'Asymptotic stabilization via control by interconnection of port-Hamiltonian systems', *Automatica*, 45, pp. 1611–1618, 2009.
63. J. Cervera, A.J. van der Schaft, A. Banos, 'Interconnection of port-Hamiltonian systems and composition of Dirac structures', *Automatica*, 4, pp. 212–225, 2007.
64. D.E. Chang, A.M. Bloch, N.E. Leonard, J.E. Marsden, C.A. Woolsey, 'The equivalence of controlled Lagrangian and controlled Hamiltonian systems', *ESAIM: Control, Optimisation and Calculus of Variations*, 8, pp. 393–422, 2002.
65. B.S. Chen, T.S. Lee, J.H. Feng, 'A nonlinear $\mathcal{H}_\infty$ control design in robotic systems under parameter perturbation and external disturbance', *Int. J. Control*, 59, pp. 439–462, 1994.
66. N. Chopra, M.W. Spong, 'Passivity-based control of multi-agent systems', pp. 107–134 in *Advances in Robot Control*, Eds. S. Kawamura, M. Svinin, Springer, Berlin, 2006.

67. F.H. Clarke, Yu. S. Ledyaev, R.J. Stern & P.R. Wolenski, 'Qualitative properties of trajectories of control systems: a survey', *Journal of Dynamical and Control Systems*, 1, pp. 1–48, 1995.

68. F.H. Clarke, Yu. S. Ledyaev, R.J. Stern & P.R. Wolenski, *Nonsmooth analysis and control theory*, Springer-Verlag, New York, 1998.

69. E. Colgate, N. Hogan, 'Robust control of dynamically interacting systems', *Int. J. Control*, 48(1), pp. 65–88, 1988.

70. J. Cortés, 'Distributed algorithms for reaching consensus on general functions', *Automatica*, 44, pp. 726–737, 2008.

71. J. Cortés, A.J. van der Schaft, P.E. Crouch, 'Characterization of gradient control systems', *SIAM J. Control Optim.*, 44(4), pp. 1192–1214, 2005.

72. T.J. Courant, 'Dirac manifolds', *Trans. American Math. Soc.*, 319, pp. 631–661, 1990.

73. P.E. Crouch, A.J. van der Schaft, *Variational and Hamiltonian Control Systems*, Lect. Notes in Control and Inf. Sciences 101, Springer-Verlag, Berlin, 1987.

74. M. Dalsmo, *Contributions to nonlinear control and mathematical description of physical systems*, PhD Dissertation, NTNU, Trondheim, 1997.

75. M. Dalsmo, O. Egeland, 'State feedback $\mathcal{H}_\infty$ control of a rigid spacecraft', in *Proc. 34th IEEE Conf. on Decision and Control*, New Orleans, pp. 3968–3973, 1995.

76. M. Dalsmo, O. Egeland, 'Tracking of rigid body motion via nonlinear $\mathcal{H}_\infty$ control', *Proc. 13th IFAC World Congress*, pp. 395–400, 1996.

77. M. Dalsmo, W.C.A. Maas, 'Singular $\mathcal{H}_\infty$ suboptimal control for a class of two-block interconnected nonlinear systems', in *Proc. 35th IEEE Conf. on Decision and Control*, Kobe, Japan, pp. 3782–3787, 1996.

78. M. Dalsmo, A.J. van der Schaft, 'On representations and integrability of mathematical structures in energy-conserving physical systems', *SIAM J. Control and Optimization*, 37, pp. 54–91, 1999.

79. S. Dashkovskiy, H. Ito, F. Wirth, 'On a small gain theorem for ISS networks in dissipative Lyapunov form', *European Journal of Control*, 17(4), pp. 357–365, 2011.

80. M.V. Day, 'On Lagrange submanifolds and viscosity solutions', *J. Mathematical Systems, Estimation and Control*, 8, 1998.

81. P.A.M. Dirac, *Lectures in Quantum Mechanics*, Belfer Graduate School of Science, Yeshiva University Press, New York, 1964; *Can. J. Math.*, 2, pp. 129–148, 1950.

82. P.A.M. Dirac, 'Generalized Hamiltonian Dynamics', *Proc. R. Soc. A*, 246, pp. 333–348, 1958.

83. C.A. Desoer, M. Vidyasagar, *Feedback Systems: Input-Output Properties*, Academic Press, New York, 1975.

84. G. Didinsky, T. Basar & P. Bernhard, 'Structural properties of minimax controllers for a class of differential games arising in nonlinear $\mathcal{H}_\infty$-control', *Systems & Control Letters*, 21, pp. 433–441, 1993.

85. I. Dorfman, *Dirac Structures and Integrability of Nonlinear Evolution Equations*, John Wiley, Chichester, 1993.

86. P.M. Dower, M.R. James, 'Dissipativity and nonlinear systems with finite power gain', *Int. J. Robust and Nonlinear Control*, 8, pp. 699–724, 1998.

87. F.J. Doyle, F. Allgöwer, M. Morari, 'A normal form approach to approximate input-output linearization for maximum phase nonlinear systems', *IEEE Trans. Aut. Contr.*, AC-41, pp. 305–309, 1996.

88. J.C. Doyle, K. Glover, P.P. Khargonekar, B.A. Francis, 'State-space solutions to standard $\mathcal{H}_2$ and $\mathcal{H}_\infty$ control problems', *IEEE Trans. Aut. Contr.*, AC-34, pp. 831–846, 1989.

89. D. Dresscher, T.J.A. de Vries, S. Stramigioli, 'A novel concept for a translational continuously variable transmission', *IEEE International Conference on Advanced Intelligent Mechatronics* (AIM), 7–11July 2015.

90. V. Duindam, G. Blankenstein, S. Stramigioli, 'Port-Based Modeling and Analysis of Snakeboard Locomotion', Proc. 16th *Int. Symp. on Mathematical Theory of Networks and Systems* (MTNS2004), Leuven, 2004.

91. V. Duindam, S. Stramigioli, 'Port-based asymptotic curve tracking for mechanical systems', *Eur. J. Control*, 10(5), pp. 411–420, 2004.

92. V. Duindam, S. Stramigioli, J.M.A. Scherpen, 'Passive compensation of nonlinear robot dynamics', *IEEE Transactions on Robotics and Automation*, 20(3), pp. 480–487, 2004.
93. V. Duindam, A. Macchelli, S. Stramigioli, H. Bruyninckx, eds., *Modeling and Control of Complex Physical Systems; the Port-Hamiltonian Approach*, Springer, Berlin, Heidelberg, 2009.
94. D. Eberard, B. Maschke, A.J. van der Schaft, 'An extension of Hamiltonian systems to the thermodynamic phase space: towards a geometry of non-reversible thermodynamics', *Reports on Mathematical Physics*, 60(2), pp. 175–198, 2007.
95. G. Escobar, A.J. van der Schaft, R. Ortega, 'A Hamiltonian viewpoint in the modelling of switching power converters', *Automatica*, Special Issue on Hybrid Systems, 35, pp. 445–452, 1999.
96. W. Feng, I. Postlethwaite, 'Robust nonlinear $\mathcal{H}_\infty$ control adaptive control of robot manipulator motion', Internal Report Dept. of Engineering, Leicester University, 1993.
97. J. Ferguson, R. H. Middleton, A. Donaire, 'Disturbance rejection via control by interconnection of port-Hamiltonian systems', *Proc. 54th IEEE Conference on Decision and Control (CDC)* Osaka, Japan, pp. 507–512, 2015.
98. Shaik Fiaz, D. Zonetti, R. Ortega, J.M.A. Scherpen, A.J. van der Schaft, 'A port-Hamiltonian approach to power network modeling and analysis', *European Journal of Control*,19, pp. 477–485, 2013.
99. W.H. Fleming, H.M. Soner, *Controlled Markov Processes and Viscosity Solutions*, Springer-Verlag, New York, 1993.
100. F. Forni, R. Sepulchre, 'On differentially dissipative dynamical systems', *Proc. 9th IFAC Symposium on Nonlinear Control Systems* (NOLCOS2013), Toulouse, France, September 4-6, 2013.
101. F. Forni, R. Sepulchre, A. J. van der Schaft, 'On differential passivity of physical systems', pp. 6580–6585 in *Proc. 52nd IEEE Conference on Decision and Control (CDC)*, 2013, Florence, Italy.
102. A.L. Fradkov, I.A. Makarov, A.S. Shiriaev, O.P. Tomchina, 'Control of oscillations in Hamiltonian systems', 4th Europ. Control Conf., Brussels, 1997.
103. B.A. Francis, *A Course in $\mathcal{H}_\infty$ Control Theory*, Lect. Notes in Control and Information Sciences 88, Springer-Verlag, Berlin, 1987.
104. B.A. Francis, 'The optimal linear quadratic time-invariant regulator with cheap control', *IEEE Transactions on Automatic Control*, AC-24(4), pp. 616–621, 1979.
105. H. Frankowska, 'Hamilton-Jacobi equations: viscosity solutions and generalized gradients', *J. Math. Anal. Appl.*, 141, pp. 21–26, 1989.
106. R.A. Freeman, P.V. Kokotivić, *Robust Nonlinear Control Design*, Birkhäuser (Systems & Control: Foundations & Applications), Boston, 1996.
107. K.Fujimoto, B.D.O. Anderson, F. de Bruyne, 'A parametrization for closed-loop identification of nonlinear systems based on differentially coprime kernel representations', *Automatica*, 37(12), pp. 1893–1907, 2001.
108. K. Fujimoto, J.M.A. Scherpen, W.S. Gray, 'Hamiltonian realizations of nonlinear adjoint operators', *Automatica*, 38, pp. 1769–1775, 2002.
109. K. Fujimoto, T. Sugie, 'Parametrization of nonlinear state feedback controllers via kernel representations', *Proc. 4th IFAC Symposium on Nonlinear Control Systems* (NOLCOS'98), Enschede, pp. 7–12, 1998.
110. K. Fujimoto, T. Sugie, 'State-space characterization of Youla parametrization for nonlinear systems based on input-to-state stability', in *Proc. 37th IEEE Conf. on Decision and Control*, Tampa, FL, pp. 2479–2484, 1998.
111. T.T. Georgiou, 'Differential stability and robust control of nonlinear systems', *Math. of Control, Signals and Systems*, 6, pp. 289–307, 1994.
112. T.T. Georgiou, M.C. Smith, 'Robustness analysis of nonlinear feedback systems: an input-output approach', *IEEE Trans. Autom. Contr.*, AC-42, pp. 1200–1221, 1997.
113. J.W. Gibbs, *Collected Works*, Volume I, Thermodynamics, Longmans, Green and Co, New York-London-Toronto, 1928.

114. C. Godsil, G. Royle, *Algebraic graph theory*, Graduate Texts in Mathematics 207, Springer, New York, 2004.

115. G. Golo, A. van der Schaft, P.C. Breedveld, B.M. Maschke, 'Hamiltonian formulation of bond graphs', *Nonlinear and Hybrid Systems in Automotive Control*, Eds. R. Johansson, A. Rantzer, pp. 351–372, Springer London, 2003.

116. F. Gómez-Estern, A.J. van der Schaft, 'Physical damping in IDA-PBC controlled underactuated mechanical systems', *European Journal on Control*, 10, Special Issue on Hamiltonian and Lagrangian Methods for Nonlinear Control, pp. 451–468, 2004.

117. M. Green, D.J.N. Limebeer, *Linear Robust Control*, Information and System Sciences Series, Prentice Hall, Englewood Cliffs, NJ, 1995.

118. J. Hamberg, 'Controlled Lagrangians, symmetries and conditions for strong matching', in *Proc. IFAC Workshop on Lagrangian and Hamiltonian Methods for Nonlinear Control*, eds. N.E. Leonard, R. Ortega, pp. 57–62, Pergamon, 2000.

119. J.W. Helton, M.R. James, *Extending $\mathcal{H}_\infty$ Control to Nonlinear Systems*, SIAM Frontiers in Applied Mathematics, 1999.

120. J.W. Helton, M.R. James, 'An information state approach to nonlinear J-inner/outer factorization', in *Proc. 33rd IEEE Conf. on Decision and Control*, Orlando, FL, pp. 2565–2571, 1994.

121. O.B. Hijab, *Minimum Energy Estimation*, Doctoral Dissertation, University of California, Berkeley, 1980.

122. D.J. Hill, 'Preliminaries on passivity and gain analysis', *IEEE CDC Tutorial Workshop on Nonlinear Controller Design using Passivity and Small-Gain Techniques* (organizers: D.J. Hill, R. Ortega & A.J. van der Schaft), 1994.

123. D.J. Hill, P.J. Moylan, 'Stability of nonlinear dissipative systems', *IEEE Trans. Aut. Contr.*, AC-21, pp. 708–711, 1976.

124. D.J. Hill, P.J. Moylan, 'Stability results for nonlinear feedback systems', *Automatica*, 13, pp. 377–382, 1977.

125. D.J. Hill, P.J. Moylan, 'Connections between finite gain and asymptotic stability', *IEEE Trans. Aut. Contr.*, AC-25, pp. 931–936, 1980.

126. D.J. Hill, P.J. Moylan, 'Dissipative dynamical systems: Basic input-output and state properties', *J. Franklin Inst.*, 309, pp. 327–357, 1980.

127. N. Hogan, Impedance control: an approach to manipulation: Part I – theory, Part II – implementation, Part III – applications, *ASME J. Dyn. Syst. Meas. Control*, 107(1), pp. 1–24, 1985.

128. N. Hogan, E. D. Fasse, Conservation Principles and Bond-Graph Junction Structures, pp. 9–13 in R. C. Rosenberg and R. Redfield (eds.), *Automated Modeling for Design*, ASME, New York, 1988.

129. T. Holzhüter, 'Optimal regulator for the inverted pendulum via Euler-Lagrange backward integration', *Automatica*, 40(9), pp. 1613–1620, 2004.

130. J. Huang, C.-F. Lin, 'Numerical approach to computing nonlinear $\mathcal{H}_\infty$ control laws', *AIAA J. Guidance, Control and Dynamics*, 18, pp. 989–994, 1995.

131. J. Imura, H. Maeda, T. Sugie, T. Yoshikawa, 'Robust stabilization of nonlinear systems by $\mathcal{H}_\infty$ state feedback', *Systems & Control Letters*, 24, pp. 103–114, 1995.

132. J.-I. Imura, T. Sugie, T. Yoshikawa, 'Characterization of the strict bounded real condition for nonlinear systems', *IEEE Trans. Aut. Contr.*, AC-42, pp. 1459–1464, 1997.

133. J.-I. Imura, T. Yoshikawa, 'Parametrization of all stabilizing controllers of nonlinear systems', *Systems & Control Letters*, 29, pp. 207–213, 1997.

134. A. Iggidr, B. Kalitine & R. Outbib, 'Semi-definite Lyapunov functions, stability and stabilization', *Math. of Control, Signals and Systems*, 9, pp. 95–106, 1996.

135. A. Isidori, *Nonlinear Control Systems* (2nd Edition), Communications and Control Engineering Series, Springer-Verlag, London, 1989, 3rd Edition, 1995.

136. A. Isidori, '$\mathcal{H}_\infty$ control via measurement feedback for affine nonlinear systems', *Int. J. Robust and Nonlinear Control*, 4, pp. 553–574, 1994.

137. A. Isidori, 'A note on almost disturbance decoupling for nonlinear minimum phase systems', *Systems & Control Letters*, 27, pp. 191–194, 1996.

138. A. Isidori, 'Global almost disturbance decoupling with stability for non minimum-phase single-input single-output nonlinear systems', *Systems & Control Letters*, 28, pp. 115–122, 1996.

139. A. Isidori, *Nonlinear Control Systems II*, Communications and Control Engineering, Springer-Verlag, London, 1999.

140. A. Isidori, A. Astolfi, 'Disturbance attenuation and $\mathcal{H}_\infty$ control via measurement feedback in nonlinear systems', *IEEE Trans. Aut. Contr.*, AC-37, pp. 1283–1293, 1992.

141. A. Isidori & C.I. Byrnes, 'Steady state response, separation principle and the output regulation of nonlinear systems', in *Proc. 28th IEEE Conf. on Decision and Control*, Tampa, FL, pp. 2247–2251, 1989.

142. A. Isidori, W. Kang, '$\mathcal{H}_\infty$ control via measurement feedback for general nonlinear systems', *IEEE Trans. Aut. Contr.*, AC-40, pp. 466–472, 1995.

143. S.M. Jalnapurkar, J.E. Marsden, 'Stabilization of relative equilibria', *IEEE Trans. Aut. Contr.* AC-45, 8, pp. 1483 –1491, 2000.

144. M.R. James, 'A partial differential inequality for nonlinear dissipative systems', *Systems & Control Letters*, 21, pp. 171–185, 1993.

145. M.R. James, J.S. Baras, 'Robust $\mathcal{H}_\infty$ output feedback control for nonlinear systems', *IEEE Trans. Aut. Contr.*, AC-40, pp. 1007–1017, 1995.

146. B. Jayawardhana, R. Ortega, E. Garcia-Canseco, F. Castanos, 'Passivity of nonlinear incremental systems: application to PI stabilization of nonlinear RLC circuits', *Systems & Control Letters*, 56(9-10), pp. 618–622, 2007.

147. D. Jeltsema, R. Ortega, J. Scherpen, 'An energy-balancing perspective of IDA-PBC of nonlinear systems', *Automatica*, 40(9), pp. 1643–1646, 2004.

148. Ying-Cun Ji, Wei-Bing Gao, 'Nonlinear $\mathcal{H}_\infty$-control and estimation of optimal $\mathcal{H}_\infty$-gain', *Systems & Control Letters*, 24, pp. 321–332, 1995.

149. Z.-P. Jiang, A.R. Teel, L. Praly, 'Small gain theorem for ISS systems and applications', *Math. of Control, Signals and Systems*, 7, pp. 95–120, 1994.

150. Z.-P. Jiang, Y. Wang, 'A generalization of the nonlinear small-gain theorem for large-scale complex systems', *Proc. of the 7th World Congress on Intelligent Control and Automation*, Chongqing, China, pp. 1188–1193, 2008.

151. A. Jokic, I. Nakic, 'On additive Lyapunov functions and existence of neutral supply rates in acyclic LTI dynamical networks', *Proc. 22nd Int. Symposium on Mathematical Theory of Networks and Systems*, pp. 345–352, Minneapolis, 2016.

152. U. Jönsson, *Lecture notes on integral quadratic constraints*, Department of Mathematics, Royal Institute of Technology (KTH), Stockholm, Sweden, 2001.

153. J. Jouffroy, J.-J. E. Slotine, 'Methodological remarks on contraction theory', in *Proc. 43rd IEEE Conference on Decision and Control*, 2004

154. R.E. Kalman, 'Lyapunov functions for the problem of Lur'e in automatic control', *Proceedings of the National Academy of Sciences*, 49 (2): 201–205, 1963.

155. R.E. Kalman, 'When is a linear control system optimal ?', *Trans. ASME Ser. D:J Basic Eng.*, 86, pp. 1–10, 1964.

156. W. Kang, 'Nonlinear $\mathcal{H}_\infty$ control and its applications to rigid spacecraft', *IEEE Trans. Aut. Contr.*, AC-40, pp. 1281–1285, 1995.

157. W. Kang, P.K. De, A. Isidori, 'Flight control in a windshear via nonlinear $\mathcal{H}_\infty$ methods', in *Proc. 31st IEEE Conf. on Decision and Control*, Tucson, AZ, pp. 1135–1142, 1992.

158. F. Kerber, A.J. van der Schaft, 'Compositional properties of passivity', *Proc. 50th IEEE Conference on Decision and Control and European Control Conference* (CDC-ECC) Orlando, FL, USA, December 12-15, pp. 4628–4633, 2011.

159. H.K. Khalil, *Nonlinear Systems*, MacMillan, New York, 1992.

160. H.K. Khalil, *Nonlinear Control*, Pearson Education, 2015.

161. P.P. Khargonekar, 'State-space $\mathcal{H}_\infty$ control theory and the *LQG*-problem', in *Mathematical System Theory - The influence of R.E. Kalman* (ed. A.C. Antoulas), Springer, Berlin, 1991.

162. P.P. Khargonekar, I.R Petersen, M.A. Rotea, '$\mathcal{H}_\infty$ optimal control with state feedback', *IEEE Trans. Automat. Contr.*, AC-33, pp. 786–788, 1988.
163. S.Z. Khong, A.J. van der Schaft, 'Converse passivity theorems', Submitted for publication, 2016.
164. G. Kirchhoff, 'Über die Auflösung der Gleichungen, auf welche man bei der Untersuchung der Linearen Verteilung galvanischer Ströme geführt wird', *Ann. Phys. Chem. 72*, pp. 497–508, 1847.
165. H. Kimura, *Chain Scattering Approach to $\mathcal{H}_\infty$-control*, Birkhäuser, Boston, 1997.
166. H.W. Knobloch, A. Isidori, D. Flockerzi, *Topics in Control Theory*, DMV Seminar Band 22, Birkhäuser Verlag, Basel, 1993.
167. I. Kolmanovsky & N.H. McClamroch, 'Developments in nonholonomic control problems', *IEEE Control Systems Magazine*, 15, pp. 20–36, 1995.
168. J.J. Koopman, D. Jeltsema, 'Casimir-based control beyond the dissipation obstacle', in *Proc. IFAC Workshop on Lagrangian and Hamiltonian Methods for Nonlinear Control*, Bertinoro, Italy, 2012.
169. H. Kraim, B. Kugelmann, H.J. Pesch, M.H. Breitner, 'Minimizing the maximum heating of a re-entering space shuttle: an optimal control problem with multiple control constraints', *Optimal Control Applications & Methods*, 17, pp. 45–69, 1996.
170. A.J. Krener, 'Necessary and sufficient conditions for nonlinear worst case ($\mathcal{H}_\infty$) control and estimation', *J. Mathematical Systems, Estimation & Control*, 5, pp. 1–25, 1995.
171. A. J. Krener, *Nonlinear Systems Toolbox V.1.0.*, MATLAB based toolbox available by request from ajkrener@ucdavis.edu, 1997.
172. A.J. Krener, C. O. Aguilar, T.W. Hunt, 'Series solutions of HJB equations', *Festschrift in Honor of Uwe Helmke on the Occasion of his Sixtieth Birthday*, Knut Húper, Jochen Trumpf, Editors, ISBN 978-1470044008, 2013.
173. M. Kristic, I. Kanellakopoulos, P.V. Kokotović, *Nonlinear and Adaptive Control Design*, Wiley, New York, 1995.
174. V. Kučera, 'A contribution to matrix quadratic equations', *IEEE Trans. Aut. Contr.*, AC-17, pp. 344–347, 1972.
175. V. Kučera, 'Algebraic Riccati equation: Hermitian and definite solutions', in *The Riccati Equation* (eds. S. Bittanti, A.J. Laub, J.C. Willems), Communications and Control Engineering Series, Springer-Verlag, Berlin, pp. 53–88, 1991.
176. A. Kugi & K. Schlacher, 'Nonlinear $\mathcal{H}_\infty$-controller design for a DC-to-DC power converter', *IEEE Trans. Control Systems Technology*, 7, pp. 230–237, 1998.
177. P. Kundur, *Power System Stability and Control*, Mc-Graw-Hill Engineering, 1993.
178. H. Kwakernaak, 'Robust control and $\mathcal{H}_\infty$-optimization-tutorial paper', *Automatica*, 29, pp. 255–273, 1993.
179. P. Libermann, C.-M. Marle, *Symplectic Geometry and Analytical Mechanics*, D. Reidel Publishing Company, Dordrecht, Holland, 1987.
180. P.Y. Li & R. Horowitz, 'Application of passive velocity field control to robot contour following problems', in *Proc. 35th IEEE Conf. Decision and Control*, Kobe, Japan, pp. 378–385, 1996.
181. N. Linard, B.D.O. Anderson, F. de Bruyne, 'Coprimeness properties of nonlinear fractional system realizations', *Systems & Control Letters*, 34, pp. 265–271, 1998.
182. T. Liu, D.J. Hill, Z.-P. Jiang, 'Lyapunov formulation of ISS cyclic-small-gain in continuous-time dynamical networks', *Automatica*, 47, pp. 2088–2093, 2011.
183. R. Lozano, 'Positive Real Transfer Functions and Passivity', handwritten notes, 1994.
184. W.-M. Lu, 'A state-space approach to parametrization of stabilizing controllers for nonlinear systems', *IEEE Trans. Aut. Contr.*, AC-40, pp. 1576–1588, 1995.
185. W.-M. Lu, J.C. Doyle, '$\mathcal{H}_\infty$-control of nonlinear systems: A class of controllers', in *Proc. 32nd IEEE Conf. on Decision and Control*, San Antonio, TX, pp. 166–171, 1993.
186. W.-M. Lu, J.C. Doyle, '$\mathcal{H}_\infty$ control of nonlinear sytems via output feedback: Controller parametrization', *IEEE Trans. Aut. Contr.*, AC-39, pp. 2517–2521, 1994.
187. D.L. Lukes, 'Optimal regulation of nonlinear dynamical systems', *SIAM J. Control*, 7, pp. 75–100, 1969.

188. W.C.A. Maas, *Nonlinear $\mathcal{H}_\infty$ Control: The Singular Case*, PhD-thesis, University of Twente, 1996.

189. W.C.A. Maas, A.J. van der Schaft, 'Singular nonlinear $\mathcal{H}_\infty$ optimal control problem', *Int. J. Robust and Nonlinear Control*, 6, pp. 669–689, 1996.

190. A. Macchelli, S. Stramigioli, A.J. van der Schaft, C. Melchiorri, 'Scattering for infinite-dimensional port Hamiltonian systems', *Proc. 41st IEEE Conf. Decision and Control*, Las Vegas, Nevada, December 2002.

191. A. Macchelli, 'Passivity-based control of implicit port-Hamiltonian systems', Proc. *European Control Conference* (ECC), 2013, Florence, Italy.

192. J. Machowski, J.W. Bialek, J.R. Bumby, *Power System Dynamics: Stability and Control*, John Wiley & Sons, Ltd, 2nd edition, 2008.

193. I.R. Manchester, J.-J.E. Slotine, 'Control contraction metrics: Convex and intrinsic criteria for nonlinear feedback design', arXiv preprint arXiv:1503.03144, 2015.

194. I.M.Y. Mareels, D.J. Hill, 'Monotone stability of nonlinear feedback systems', *J. of Mathematical Systems, Estimation and Control*, 2, pp. 275–291, 1992.

195. R. Marino, P. Tomei, *Nonlinear Control Design*, Prentice Hall International (UK), 1995.

196. R. Marino, W. Respondek, A.J. van der Schaft & P. Tomei, 'Nonlinear $\mathcal{H}_\infty$ almost disturbance decoupling', *Systems & Control Letters*, 23, pp. 159–168, 1994.

197. J.E. Marsden, T.S. Ratiu, *Introduction to Mechanics and Symmetry*, Texts in Applied Mathematics 17, Springer-Verlag, New York, 1994.

198. B.M. Maschke, R. Ortega, A.J. van der Schaft, 'Energy-based Lyapunov functions for forced Hamiltonian systems with dissipation', in *Proc. 37th IEEE Conference on Decision and Control*, Tampa, FL, pp. 3599–3604, 1998.

199. B.M. Maschke, R. Ortega, A.J. van der Schaft, 'Energy-based Lyapunov functions for forced Hamiltonian systems with dissipation', *IEEE Transactions on Automatic Control*, 45, pp. 1498–1502, 2000.

200. B.M. Maschke, R. Ortega, A.J. van der Schaft, G. Escobar, 'An energy-based derivation of Lyapunov functions for forced systems with application to stabilizing control', in *Proc. 14th IFAC World Congress*, Beijing, Vol. E, pp. 409–414, 1999.

201. B.M. Maschke & A.J. van der Schaft, 'Port-controlled Hamiltonian systems: Modelling origins and system-theoretic properties', in *Proc. 2nd IFAC NOLCOS*, Bordeaux, pp. 282–288, 1992.

202. B.M. Maschke, A.J. van der Schaft, 'System-theoretic properties of port-controlled Hamiltonian systems', in *Systems and Networks: Mathematical Theory and Applications*, Vol. II, Akademie-Verlag, Berlin, pp. 349–352, 1994.

203. B.M. Maschke, A.J. van der Schaft, P.C. Breedveld, 'An intrinsic Hamiltonian formulation of network dynamics: Non-standard Poisson structures and gyrators', *J. Franklin Inst.*, 329, pp. 923–966, 1992.

204. B.M. Maschke, A.J. van der Schaft, 'A Hamiltonian approach to stabilization of nonholonomic mechanical systems', in *Proc. 33rd IEEE Conf. on Decision and Control*, Orlando, FL, pp. 2950–2954, 1994.

205. B.M. Maschke, A.J. van der Schaft, P.C. Breedveld, 'An intrinsic Hamiltonian formulation of the dynamics of LC-circuits', *IEEE Trans. Circ. and Syst.*, CAS-42, pp. 73–82, 1995.

206. B.M. Maschke, A.J. van der Schaft, 'Interconnected Mechanical Systems, Part II: The Dynamics of Spatial Mechanical Networks', in *Modelling and Control of Mechanical Systems*, (eds. A. Astolfi, D.J.N. Limebeer, C. Melchiorri, A. Tornambe, R.B. Vinter), pp. 17–30, Imperial College Press, London, 1997.

207. B.M. Maschke, A.J. van der Schaft, 'Note on the dynamics of LC circuits with elements in excess', Memorandum no. 1426, University of Twente, Faculty of Mathematical Sciences, January 1998.

208. B.M. Maschke, A.J. van der Schaft, 'Scattering representation of Dirac structures and interconnection in network models', in *Mathematical Theory of Networks and Systems*, eds. A. Beghi, L. Finesso, G. Picci, Il Poligrafo, Padua, pp. 305–308, 1998.

209. W.M. McEneaney, *Max-Plus Methods for Nonlinear Control and Estimation*, Birkhäuser Systems and Control Series, 2006.
210. W.M. McEneaney, K.D. Mease, 'Nonlinear $\mathcal{H}_\infty$ control of aerospace plane ascent', in *Proc. 34th IEEE Conf. on Decision and Control*, New Orleans, pp. 3994–3995, 1995.
211. A.G.J. MacFarlane, *Dynamical System Models*, G.G. Harrap & Co., London, 1970.
212. D.G. McFarlane, K. Glover, *Robust Controller Design using Normalized Coprime Factor Plant Descriptions*, Lect. Notes Contr. and Inf. Sciences 138, Springer-Verlag, Berlin, 1990.
213. R.I. McLachlan, G.R.W. Quispel, N. Robidoux, 'A unified approach to Hamiltonian systems, Poisson systems, gradient systems and systems with Lyapunov functions and/or first integrals', *Phys. Rev. Lett.* 81, pp. 2399–2403, 1998.
214. J.C. van der Meer, *The Hamiltonian Hopf Bifurcation*, Lect. Notes in Mathematics 1160, Springer-Verlag, Berlin, 1985.
215. J.A. Megretski, A. Rantzer, 'System analysis via integral quadratic constraints', *IEEE Trans. Aut. Contr.*, AC-42, pp. 819–830, 1997.
216. G. Meinsma, H. Zwart, 'On H^∞ control for dead-time systems', *IEEE Transactions on Automatic Control*, vol. 45, pp. 272–285, 2000.
217. C. Meissen, L. Lessard, M. Arcak, A.K. Packard, 'Compositional performance certification of interconnected systems using ADMM', *Automatica*, 61, 55–63, 2015.
218. J. Merker, 'On the geometric structure of Hamiltonian systems with ports', *J. Nonlinear Sci.*, 19, pp. 717–738, 2009.
219. A.N. Michel, R.K. Miller, *Qualitative Analysis of Large Scale Dynamical Systems*, Academic Press, New York, 1977.
220. I. Mirzaev, J. Gunawardena, 'Laplacian dynamics on general graphs', *Bull. Math. Biol.*, 75, pp. 2118–2149, 2013.
221. B.P. Molinari, 'The time-invariant linear-quadratic optimal control problem', *Automatica*, 13, pp. 347–357, 1977.
222. M. Morari, E. Zafiriou, *Robust Process Control*, Prentice Hall, Englewood Cliffs, NJ, 1989.
223. P.J. Morrison, 'A paradigm for joined Hamiltonian and dissipative systems', *Physica D*, pp. 410–419, 1985.
224. R.E. Mortensen, 'Maximum likelihood recursive nonlinear filtering', *J. Optimization Theory and Appl,*, 2, pp. 386–394, 1968.
225. P. Moylan, *Dissipative Systems and Stability*, Lecture Notes, 2014.
226. P.J. Moylan, B.D.O. Anderson, 'Nonlinear regulator theory and an inverse optimal control problem', *IEEE Trans. Aut. Contr.*, AC-18, pp. 460–465, 1973.
227. P.J. Moylan, D.J. Hill, 'Stability criteria for large-scale systems', *IEEE Trans. Aut. Contr.*, AC-23, pp. 143–49, 1978.
228. H. Narayanan, 'Some applications of an implicit duality theorem to connections of structures of special types including Dirac and reciprocal structures', *Systems & Control Letters*, 45, pp. 87–96, 2002.
229. C. Navasca, A.J. Krener, 'Patchy solutions of Hamilton-Jacobi-Bellman partial differential equations', in Lecture Notes in Control and Information Sciences, 364, pp. 251–270, 2007.
230. J.I. Neimark, N.A. Fufaev, *Dynamics of Nonholonomic Systems*, Vol. 33 of Translations of Mathematical Monographs, American Mathematical Society, Providence, Rhode Island, 1972.
231. S.K. Nguang, M. Fu, '$\mathcal{H}_\infty$ filtering for known and uncertain nonlinear systems', in *Proc. IFAC Symp. Robust Control Design*, Rio de Janeiro, Brazil, 1994.
232. S.K. Nguang, 'Robust nonlinear $\mathcal{H}_\infty$ output feedback control', *IEEE Trans. Aut. Contr.*, pp. 1003–1007, 1996.
233. H. Nijmeijer, A.J. van der Schaft, *Nonlinear Dynamical Control Systems*, Springer-Verlag, New York, 1990, corrected printing 2016.
234. H. Nijmeijer, R. Ortega, A.C. Ruiz, A.J. van der Schaft, 'On passive systems: From linearity to nonlinearity', in *Proc. 2nd IFAC NOLCOS*, Bordeaux, pp. 214–219, 1992.
235. P.J. Olver, *Applications of Lie groups to differential equations*, Springer-Verlag, New York, 1986.

236. J.-P. Ortega, V. Planas-Bielsa, 'Dynamics on Leibniz manifolds', *Journal of Geometry and Physics*, 52, p. 127, 2004.

237. R. Ortega, 'Passivity properties for stabilization of cascaded nonlinear systems', *Automatica*, 27, pp. 423–424, 1989.

238. R. Ortega, E. Garcia-Canseco, 'Interconnection and damping assignment passivity-based control: a survey', *European Journal of Control*, 10, pp. 432–450, 2004.

239. R. Ortega, A. Loria, P.J. Nicklasson & H. Sira-Ramirez, *Passivity-based Control of Euler-Lagrange Systems*, Springer-Verlag, London, 1998.

240. R. Ortega, A.J. van der Schaft, F. Castaños, A. Astolfi, 'Control by interconnection and standard passivity-based control of port-Hamiltonian systems', *IEEE Transactions on Automatic Control*, AC-53, pp. 2527–2542, 2008.

241. R. Ortega, A.J. van der Schaft, I. Mareels, & B.M. Maschke, 'Putting energy back in control', *Control Systems Magazine*, 21, pp. 18–33, 2001.

242. R. Ortega, A.J. van der Schaft, B.M. Maschke & G. Escobar, 'Interconnection and damping assignment passivity-based control of port-controlled Hamiltonian systems', *Automatica*, 38, pp. 585–596, 2002.

243. R. Ortega, M.W. Spong, 'Adaptive motion control of rigid robots: A tutorial', *Automatica*, 25, pp. 877–888, 1989.

244. R. Ortega, M.W. Spong, F. Gómez-Estern, G. Blankenstein, 'Stabilization of underactuated mechanical systems via interconnection and damping assignment', *IEEE Trans. Automatic Control*, AC-47(8), pp. 1218–1233, 2002.

245. H. M. Osinga, J. Hauser, 'The geometry of the solution set of nonlinear optimal control problems', *Journal of Dynamics and Differential Equations*, 18(4), pp. 881–900, 2006.

246. H.C Öttinger, *Beyond Equilibrium Thermodynamics*, Wiley, Hoboken, 2004.

247. A.D.B. Paice, A.J. van der Schaft, 'Stable kernel representations and the Youla parametrization for nonlinear systems', in *Proc. IEEE 33rd Conf. on Decision and Control*, Orlando, FL, pp. 781–786, 1994.

248. A.D.B. Paice, A.J. van der Schaft, 'On the parametrization and construction of nonlinear stabilizing controllers', in *Proc. 3rd IFAC NOLCOS*, Tahoe City, CA, pp. 509–513, 1995.

249. A.D.B. Paice, A.J. van der Schaft, 'The class of stabilizing nonlinear plant-controller pairs', *IEEE Trans. Aut. Contr.*, AC-41, pp. 634–645, 1996.

250. Z. Pan, T. Basar, 'Adaptive controller design for tracking and disturbance attenuation in parametric-strict-feedback nonlinear systems', *IEEE Trans. Aut. Contr.*, AC-43, pp. 1066–1084, 1998.

251. R. Pasumarthy, A.J. van der Schaft, 'Achievable Casimirs and its implications on control of port-Hamiltonian systems', *Int. J. Control*, 80(9), pp. 1421–1438, 2007.

252. L. Pavel, F.W. Fairman, 'Robust stabilization of nonlinear plants - an L_2-approach', *Int. J. Robust and Nonlinear Control*, 6, pp. 691–726, 1996.

253. L. Pavel & F.W. Fairman, 'Nonlinear $\mathcal{H}_\infty$ control: A J-dissipative approach', *IEEE Trans. Aut. Contr.*, AC-42, pp. 1636–1653, 1997.

254. A. Pavlov, N. van de Wouw, H. Nijmeijer, 'Frequency response functions for nonlinear convergent systems', *IEEE Trans. Autom. Control* 52(6), pp. 1159–1165, 2007.

255. V.M. Popov, *Hyperstability of Control Systems*, Springer-Verlag, New York, 1973.

256. S. Prajna, A.J. van der Schaft, G. Meinsma. 'An LMI approach to stabilization of linear port-controlled Hamiltonian systems', *Systems & Control Letters*, 45, pp. 371–385, 2002.

257. A. Rantzer, 'On the Kalman-Yakubovich-Popov Lemma', *Systems & Control Letters*, 28, pp. 7–10, 1996.

258. S. Rao, A.J. van der Schaft, B. Jayawardhana, 'A graph-theoretical approach for the analysis and model reduction of complex-balanced chemical reaction networks', *J Math Chem*, 51, pp. 2401–2422, 2013.

259. R.M. Redheffer, 'On a certain linear fractional transformation', *J. Math. and Physics*, 39, pp. 269–286, 1960.

260. E.A. Robinson, 'External representation of stationary stochastic processes', *Arkiv für Mathematik*, 4, pp. 379–384, 1962.

261. B. Rüffer, *Monotone systems, graphs, and stability of large-scale interconnected systems*, Ph.D. thesis, University of Bremen, Germany, 2007.
262. M.G. Safonov, *Stability and Robustness of Multivariable Feedback Systems*, The MIT Press, Cambridge, MA, 1980.
263. N. Sakamoto, 'Analysis of the Hamilton-Jacobi equation in nonlinear control theory by symplectic geometry', *SIAM J. Contr. Optimization*, 40(6), pp. 1924–1937, 2002.
264. N. Sakamoto, 'Case studies on the application of the stable manifold approach for nonlinear optimal control design', *Automatica*, 49(2), pp. 568–576, 2013.
265. N. Sakamoto, A.J. van der Schaft, 'An approximation method for the stabilizing solution of the Hamilton-Jacobi equation for integrable systems, a Hamiltonian perturbation approach', *Transactions of the Society of Instrument and Control Engineers* (SICE), vol. 43, pp. 572–580, 2007.
266. N. Sakamoto, A.J. van der Schaft, 'Analytical approximation methods for the stabilizing solution of the Hamilton-Jacobi equation', *IEEE Transactions on Automatic Control*, vol. 53, pp. 2335–2350, 2008.
267. S. Sastry, *Nonlinear Systems: Analysis, Stability, and Control*, Interdisciplinary Applied Mathematics, Springer, New York, 1999.
268. M. Sassano, A. Astolfi, 'Dynamic approximate solutions of the HJ inequality and of the HJB equation for input-affine nonlinear systems', *IEEE Transactions on Automatic Control*, AC-57(10), pp. 2490–2503, 2012.
269. A.J. van der Schaft, *System theoretic properties of physical systems*, CWI Tract 3, CWI, Amsterdam, 1984.
270. A.J. van der Schaft, 'Stabilization of Hamiltonian systems', *Nonl. An. Th. Math. Appl.*, 10, pp. 1021–1035, 1986.
271. A.J. van der Schaft, 'Equations of motion for Hamiltonian systems with constraints', *J. Physics A: Math. Gen.*, 20, pp. 3271–3277, 1987.
272. A.J. van der Schaft, 'On a state space approach to nonlinear $\mathcal{H}_\infty$ control', *Systems & Control Letters*, 16, pp. 1–8, 1991.
273. A.J. van der Schaft, 'On the Hamilton-Jacobi equation of nonlinear $\mathcal{H}_\infty$ optimal control', in *Proc. 1st Europ. Control Conf.*, Grenoble, July 1991, Hermes, Paris, pp. 649–654, 1991.
274. A.J. van der Schaft, 'Relations between ($\mathcal{H}_\infty$-) optimal control of a nonlinear system and its linearization', in *Proc. 30th IEEE Conf. on Decision and Control*, Brighton, UK, pp. 1807–1808, 1991.
275. A.J. van der Schaft, 'L_2-gain analysis of nonlinear systems and nonlinear state feedback $\mathcal{H}_\infty$ control', *IEEE Trans. Autom. Contr.*, AC-37, pp. 770–784, 1992.
276. A.J. van der Schaft, 'Nonlinear $\mathcal{H}_\infty$ control and Hamilton-Jacobi inequalities', in *Proc. 2nd IFAC NOLCOS* (ed. M. Fliess), Bordeaux, pp. 130–135, 1992.
277. A.J. van der Schaft, 'Complements to nonlinear $\mathcal{H}_\infty$ optimal control by state feedback', *IMA J. Math. Contr. Inf.*, 9, pp. 245–254, 1992.
278. A.J. van der Schaft, 'Nonlinear state space $\mathcal{H}_\infty$ control theory', pp. 153–190 in *Essays on Control: Perspectives in the Theory and its Applications* (Eds. H.L. Trentelman, J.C. Willems), PSCT14, Birkhäuser, Basel, 1993.
279. A.J. van der Schaft, 'Robust stabilization of nonlinear systems via stable kernel representations with L_2-gain bounded uncertainty', *Systems & Control Letters*, 24, pp. 75–81, 1995.
280. A.J. van der Schaft, 'Implicit Hamiltonian systems with symmetry', *Rep. Math. Phys.*, 41, pp. 203–221, 1998.
281. A.J. van der Schaft, 'Interconnection and geometry', in *The Mathematics of Systems and Control, From Intelligent Control to Behavioral Systems* (eds. J.W. Polderman, H.L. Trentelman), Groningen, 1999.
282. A.J. van der Schaft, 'Port-Hamiltonian systems: an introductory survey', pp. 1339–1365 in *Proceedings of the International Congress of Mathematicians*, Volume III, Invited Lectures, eds. Marta Sanz-Sole, Javier Soria, Juan Luis Verona, Joan Verdura, Madrid, Spain, 2006.
283. A.J. van der Schaft, 'Balancing of lossless and passive systems', *IEEE Transactions on Automatic Control*, AC-53, pp. 2153–2157, 2008.

284. A.J. van der Schaft, 'On the relation between port-Hamiltonian and gradient systems', *Preprints of the 18th IFAC World Congress*, Milano (Italy), pp. 3321–3326, 2011.

285. A. J. van der Schaft, 'On differential passivity', pp. 21–25 in *Proc. 9th IFAC Symposium on Nonlinear Control Systems* (NOLCOS2013), Toulouse, France, September 4-6, 2013.

286. A.J. van der Schaft, 'Port-Hamiltonian differential-algebraic systems', pp. 173–226 in *Surveys in Differential-Algebraic Equations I* (eds. A. Ilchmann, T. Reis), Differential- Algebraic Equations Forum, Springer, 2013.

287. A.J. van der Schaft, Modeling of physical network systems, *Systems & Control Letters*, 2015.

288. A.J. van der Schaft, J.A. Ball, 'Inner-outer factorization of nonlinear state space systems', in *Systems and Networks: Mathematical Theory and Applications*, Vol. II, Akademie-Verlag, Berlin, pp. 529–532, 1994.

289. A.J. van der Schaft, J.A. Ball, 'Nonlinear inner-outer factorization', in *Proc. IEEE 33rd Conf. on Decision and Control*, Orlando, FL, pp. 2549–2552, 1994.

290. A.J. van der Schaft, M.K. Camlibel, 'A state transfer principle for switching port-Hamiltonian systems', pp. 45–50 in *Proc. 48th IEEE Conf. on Decision and Control*, Shanghai, China, December 16-18, 2009.

291. A.J. van der Schaft, D. Jeltsema, 'Port-Hamiltonian Systems Theory: An Introductory Overview', *Foundations and Trends in Systems and Control*, 1(2-3), pp. 173–378, 2014.

292. A.J. van der Schaft, B.M. Maschke, 'Interconnected Mechanical Systems, Part I: Geometry of Interconnection and implicit Hamiltonian Systems', in *Modelling and Control of Mechanical Systems*, (eds. A. Astolfi, D.J.N. Limebeer, C. Melchiorri, A. Tornambe, R.B. Vinter), pp. 1–15, Imperial College Press, London, 1997.

293. A.J. van der Schaft, B.M. Maschke, 'On the Hamiltonian formulation of nonholonomic mechanical systems', *Rep. Math. Phys.*, 34, pp. 225–233, 1994.

294. A.J. van der Schaft, B.M. Maschke, 'The Hamiltonian formulation of energy conserving physical systems with external ports', *Archiv für Elektronik und Übertragungstechnik*, 49, pp. 362–371, 1995.

295. A.J. van der Schaft, B.M. Maschke, 'Mathematical modeling of constrained Hamiltonian systems', in *Proc. 3rd IFAC NOLCOS '95*, Tahoe City, CA, pp. 678–683, 1995.

296. A.J. van der Schaft, B.M. Maschke, 'Hamiltonian formulation of distributed-parameter systems with boundary energy flow', *Journal of Geometry and Physics*, 42, pp. 166–194, 2002.

297. A.J. van der Schaft, B.M. Maschke, 'Conservation laws and lumped system dynamics', pp 31–48 in *Model-Based Control: Bridging Rigorous Theory and Advanced Technology*, P.M.J. Van den Hof, C. Scherer, P.S.C. Heuberger, eds., Springer, Berlin-Heidelberg, 2009.

298. A.J. van der Schaft, B. Maschke, 'Port-Hamiltonian systems on graphs', *SIAM J. Control Optimization*, 51(2), pp. 906–937, 2013.

299. A.J. van der Schaft, S. Rao, B. Jayawardhana, 'On the mathematical structure of balanced chemical reaction networks governed by mass action kinetics', *SIAM J. Appl. Math.*, 73(2), pp. 953–973, 2013.

300. A.J. van der Schaft, S. Rao, B. Jayawardhana, 'On the network thermodynamics of mass action chemical reaction networks', *Proc. 1st IFAC Workshop on Thermodynamic Foundations of Mathematical Systems Theory*, Lyon, France, July 2013.

301. A.J. van der Schaft, S. Rao, B. Jayawardhana, 'Complex and detailed balancing of chemical reaction networks revisited', *J. of Mathematical Chemistry*, 53(6), pp. 1445–1458, 2015.

302. A.J. van der Schaft, J.M. Schumacher, *An Introduction to Hybrid Dynamical Systems*, Springer Lect. Notes in Control and Information Sciences, Vol. 251, Springer-Verlag, London, 2000, p. xiv+174.

303. A.J. van der Schaft, T.W. Stegink, 'Perspectives in modelling for control of power networks', *Annual Reviews in Control*, 41, pp. 119–132, 2016.

304. C. Scherer, '$\mathcal{H}_\infty$-control by state feedback: An iterative algorithm and characterization of high-gain occurrence', *Systems & Control Letters*, 12, pp. 383–391, 1989.

305. C. Scherer, *The Riccati inequality and state space $\mathcal{H}_\infty$ control theory*, Doctoral Dissertation, University of Würzburg, Germany, 1991.

306. C.W. Scherer, P. Gahinet, M. Chilali, 'Multi-objective control by LMI optimiza- tion', *IEEE Trans. Automat. Control*, 42, pp. 896–911, 1997.

307. C.W. Scherer, 'LPV control and full block multipliers', *Automatica*, 37, pp. 363–373, 2001.

308. J.M.A. Scherpen, 'Balancing for nonlinear systems', *Systems & Control Letters*, 21, pp. 143–153, 1993.

309. J.M.A. Scherpen, W.S. Gray, 'Nonlinear Hilbert adjoints: properties and applications to Hankel singular value analysis', *Nonlinear Analysis*, 51, pp. 883–901, 2002.

310. J.M.A. Scherpen, A.J. van der Schaft, 'Normalized coprime factorizations and balancing for unstable nonlinear systems', *Int. J. Control*, 60, pp. 1193–1222, 1994.

311. J.M. Schumacher, 'The role of the dissipation matrix in singular optimal control', *Systems & Control Letters*, 2, pp. 262–266, 1983.

312. R. Sepulchre, M. Janković, P. Kokotović, *Constructive Nonlinear Control*, Communications and Control Engineering Series, Springer-Verlag London, 1997.

313. T. Shen, K. Tamura, 'Robust $\mathcal{H}_\infty$ control of uncertain nonlinear systems via state feedback', *IEEE Trans. Aut. Contr.*, AC-40, pp. 766–768, 1995.

314. S.L. Shishkin, D.J. Hill, 'Dissipativity and global stabilizability of nonlinear systems', in *Proc. 34th IEEE Conf. on Decision and Control*, New Orleans, pp. 2227–2232, 1995.

315. D.D. Siljak, *Large-scale dynamic systems: stability and structure*, North Holland, Amsterdam, 1978.

316. J.-J. Slotine, W. Li, 'Adaptive manipulator control: a case study', *IEEE Trans. Aut. Contr.*, 33, pp. 995–1003, 1988.

317. E.D. Sontag, *Mathematical Control Theory*, Texts in Applied Mathematics 6, Springer-Verlag, New York, 1990.

318. E.D. Sontag, 'On the input-to-state stability property', *Eur. J. Control*, 1, pp. 24–36, 1995.

319. E.D. Sontag, 'Comments on integral variants of ISS', *Systems & Control Letters*, 34, pp. 93–100, 1998.

320. E.D. Sontag, 'Structure and stability of certain chemical networks and applications to the kinetic proofreading model of T-cell receptor signal transduction', *IEEE Trans. Autom. Control*, 46(7), pp. 1028–1047, 2001.

321. E.D. Sontag, 'Input to state stability: basic concepts and results', P. Nistri & G. Stefani (eds.), *Nonlinear and Optimal Control Theory*, pp. 163–220, Springer-Verlag, Berlin, 2006.

322. E.D. Sontag, A. Teel, 'Changing supply functions in input/state stable systems', *IEEE Trans. Automatic Control*, 40, pp. 1476–1478, 1995.

323. E.D. Sontag, Y. Wang, 'On characterizations of the input-to-state stability property', *Systems & Control Letters*, 24, pp. 351–359, 1995.

324. E.D. Sontag, Y. Wang, 'New characterizations of the input-to-state stability property', *IEEE Trans. Autom. Contr.*, AC-41, pp. 1283–1294, 1996.

325. P. Soravia, '$\mathcal{H}_\infty$ control of nonlinear systems: Differential games and viscosity solutions', *SIAM J. Control and Optimization*, 34, pp. 1071–1097, 1996.

326. J. Steigenberger, 'Classical framework for nonholonomic mechanical control systems', *Int. J. Robust and Nonlinear Control*, 5, pp. 331–342, 1995.

327. A.A. Stoorvogel, *The $\mathcal{H}_\infty$ control problem: a state space approach*, Prentice Hall, Englewood Cliffs, 1992.

328. S. Stramigioli, *From Differentiable Manifolds to Interactive Robot Control*, PhD Dissertation, University of Delft, 1998.

329. S. Stramigioli, Energy-aware robotics, pp. 37–50 in *Mathematical Control Theory I, Nonlinear and Hybrid Control Systems*, Eds. M.K. Camlibel, A.A. Julius, R. Pasumarthy, J.M.A. Scherpen, Lect. Notes in Control and Information Sciences 461, Springer, 2015.

330. S. Stramigioli, B.M. Maschke, A.J. van der Schaft, 'Passive output feedback and port interconnection', in *Proc. 4th IFAC NOLCOS*, Enschede, pp. 613–618, 1998.

331. S. Stramigioli, A.J. van der Schaft, B.M. Maschke, C. Melchiorri, 'Geometric scattering in robotic manipulation', *IEEE Transactions on Robotics and Automation*, 18, pp. 588–596, 2002.

332. H.J. Sussmann, P.V. Kokotović, 'The peaking phenomenon and the global stabilization of nonlinear systems', *IEEE Trans. Aut. Contr.*, 36, pp. 424–439, 1991.
333. G. Tadmor, 'Worst-case design in the time domain: the maximum principle and the standard $\mathcal{H}_\infty$ problem', *Math. of Control, Signals and Systems*, 3, pp. 301–324, 1990.
334. M. Takegaki, S. Arimoto, 'A new feedback method for dynamic control of manipulators', *Trans. ASME, J. Dyn. Systems, Meas. Control*, 103, pp. 119–125, 1981.
335. K. Tchon, 'On some applications of transversality to system theory', *Systems & Control Letters*, 4, pp. 149–156, 1984.
336. A.R. Teel, 'A nonlinear small gain theorem for the analysis of control systems with saturation', *IEEE Trans. Autom. Contr.*, AC-41, pp. 1256–1271, 1996.
337. A.R. Teel, 'On graphs, conic relations and input-output stability of nonlinear feedback systems', *IEEE Trans. Autom. Contr.*, AC-41, pp. 702–709, 1996.
338. H.L. Trentelman, J.C. Willems, 'The dissipation inequality and the algebraic Riccati equation', in *The Riccati equation* (eds. S. Bittanti, A.J. Laub & J.C. Willems), pp. 197–242, Communication and Control Engineering Series, Springer-Verlag, Berlin, 1991.
339. H.L. Trentelman, J.C. Willems, 'Synthesis of dissipative systems using quadratic differential forms: Part II', *IEEE Trans. Aut. Contr.*, AC-47(1), pp. 70–86, 2002.
340. C. Valentin, M. Magos, B. Maschke, 'A port-Hamiltonian formulation of physical switching systems with varying constraints', *Automatica*, 43(7), pp. 1125–1133, 2007.
341. J. Veenman, C.W. Scherer, 'Stability analysis with integral quadratic constraints: a dissipativity based proof', *52nd IEEE Conference on Decision and Control*, Florence, Italy, pp. 3770–3775, 2013.
342. A. Venkatraman, A.J. van der Schaft, 'Energy shaping of port-Hamiltonian systems by using alternate passive input-output pairs', *European Journal of Control*, 6, pp. 1–13, 2010.
343. M. Vidyasagar, *Nonlinear Systems Analysis* (2nd Edition), Prentice Hall, London, 1993 (1st Edition, 1978).
344. W. Wang, J.-J. E. Slotine, 'On partial contraction analysis for coupled nonlinear oscillators', *Biological cybernetics*, 92(1), pp. 38–53, 2005.
345. N. Viswanadham, M. Vidyasagar, 'Stabilization of linear and nonlinear dynamical systems using an observer-controller configuration', *Systems & Control Letters*, 1, pp. 87–91, 1981.
346. S. Weiland, J.C. Willems, 'Dissipative dynamical systems in a behavioral context', *Mathematical Models and Methods in Applied Sciences*, 1, pp. 1–25, 1991.
347. A. Weinstein, 'The local structure of Poisson manifolds', *J. Differential Geometry*, 18, pp. 523–557, 1983.
348. J.C. Willems, *The Analysis of Feedback Systems*, The MIT Press, Cambridge, Massachusetts, USA, 1971.
349. J.C. Willems, 'Least-squares stationary optimal control and the algebraic Riccati equation', *IEEE Trans. Aut. Contr.*, AC-16, pp. 621–634, 1971.
350. J.C. Willems, 'Dissipative dynamical systems - Part I: General Theory', *Archive for Rational Mechanics and Analysis*, 45, pp. 321–351, 1972.
351. J.C. Willems, 'Dissipative dynamical systems - Part II: Linear systems with quadratic supply rates', *Archive for Rational Mechanics and Analysis*, 45, pp. 352–393, 1972.
352. J.C. Willems, 'Mechanisms for the stability and instability in feedback systems', *Proc. IEEE*, 63, pp. 24–35, 1976.
353. J.C. Willems, H.L. Trentelman, 'Synthesis of dissipative systems using quadratic differential forms: Part I', *IEEE Trans. Aut. Contr.*, AC-47, 1, pp. 53–69, 2002.
354. S. Wolf, G. Grioli, O. Eiberger, W. Friedl, M. Grebenstein, H. Hppner, E. Burdet, D. G. Caldwell, R. Carloni, M. G. Catalano, D. Lefeber, S. Stramigioli, N. Tsagarakis, M. Van Damme, R. Van Ham, B. Vanderborght, L. C. Visser, A. Bicchi, A. Albu-Schäffer, 'Variable Stiffness Actuators: Review on Design and Components', *IEEE/ASME Transactions on Mechatronics*, 21(5), pp. 2426–2430, 2016.
355. J. Won, N. Hogan, 'Coupled stability of non-nodic physical systems', in *Proc. 4th IFAC NOLCOS*, Enschede, pp. 595–600, 1998.

356. W.M. Wonham, *Linear Multivariable Control: a Geometric Approach* (2nd edition), Applications of Mathematics, 10, Springer-Verlag, New-York, 1979.

357. R.A. Wright, C. Kravaris, 'Nonminimum-phase compensation for nonlinear processes', *AIChEj*, 38, pp. 26–40, 1992.

358. L. Xie, C.E. de Souza, 'Robust $\mathcal{H}_\infty$ control for linear systems with norm-bounded time-varying uncertainty', *IEEE Trans. Aut. Contr.*, AC-37, pp. 1188–1191, 1992.

359. V.A. Yakubovich, 'The method of matrix inequalities in the theory of stability of nonlinear control systems', Part I-III, *Automatika i Telemechanika* 25(7): 1017–1029, 26(4): 577–599, 1964, 26(5): 753–763, 1965 (English translation in *Automation and Remote Control*).

360. S. Yuliar, M.R. James, 'General dissipative output feedback control for nonlinear systems', in *Proc. 34th IEEE Conf. on Decision and Control*, New Orleans, LA, pp. 2221–2226, 1995.

361. S. Yuliar, M.R. James, J.W. Helton, 'State feedback dissipative control synthesis', *J. Math. Control, Signals and Systems*, 11, pp. 335–356, 1998.

362. G. Zames, 'On the input-output stability of time-varying nonlinear feedback systems, Part I: Conditions derived using concepts of loop gain, conicity, and positivity', *IEEE Transactions on Automatic Control*, AC-11, pp. 228–238, 1966.

363. G. Zames, 'On the input-output stability of time-varying nonlinear feedback systems, Part II: Conditions involving circles in the frequency plane and sector nonlinearities', *IEEE Transactions on Automatic Control*, AC-11, pp. 465–476, 1966.

364. H. Zhang, F.L. Lewis, Z. Qu, 'Lyapunov, adaptive, and optimal design techniques for cooperative systems on directed communication graphs', *IEEE Trans. Industrial Electronics*, 59(7), pp. 3026–3041, 2012.

Index

© Springer International Publishing AG 2017
A. van der Schaft, *L₂-Gain and Passivity Techniques in Nonlinear Control*,
Communications and Control Engineering, DOI 10.1007/978-3-319-49992-5